UI

research projects as Principal and Co-Principal Investigator. He is an elected Fellow of the WB Academy of Science and Technology (WAST) and a Fellow of the International Society for Research and Development (ISRD, UK), the Society for Applied Biotechnology (SAB), International Scientific Research Organisation for Science, Engineering and Technology (ISROSET), Academy of Environment and Life Sciences (AELS), the Scientific Society of Advanced Research and Social Change (SSARSC), and Society of Pharmacognosy and Phytochemistry (SPP), etc. He is an active member of many other science academies and societies, including the New York Academy of Science (NYAS), World Academy of Science, Engineering and Technology (WASET), African Forest Forum (AFF), Association for Tropical Biology & Conservation (ATBC, USA) to name a few.

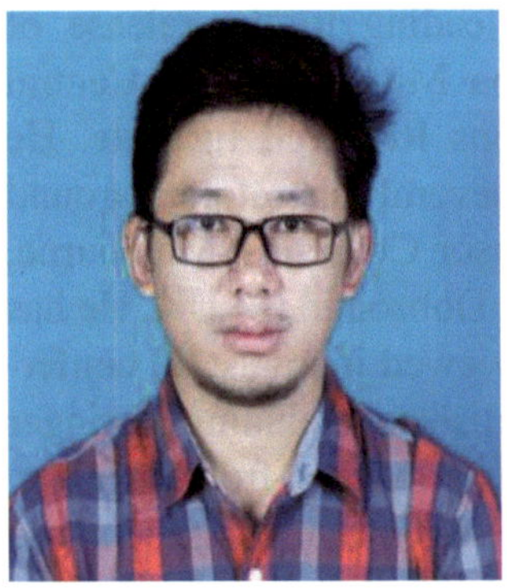

Pemba Hissay Bhutia is pursuing his PhD in Horticulture (spices and plantation crops) at Bidhan Chandra Krishi Viswavidyalaya, Mohanpur, Nadia, West Bengal, India. He has qualified for the ICAR-NET examination in spices, plantation, and medicinal and aromatic plants in 2015. He completed his postgraduate studies in spices and plantation crops from Bidhan Chandra Krishi Viswavidyalaya, Mohanpur, Nadia, India.

Akkabathula Chandini Raj is presently continuing her PhD program at Bidhan Chandra Krishi Viswavidyalaya, Mohanpur, West Bengal, India. She completed her BSc (Hons) in Horticulture at the Horticulture College and Research Station, Dr. Y. S. R. Horticultural University, Venkataramannagudem in 2012. She pursued her postgraduate degree from Bidhan Chandra Krishi Viswavidyalaya, Mohanpur, West Bengal, in spices and plantation crops during 2012–2014, where she stood first in her class She has been awarded a Rajiv Gandhi National Fellowship for her research

UNDEREXPLOITED SPICE CROPS

Present Status, Agrotechnology, and Future Research Directions

Innovations in Horticultural Science

UNDEREXPLOITED SPICE CROPS

Present Status, Agrotechnology, and Future Research Directions

Amit Baran Sharangi, PhD
Pemba H. Bhutia
Akkabathula Chandini Raj
Majjiga Sreenivas

APPLE ACADEMIC PRESS

Apple Academic Press Inc.
3333 Mistwell Crescent
Oakville, ON L6L 0A2 Canada

Apple Academic Press Inc.
9 Spinnaker Way
Waretown, NJ 08758 USA

International Standard Book Number-13: 978-1-77188-697-0 (Hardcover)
International Standard Book Number-13: 978-1-351-13646-4 (eBook)

Library and Archives Canada Cataloguing in Publication

Sharangi, A. B. (Amit Baran), author
Underexploited spice crops : present status, agrotechnology, and future research directions /
Amit Baran Sharangi, PhD, Pemba H. Bhutia, Akkabathula Chandini Raj, Majjiga Sreenivas.

(Innovations in horticultural science)
Includes bibliographical references and index.
Issued in print and electronic formats.
ISBN 978-1-77188-697-0 (hardcover).--ISBN 978-1-351-13646-4 (PDF)

1. Spice plants. 2. Spices. I. Bhutia, Pemba H., author
II. Raj, Akkabathula Chandini, author III. Sreenivas, Majjiga, author
IV. Title. V. Series: Innovations in horticultural science

SB305.S53 2018 633.8'3 C2018-903487-4 C2018-903488-2

Library of Congress Cataloging-in-Publication Data

Names: Sharangi, A. B. (Amit Baran), author.

Title: Underexploited spice crops : present status, agrotechnology, and future research directions / authors: Amit Baran Sharangi, Pemba H. Bhutia, Akkabathula Chandini Raj, Majjiga Sreenivas.

Description: Waretown, NJ : Apple Academic Press, 2018. | Series: Innovations in horticultural science | Includes bibliographical references and index.

Identifiers: LCCN 2018026728 (print) | LCCN 2018028696 (ebook) | ISBN 9781351136464 (ebook) | ISBN 9781771886970 (hardcover : alk. paper)

Subjects: LCSH: Spice plants. | Spices.

Classification: LCC SB305 (ebook) | LCC SB305 .S537 2018 (print) | DDC 633.8/3--dc23

LC record available at https://lccn.loc.gov/2018026728

Apple Academic Press also publishes its books in a variety of electronic formats. Some content that appears in print may not be available in electronic format. For information about Apple Academic Press products, visit our website at **www.appleacademicpress.com** and the CRC Press website at **www.crcpress.com**

CONTENTS

ABOUT THE AUTHORS

Amit Baran Sharangi, PhD, is Professor in Horticultural Science and Head of the Department of Spices and Plantation Crops in the Faculty of Horticulture at Bidhan Chandra Krishi Viswavidyalaya (Agricultural University), India. He has been teaching for twenty years and was instrumental in the process of coconut improvement leading to the release of a variety Kalpa Mitra from the Central Plantation Crops Research Institute. He spent time at several laboratories around the world, including the laboratories of Professor Cousen in Melbourne, Australia; Professor Picha in the USA; and Dr. Dobson in the UK. He has published about 65 research papers in peer-reviewed journals, 60 conference papers, and 16 books with reputed publishers including Springer Nature. He has also published chapters in books published from Springer, CRC Press, Nova Science Publishers, and others. One of his papers was ranked among the top 25 articles in ScienceDirect. Presently he is associated with 40 international and national journals in a variety of roles, including editor-in-chief, regional editor, technical editor, editorial board member, and reviewer.

Professor Sharangi has visited abroad extensively on academic missions and has received several international awards, such as the Endeavour Postdoctoral Award (Australia), INSA-RSE (Indian National Science Academy) Visiting Scientist Fellowship (UK), Fulbright Visiting Faculty Fellowship (USA), Achiever's Award (Society for Advancement of Human and Nature, Man of the Year—2015 (Cambridge, UK), Outstanding Scientist, Higher Education Leadership Award (Venus International Foundation), etc. He has delivered invited lectures in the UK, USA, Australia, Thailand, Israel, and Bangladesh on several aspects of herbs and spices. Professor Sharangi is associated with a number of research projects as Principal and Co-Principal Investigator. He is an elected Fellow of the WB Academy of Science and Technology (WAST) and a Fellow of the International Society for Research and Development (ISRD, UK), the Society for Applied Biotechnology (SAB), International Scientific Research Organisation for Science, Engineering and Technology (ISROSET), Academy of Environment and Life Sciences (AELS), the Scientific Society of Advanced Research and Social Change (SSARSC), and Society of Pharmacognosy and Phytochemistry (SPP), etc. He is an active member of many other science academies and societies, including the New York Academy of Science (NYAS), World Academy of Science, Engineering and Technology (WASET), African Forest Forum (AFF), Association for Tropical Biology & Conservation (ATBC, USA) to name a few.

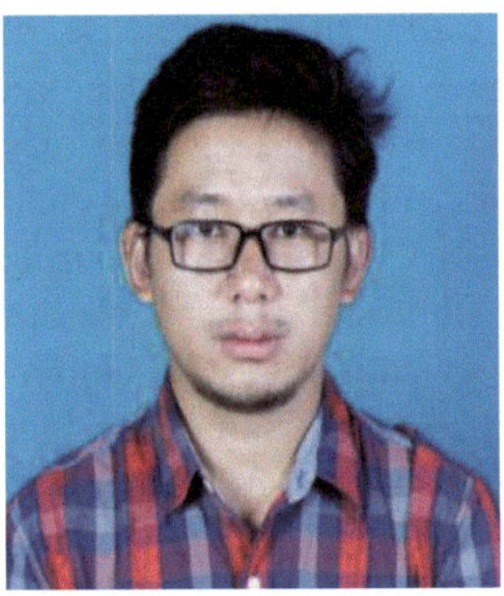

Pemba Hissay Bhutia is pursuing his PhD in Horticulture (spices and plantation crops) at Bidhan Chandra Krishi Viswavidyalaya, Mohanpur, Nadia, West Bengal, India. He has qualified for the ICAR-NET examination in spices, plantation, and medicinal and aromatic plants in 2015. He completed his postgraduate studies in spices and plantation crops from Bidhan Chandra Krishi Viswavidyalaya, Mohanpur, Nadia, India.

Akkabathula Chandini Raj is presently continuing her PhD program at Bidhan Chandra Krishi Viswavidyalaya, Mohanpur, West Bengal, India. She completed her BSc (Hons) in Horticulture at the Horticulture College and Research Station, Dr. Y. S. R. Horticultural University, Venkataramannagudem in 2012. She pursued her postgraduate degree from Bidhan Chandra Krishi Viswavidyalaya, Mohanpur, West Bengal, in spices and plantation crops during 2012–2014, where she stood first in her class She has been awarded a Rajiv Gandhi National Fellowship for her research work and has qualified for the ICAR-NET (Indian Council of Agricultural Research) in spices, plantation, medicinal, and aromatic plants in 2015 and is one among the ten students who appeared for ARS Viva-voce of Agricultural Scientists Recruitment Board (ASRB) in the same year. She earned her postgraduate degree in spices and plantation crops.

Majjiga Sreenivas is now pursuing a PhD in spices and plantation crops at Bidhan Chandra Krishi Viswavidyalaya, Mohanpur, West Bengal, India, and is engaged as a Senior Research Fellow on a project associated with his research. He graduated with BSc (Hons.) in Horticulture from the College of Horticulture, Dr. Y. S. R. Horticultural University, Mojerla, Telangana state in 2012. He completed postgraduate studies at the Horticulture College and Research Institute, Venkataramannagudem in Plantation Spices Medicinal and Aromatic crops in 2014. He qualified for ICAR-NET (Indian Council of Agricultural Research) in spices, plantation, medicinal, and aromatic plants in 2015. He completed his postgraduate program at the Horticulture College and Research Institute, Venkataramannagudem, India, in plantation spices medicinal and aromatic crops.

INNOVATIONS IN HORTICULTURAL SCIENCE

About the Series

Editor-in-Chief:
Dr. Mohammed Wasim Siddiqui Assistant Professor-cum- Scientist
Bihar Agricultural University | www.bausabour.ac.in
Department of Food Science and Postharvest Technology
Sabour | Bhagalpur | Bihar | P. O. Box 813210 | INDIA
Contacts: (91) 9835502897
Email: wasim_serene@yahoo.com | wasim@appleacademicpress.com

The horticulture sector is considered as the most dynamic and sustainable segment of agriculture all over the world. It covers pre- and postharvest management of a wide spectrum of crops, including fruits and nuts, vegetables (including potatoes), flowering and aromatic plants, tuber crops, mushrooms, spices, plantation crops, edible bamboos etc. Shifting food pattern in wake of increasing income and health awareness of the populace has transformed horticulture into a vibrant commercial venture for the farming community all over the world.

It is a well-established fact that horticulture is one of the best options for improving the productivity of land, ensuring nutritional security for mankind and for sustaining the livelihood of the farming community worldwide. The world's populace is projected to be 9 billion by the year 2030, and the largest increase will be confined to the developing countries, where chronic food shortages and malnutrition already persist. This projected increase of population will certainly reduce the per capita availability of natural resources and may hinder the equilibrium and sustainability of agricultural systems due to overexploitation of natural resources, which will ultimately lead to more poverty, starvation, malnutrition, and higher food prices. The judicious utilization of natural resources is thus needed and must be addressed immediately.

Climate change is emerging as a major threat to the agriculture throughout the world as well. Surface temperatures of the earth have risen significantly over the past century, and the impact is most significant on agriculture. The rise in temperature enhances the rate of respiration, reduces cropping periods, advances ripening, and hastens crop maturity, which adversely affects crop productivity. Several climatic extremes such as droughts, floods, tropical cyclones, heavy precipitation events, hot extremes, and heat waves cause a negative impact on agriculture and are mainly caused and triggered by climate change.

In order to optimize the use of resources, hi-tech interventions like precision farming, which comprises temporal and spatial management of resources in horticulture, is essentially required. Infusion of technology for an efficient utilization of resources is intended for deriving higher crop productivity per unit of inputs. This would be possible only through deployment of modern hi-tech applications and precision farming methods. For improvement in crop production and returns to farmers, these technologies have to be widely spread and adopted. Considering the above-mentioned challenges of horticulturist and their expected role in ensuring food and nutritional security to mankind, a compilation of hi-tech cultivation techniques and postharvest management of horticultural crops is needed.

This book series, Innovations in Horticultural Science, is designed to address the need for advance knowledge for horticulture researchers and students. Moreover, the major advancements and developments in this subject area to be covered in this series would be beneficial to mankind.

Topics of interest include:

1. Importance of horticultural crops for livelihood
2. Dynamics in sustainable horticulture production
3. Precision horticulture for sustainability
4. Protected horticulture for sustainability
5. Classification of fruit, vegetables, flowers, and other horticultural crops
6. Nursery and orchard management
7. Propagation of horticultural crops
8. Rootstocks in fruit and vegetable production
9. Growth and development of horticultural crops
10. Horticultural plant physiology
11. Role of plant growth regulator in horticultural production
12. Nutrient and irrigation management
13. Fertigation in fruit and vegetables crops
14. High-density planting of fruit crops
15. Training and pruning of plants
16. Pollination management in horticultural crops
17. Organic crop production
18. Pest management dynamics for sustainable horticulture
19. Physiological disorders and their management
20. Biotic and abiotic stress management of fruit crops
21. Postharvest management of horticultural crops
22. Marketing strategies for horticultural crops
23. Climate change and sustainable horticulture
24. Molecular markers in horticultural science
25. Conventional and modern breeding approaches for quality improvement
26. Mushroom, bamboo, spices, medicinal, and plantation crop production

BOOKS IN THE SERIES

- Spices: Agrotechniques for Quality Produce
  Amit Baran Sharangi, PhD, S. Datta, PhD, and Prahlad Deb, PhD
- Sustainable Horticulture, Volume 1: Diversity, Production, and Crop Improvement
  Editors: Debashis Mandal, PhD, Amritesh C. Shukla, PhD, and Mohammed Wasim Siddiqui, PhD
- Sustainable Horticulture, Volume 2: Food, Health, and Nutrition
  Editors: Debashis Mandal, PhD, Amritesh C. Shukla, PhD, and Mohammed Wasim Siddiqui, PhD
- Underexploited Spice Crops: Present Status, Agrotechnology, and Future Research Directions
  Amit Baran Sharangi, PhD, Pemba H. Bhutia, Akkabathula Chandini Raj, and Majjiga Sreenivas

LIST OF ACRONYMS

CAGR	compound annual growth rate
CFF	crops for the future
EU	European Union
FAO	Food and Agriculture Organization
FYM	farm yard manure
GAP	good agricultural practices
GC–MS	gas chromatography–mass spectrometry
GI	geographical indication
GIS	geographic information system
GMP	good manufacturing practices
HACCP	hazard analysis critical control point
IAA	indole-3-acetic acid
IDM	integrated disease management
IFAD	International Fund for Agricultural Development
IISR	Indian Institute of Spices Research
IPM	integrated pest management
ISO	International Organization for Standardization
LC–MS	liquid chromatography–mass spectrometry
MAS	marker aided selection
NER	north-eastern region
NMR	nuclear magnetic resonance
NRCSS	National Research Centre on Seed Spices
PCC	pure Ceylon cinnamon
PCR	polymerase chain reaction
TIS	temporary immersion system
UAE	United Arab Emirates
USSR	Union of Soviet Socialist Republics

FOREWORD

Both overpopulation and underutilization of genetic resources, gifted by nature and blessed by the good earth, have had a negative impact on the growth and prosperity of agriculture. Although spices are embedded in the text and dialogue of Indian history, tradition, culture, and praxis along with economic and ecological pursuits, there has been an untold story of underutilized spices. Over centuries spices have been offering livelihood and ecological deliverables to mankind. Then also their proper utilization, protection and stewardship are not in proper place or in proper management. Their rare biochemical properties and ecological suitability have opened up vistas for community enterprise and livelihood, nutraceuticals and ethnomedicines, commerce, and cultural aroma.

This book deals with botanical, cultural, and technological aspects of 24 underutilized or poorly noticed, nondescript spice crops that could be a boon to spice-related research and application. This is more so since we have to combat and mitigate the brunt of climate changes. The inquisitive scholastic minds from across the disciplines along with faculties and professionals will be well drenched by this pool of knowledge and information on this previously neglected area of research and intervention.

—M. S. Swaminathan
President, International Union for the Conservation of Nature and Natural Resources;
Former Director General, Indian Council of Agricultural Research;
Former Secretary, Ministry of Agriculture;
Former Director General, International Rice Research Institute

PREFACE

Changing food habits, the demographic scenario, the human need to move from one country to another, attraction to various ethnic foods, and a liberalized economy, all these have contributed to the emerging popularity of spices across the globe. As such, spices are not food, as these provide us nutraceutical value rather than nutritional value. The connotation speaks for itself and, thus, is self-explanatory. It is very difficult to single out any spice that does not enriched human health. But the unraveling process is rather slower than what has been expected, in spite of many research works, the numerous tools for identification and quantification, and the rapid fine-tuning, redefining and transforming of traditional wisdom in the light of scientific vision. We are now aware that apart from the popular and major spices, there are several other spices that, in spite of their enormous contribution to human health, remain underexploited or underutilized.

The book, *Underexploited Spice Crops: Present Status, Agrotechnology, and Future Research Directions,* with its comprehensive coverage, is a rich compilation of agrotechniques coupled with background information, research work, and scientific discussion on basic and applied aspects on the subject. The main feature of the book is an in-depth narration of the scope of quality underexploited spice crops as a product in influencing present-day global export market arising out of renewed interest in these crops throughout the world.

This book deals with the scientific approach of growing underexploited spices with the intention to popularizing them. It is a compilation of information on various aspects of these spices. Separate chapters on the importance of each spice, methods of growing and harvesting, recent research going on around the world with future strategies, etc. are supplemented with tables, chemical structures, and illustrations wherever necessary.

This book will be a valuable resource for the higher-level undergraduate and graduate courses in the area of spices. It will fill the gap of a long-felt need for technological information on underexploited spices. Moreover, it will cater to the need of the inquisitive common readers toward knowing different aspects of such spices that grow either in wild or which have grown in an unorganized way since antiquity, inviting due attention and scientific intervention to go hand in hand with the so-called popular spices with similar fame.

—Prof. A. B. Sharangi
Mr. Pemba H. Bhutia
Miss. A. Chandini Raj
Mr. M. Sreenivas
Kalyani, India

INTRODUCTION

In the Middle Ages, usage of spices at a meal was one of the determining index of the social status of a family. The extent and variety of aroma, taste and color provided by the spices adorning a lunch or dinner plate were indicative of wealth. Beyond being cooking ingredients, spices were also regarded for their magical medicinal remedies and protective action.

During the Middle Ages, Arab merchants dominated the spice trade with huge commercial shipments that moved from South and Southeast Asia to Europe. It is an age-old matter of debate to mark the exact beginning of the Modern Age, and many people count the event of Vasco da Gama's Portuguese expedition to find the Spice Islands to be the beginning of the same. This is because of all the excitement and thrill of having a direct trade route to Asia where Europeans encountered thriving commerce and where 'cosmopolitan' intellectuals were driven by an intense desire to control the spice trade. The Europeans ultimately colonized the world that irrevocablly changed the world spice trade map by their global exploration.

The phytochemicals in spices protect the human body from a wide array of diseases. Spices, especially the so-called underutilized ones, are well adapted to existing and adverse environmental conditions and generally resistant to pests and diseases. Furthermore, these crops have long been a traditional part of cropping systems. Their cultivation, utilization, and acceptability should not be a real problem.

This book comprises 24 underutilized spice crops with fundamental and practical aspects of qualitative production. Comprehensive information has been given about the composition and potential utilization of different underutilized crops. Discussing the fascinating profile of underutilized spice crops and their potential for supplementing the major ones, *Underexploited Spice Crops* provides a roadmap for understanding the broad sweep of agricultural, botanical, pharmacological, sociological, and policy issues that intermingle and intertwine.

Meticulously elucidated with numerous references, this book essentially explores key technologies, including agronomy, horticulture, postharvest technologies, biotechnology, biochemistry, economics, and sociology of minor spices. With so much exhaustive information scattered throughout the literature, it is often difficult to make sense of what is rational and what really or apparently has no scientific credibility. This book also provides the historical perspective of traditional herbs and spices on which to anchor the information and outlines the strengths and constraints of the different underutilized spices to transform them into utilizable ones.

CHAPTER 1

UNDEREXPLOITED SPICE CROPS: INTRODUCTION, SIGNIFICANCE AND USES

CONTENTS

1.1 INTRODUCTION

If fire is the first greatest discovery of prehistoric men to take credit in turning their raw animal flesh and plant-based foods to their tasty counterparts, then the second thing to be credited for adding taste, aroma, colour and pungency in the same food are definitely the spices. The uses of spices spread abundantly when the Caveman cooked the meat and consciously applied them initially for seasoning and later for taste and aroma as well. Gradually, the drinks were also flavoured with aromatic spices along with cooked food although their curative power was still unknown and hidden. However, slowly men realized that spices not only add flavour and taste to our food but also enhance quality and medicinal values of food. The demand for this magical product was so powerful that it led to the voyages of discovery and colonial conquest during Middle Ages. These voyages were intended to find direct routes to the spices of India and the East. Spices, thus, have a profound influence on the course of human civilization.

Approximately 2600–2100 B.C. ago during the Pyramid Age in Egypt, the first authentic records of the use of spices and herbs were found. The use of spices in daily life is known from a variety of texts, including recipe books and medicinal treatises (e.g. Dioscorides, 2000; Pliny, 1945; Dalby, 2000a, 2000b; Rodinson et al., 2001; Zaouali, 2007; Dietrich, 2008; Freedman, 2008). Out of the 109 spices listed by the International Organization for Standardization, India, as the 'Land of Spices', produces as many as 63 owing to its varied agroclimatic regions.

1.2 DEFINITION

Normally, the plants which are useful to human beings and are grown by them are called crops. However, there are certain crop species that are not used to their full potential and are accordingly given numerous connotations, namely underutilized, neglected, orphan, minor, under-researched, forgotten, traditional or indigenous crops/plant species. Moreover, wild relatives of major crops are also included as well. Some traditional crops are not indigenous, some indigenous crops are well established, and crops neglected in one country or part of a country may be overused elsewhere. For clarity, Jaenicke and Höschle-Zeledon (2006) defined underutilized crops/plant species as 'those species with under-exploited potential for contributing to food security, health (nutritional/medicinal), income generation, and environmental services'.

Past few centuries has witnessed shrinking land resources coupled with the abnormal rate of growth in world population resulting with increased concern about food and nutritional security in emerging economies. So far as the question of food security is concerned, the very scenario has invariably emerged to be as one of the major issues that needs to be immediately addressed by agricultural research. The inevitable and, till date, unpredictable effects of climate change on agricultural production, especially in flood-prone and drought-affected areas, accelerate the potential of underutilized crops to provide a buffer to climate change situations in spite of many uncertainties spinning to the issues. The very fact that many of these crops have adapted to the local climate to tackle the pest and disease incidence and survive without intensive external inputs as well as breeding is believed to make them more durable under climatic stresses. Apart from this, use of local landraces is getting additional fillip for their increased diversity towards achieving a safety net against climatic vagaries. The debate of underutilized crops to world food security has been reopened again after being sparked by the consequences of 2007 food crisis (Jaenicke, 2009; Parker, 2010).

Malnutrition, ill health and hunger still contribute a principal part in the world's greatest challenges (Baldermann et al., 2016). While major crops will take care of the food towards mitigating food scarcity and hunger, minor crops including minor spices have no less major role in supplying the nutraceuticals, phytopharmaceuticals and bioactive compounds for protection from diseases, proper physical and mental growth, and impairment of acute deficiencies. With malnutrition and disease going hand in hand, alternative food strategies combined with health and nutrition awareness programmes have to be initiated to better manage the chronic diseases such as cancer, cardiovascular diseases, diabetes, obesity, cognitive disorders, and so forth. Many underutilized and so-called neglected spice crops can be remedies for such diseases.

1.3 CLASSIFICATION

Very few efforts have been made to classify the so-called underutilized or minor spices. However, a very simple and general form is to classify them into three broad categories as follows:

a) Tree spice crops, namely allspice, cambodge, cassia, clove, Indian cassia, juniper, mango, tamarind, and so forth. These spices can be planted along roadsides. A few others can be utilized in the home gardening system.
b) Shrubs, namely curry leaf, hyssop, mace, salvia, star anise, vanilla, and so forth. Star anise can be planted in home gardens with trees.
c) Herbs, namely aniseed, basil, caraway, rosemary, and so forth.. These are also excellent pot plants. They not only satisfy our requirements as spices or condiments but also maintain the ecological balance.

A few underutilized spice crops are tabulated in Table 1.1.

1.4 BASIC USES OF HERBS AND SPICES

Herbs and spices are used since time immemorial and have various specific functions to impart in the prepared food, namely flavouring, deodorizing/masking, adding pungency, adding colour, and so forth.

1.5 MEDICINAL PROPERTIES OF SELECTED UNDERUTILIZED SPICES

Large cardamom: Hypnotic, appetizer, astringent to bowels, tonic to heart and liver.

Celery: Stimulant, tonic, diuretic, carminative, anti-inflammatory.

Aniseed: mild expectorant, stimulating, carminative, diuretic, diaphoretic.

Caraway: Stomachic, carminative, anthelmintic, lactagogue, corrective for nauseating and griping effects of medicines.

Dill: Carminative, stomachic, antipyretic.

Cinnamon: Astringent, diuretic, carminative, aphrodisiac, deodorant, expectorant, febrifuge, stomachic.

Cassia: Astringent, stimulant, carminative, germicidal and for checking nausea and vomiting.

Curry leaf: Astringent, anthelmintic, febrifuge, stomachic, appetizing, carminative, constipating, anti-inflammatory, antiseptic.

TABLE 1.1 Underutilized Spice Crops.

Sl. No.	Common name	Botanical name	Family	Habit	Edible part(s)
1	Ajowan/Bishop's weed	*Trachyspermum ammi* L.	Apiaceae	Herb	Fruit
2	Aniseed	*Pimpinella anisum* L.	Apiaceae	Herb	Fruit
3	Asafoetida	*Ferula asafoetida* L.	Apiaceae	Herb	Dried latex (gum oleoresin) exuded from the rhizome or tap root
4	Balm	*Melissa officinalis* L.	Lamiaceae	Herb	Leaf
5	Basil	*Ocimum basilicum* L.	Lamiaceae	Herb	Leaf
6	Bay Leaf	*Laurus nobilis* L.	Lauraceae	Tree	Leaf
7	Black cumin	*Nigella sativa* L.	Ranunculaceae	Herb	Fruit
8	Black mustard	*Brassica nigra* Koch.	Brassicaceae	Herb	Seed
9	Cambodge	*Garcinia cambodgia* Desr.	Clusiaceae	Tree	Pericarp lobes
10	Caper	*Capparis spinosa* L.	Capparaceae	Shrub	Unopened flower bud
11	Caraway	*Carum carvi* L.	Apiaceae	Herb	Fruit
12	Cardamom (large)	*Amomum subulatum* Roxb.	Zingiberaceae	Shrub	Seed
13	Cassia	*Cinnamomum cassia* Blume	Lauraceae	Tree	Bark
14	Celery	*Apium graveolens* L.	Apiaceae		Seed
15	Cinnamon	*Cinnamomum zeylanicum* Breyn	Lauraceae	Tree	Bark
16	Common juniper	*Juniperus communis* L.	Cupressaceae	Tree	Fruit
17	Curry leaf	*Bergera koenigii*, Syn.: *Murraya koenigii* (L) Sprengel	Rutaceae	Tree	Leaf
18	Dill	*Anethum graveolens* L.	Apiaceae	Herb	Fruit
19	Greater galangal	*Alpinia galangal Wild.*		Herb	Rhizome
20	Horseradish	*Armoracia rusticana* Gaertn.	Brassicaceae	Herb	Root
21	Hyssop	*Hyssopus officinalis* L.	Lamiaceae	Shrub	Leaves
22	Indian cassia	*Cinnamomum tamala* Nees and Eberum	Lauraceae	Tree	Bark, leaf

TABLE 1.1 *(Continued)*

Sl. No.	Common name	Botanical name	Family	Habit	Edible part(s)
23	Indian mustard	*Brassica juncea* Czen and Cross.	Brassicaceae	Herb	Seed
24	Kokum	*Garcinia indica* Choisy	Clusiaceae	Tree	Peel of fruit
25	Lovage	*Levisticum officinale* Koch.	Apiaceae	Herb	Fruit, root
26	Mango	*Mangifera indicia* L.	Anacardiaceae	Tree	Unriped fruit
27	Mint	*Mentha piperita* L.	Lamiaceae	Herb	Leaves
28	Nutmeg and mace	*Myristica fragrans* Houtt.	Myristicaceae	Tree	Seed and aril
29	Oregano	*Origanum vulgare*	Lamiaceae	Herb	Leaf, flowering top
30	Parsley	*Petroselinum crispum* Mill.	Apiaceae	Herb	Leaf
31	Pepper long	*Piper longum* L.	Piperaceae	Shrub	Berry
32	Pomegranate	*Punica granatum* L.	Punicaceae	Tree	Seed
33	Poppy seed	*Papaver somniferum* L.	Papaveraceae	Herb	Seed
34	Rosemary	*Rosmarinus officinalis* L.	Lamiaceae	Herb	Leaves
35	Saffron	*Crocus sativus* L.	Iridaceae	Trees/ shrubs	Stamen
36	Sage	*Salvia officinalis* L.	Lamiaceae	Shrub	Leaf
37	Star anise	*Illicium verum* Hook.	Illiciaceae	Tree	Fruit
38	Summer savoury	*Satureja hortensis* L.	Lamiaceae	Herb	Stem, leaf, flowering top
39	Sweet flag	*Acorus calamus* L.	Araceae	Herb	Roots
40	Sweet marjoram	*Marjorana hortensis* Moench.	Lamiaceae/ Labiatae	Herb	Leaf, flowering top
41	Tamarind	*Tamarindus indica* L.	Fabaceae	Tree	Pulp, seed
42	Thyme	*Thymus vulgaris* L.	Lamiaceae	Herb	Leaves, flowers
43	Vanilla	*Vanilla planifolia* Andr.	Orchidaceae	Vine	Pod

Kokum: Anthelmintic, cardiotonic, astringent, emollient, useful in piles, dysentery, heart complaints.

Mint: Stimulant, stomachic, carminative, antiseptic, digestive, antispasmodic, contraceptive, used in vomiting, skin diseases, amenorrhoea, dental caries.

Parsley: Stimulant, diuretic, carminative, antipyretic, anti-inflammatory, emetic, aphrodisiac, alexipharmic, refrigerant.

Saffron: Stimulant, tonic, stomachic, aphrodisiac, antispasmodic, emmenagogue, diuretic, laxative, used in bronchitis, fever, epilepsy, skin diseases, decolouration of skin.

Vanilla: Aphrodisiac.

Pepper long: Expectorant, thermogenic, diuretic, tonic, purgative, stomachic, digestive, emollient, antiseptic, used in bronchitis, fever, asthma.

Star anise: Astringent, carminative, deodorant, expectorant, digestive.

Sweet flag: Thermogenic, constipating, emmenagogue, intellect promoting, emetic, carminative, stomachic, expectorant, antipyretic, resusticative, tranquilizing, sedative, nervine tonic.

Horseradish: Appetizing, digestive, stomachic, laxative, anti-inflammatory, refreshing, antibacterial.

Sweet basil: Stomachic, anthelmintic, diaphoretic, expectorant, antipyretic, carminative, stimulant, diuretic, demulcent, bay leaves (laurel): stimulant in sprains, narcotic and in veterinary medicine.

Poppy seeds: Expectorant, sudorific, sedative, nervine tonic, constipating, aphrodisiac, used in internal haemorrhages, diarrhoea, dysentery.

Thyme: Antispasmodic, carminative, emmenagogue, anthelmintic, spasmodic, laxative, stomachic, tonic.

Rosemary: Astringent, nervine tonic, stomachic, antibacterial, protistocidal, rubefacient, used in headaches and hardy menstruation.

Allspice: Stimulant, digestive, carminative.

Bishop's weed (ajowan): Digestive, antispasmodic, stimulant, carminative, expectorant.

Caper: Diuretic, aspirant, expectorant, emmenagogue, tonic and used in scurvy, rheumatism, gout.

Cambodge: Astringent, digestive, thermogenic, constipating, used in haemorrhoids, diarrhoea, and to control obesity.

Clove: Refrigerant, ophthalmic, digestive, carminative, stomachic, stimulant, antispasmodic, antibacterial, expectorant, rubefacient, aphrodisiac, appetizer, emollient.

Coriander: Carminative, diuretic, tonic, stimulant, stomachic, refrigerant, aphrodisiac, analgesic, anti-inflammatory.

Cumin: Digestive, carminative, astringent, anti-inflammatory, constipating, diuretic, revulsive, galactagogue, uterine and nerve stimulant.

Fennel: Stimulant, carminative, stomachic, emmenagogue, refrigerant, cardiac stimulant, anti-emetic, aphrodisiac, anthelmintic.

Fenugreek: Carminative, tonic, aphrodisiac, emollient, antibacterial, used in fever, anorexia, colonitis.

Greater galangal: Carminative, expectorant, digestive, febrifuge, stimulant, depurative, used in skin diseases, rheumatism, asthma, wounds, fever, haemorrhoids.

Hyssop: Stimulant, carminative, used in nervous disorders, toothache, pulmonary and uterine troubles.

Juniper berry: Carminative, stimulant, diuretic, useful in dropsy, leucorrhoea, urino-genital disorders.

Marjoram: Carminative, expectorant, tonic, astringent.

Oregano: Stimulant, carminative, stomachic, diuretic.

Pomegranate: Astringent, cooling, tonic, aphrodisiac, laxative, diuretic, cardiotonic, used in dysentery, diarrhoea, vomiting.

Sage: Mild tonic, astringent, carminative, deodorant, insecticidal, antipyretic, used in mouthwash, gargles.

Savoury: Antispasmodic, astringent, carminative, laxative, diuretic, stomachic, vermifuge.

Tamarind: Retrigerant, digestive, carminative, laxative, febrifuge, ophthalmic useful in gastropathy, datura poisoning, alcoholic intoxication, scabies, constipation.

Tarragon: Aperient, stomachic, stimulant, febrifuge.

Tejpat: Carminative, used in colic, diarrhoea.

Important constituents of some spices with the chemical structures have been given below (Table 1.2):

TABLE 1.2 Important Constituents of Some Spices with the Chemical Structures.

Spices	Important constituents	Structures		
Celery	Myrcene, limonene, α-pinene	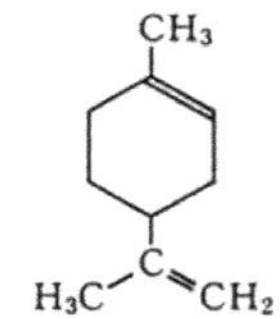 limonene	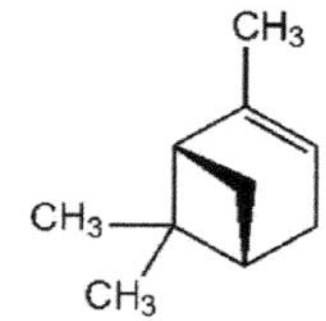 α- pinene	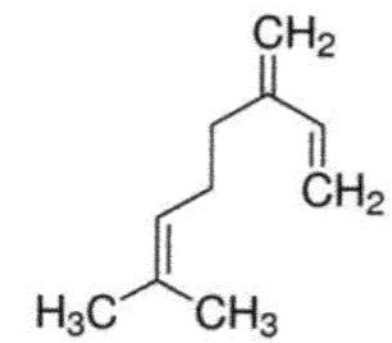 myrcene
Ajowan	Thymol, γ-terpinene	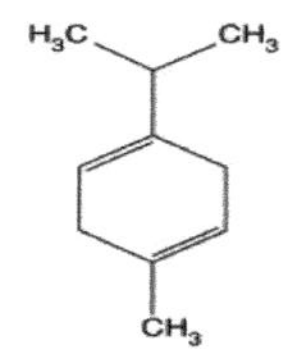 γ-Terpinene	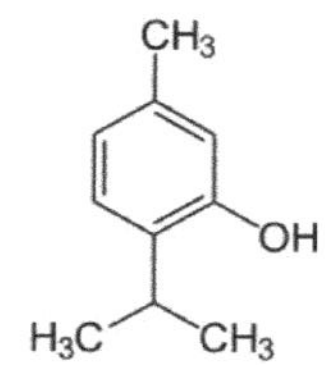 thymol	
Dill	d-carvone	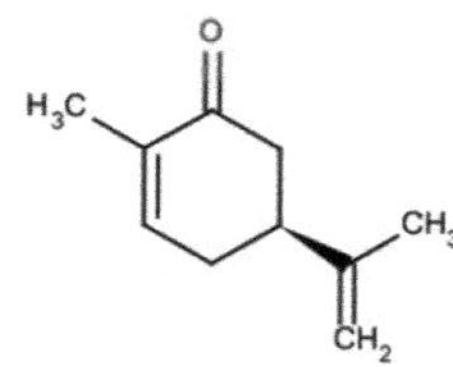	d-carvone	

TABLE 1.2 *(Continued)*

Spices	Important constituents	Structures
Anise	(E)-anethole, methyl chavicol	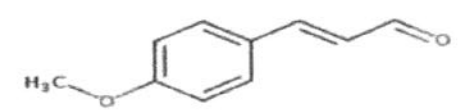 (E)-anethole 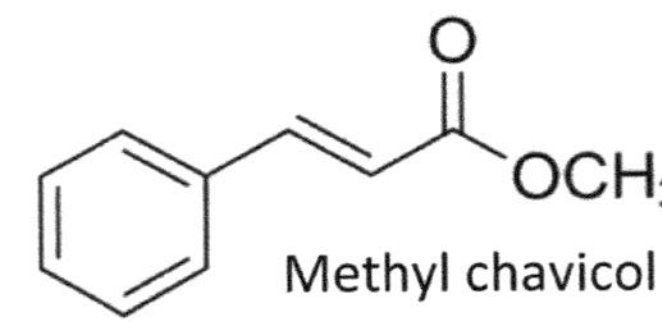 Methyl chavicol
Cassia	Eugenol, benzyl benzoate, cinnamaldehyde	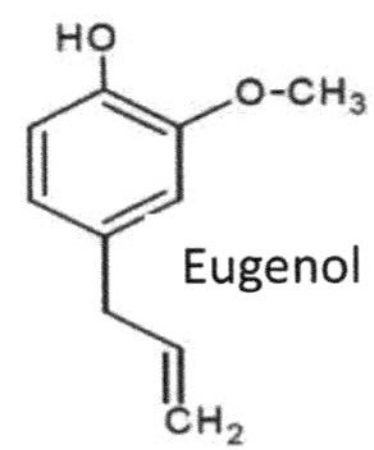 Eugenol 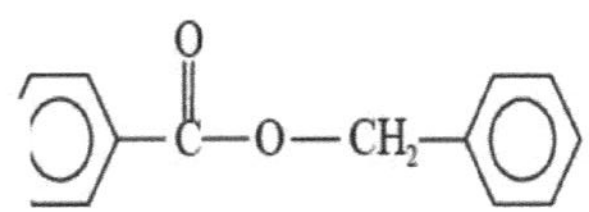 Benzyl Benzoate 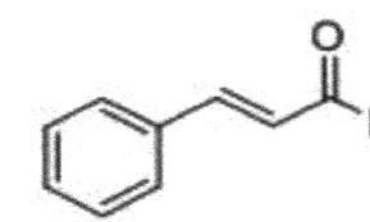 cinnamaldehyde
Kokum	α-humulene, valencene, β-caryophyllene	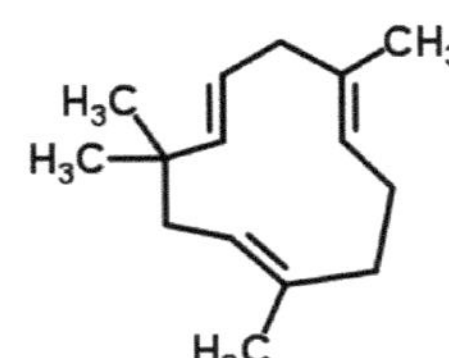 α-humelene 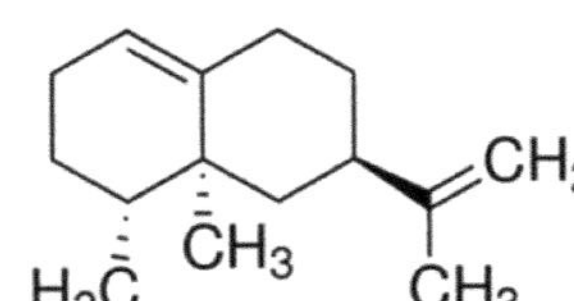 Valencene 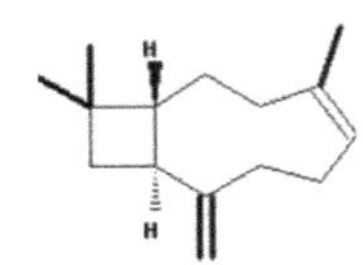 β-caryophyllene

TABLE 1.2 *(Continued)*

Spices	Important constituents	Structures
Parsley	1,3,8-p-menthatriene, β-phellandrene, myristicin	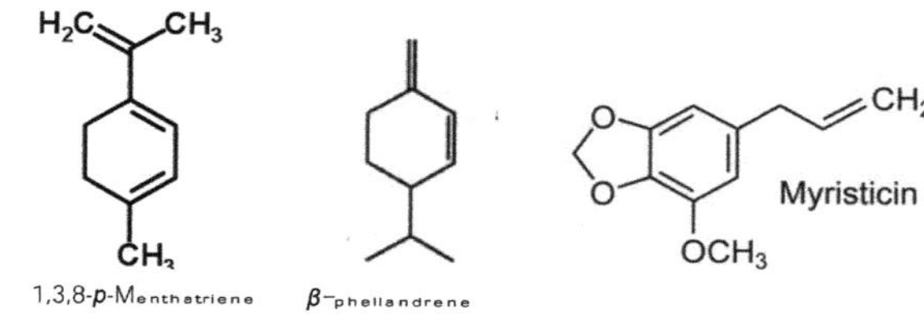
Vanilla	Vanillin,	
Long pepper	Piperine, β-caryophyllene	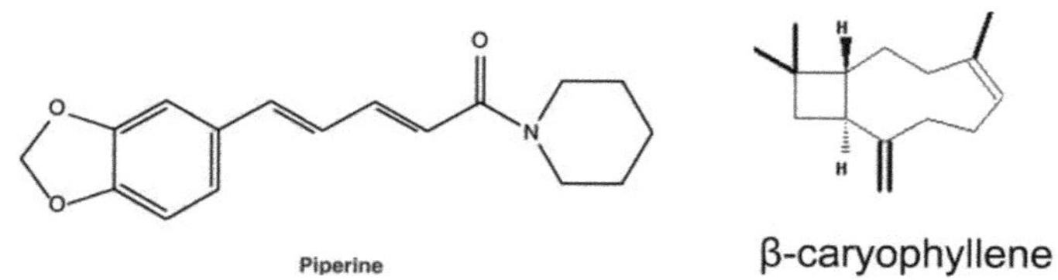
Sweet flag	Asarone, α-pinene, β-asarone, eugenol	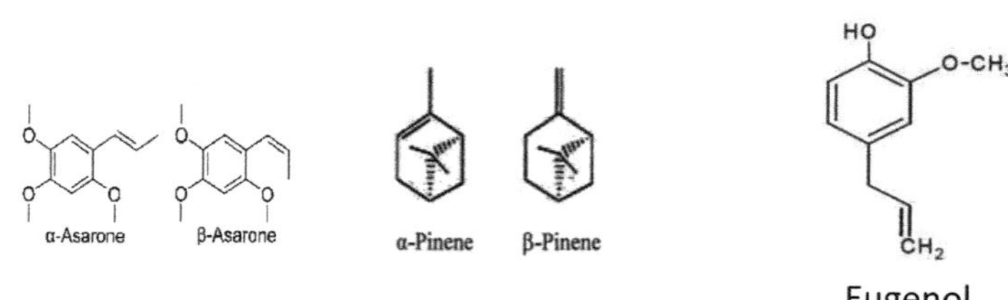

TABLE 1.2 *(Continued)*

Spices	Important constituents	Structures
Asafoetida	Ferulic acid, umbelliferone, 2-butyl propenyl disulphide	2-butyl propenyl 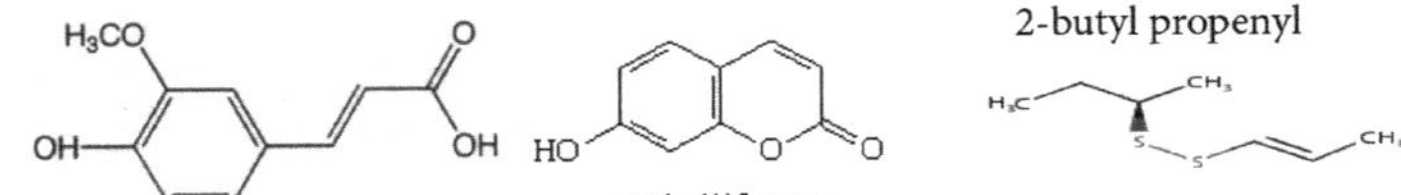 Ferulic Acid; umbelliferone
Bay leaf	1,8-cineole, linalool, α-terpinyl acetate, methyl eugenol	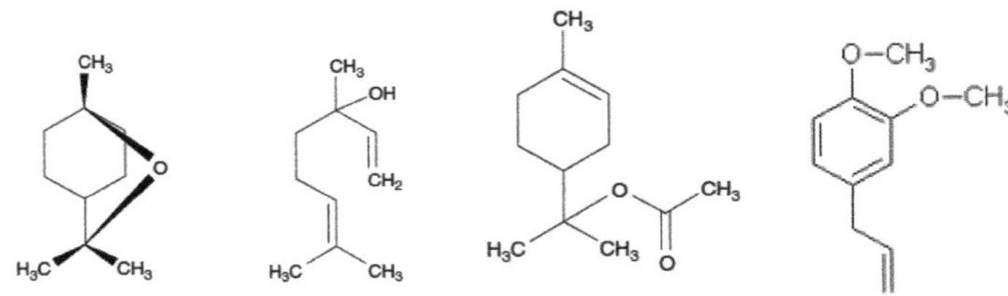 1,8-Cineole; Linalool; α-Terpinyl acetate; Methyl Eugenol
Basil	Methyl chavicol, linalool, methyl eugenol	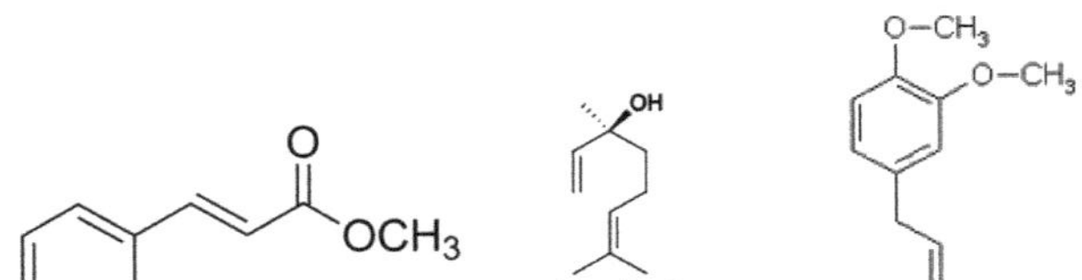 Methyl chavicol; Linalool; Methyl Eugenol
Rosemary	Verbenone, 1,8-cineole, camphor, linanool	 verbenone; 1,8-cineole; camphor; Linalool

TABLE 1.2 *(Continued)*

Spices	Important constituents	Structures
Thyme	Thymol, γ-terpinene	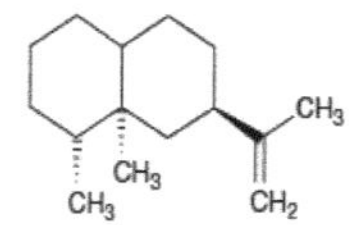 Valencene 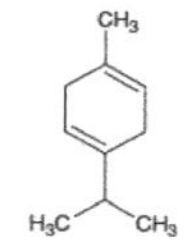 γ-Terpenene 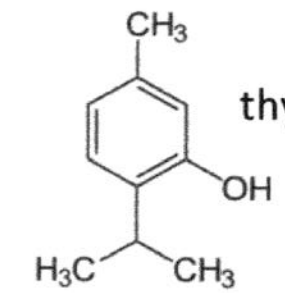 thymol
Sage	Thujone, 1,8-cineole, camphor	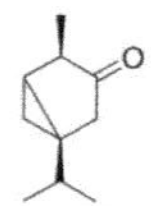 α-thujone 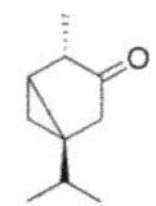 β-thujone 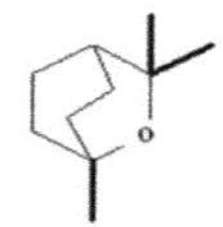 1,8-cineole 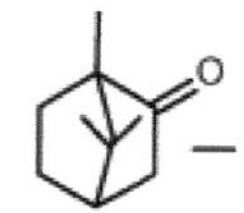 camphor
Oregano	Carvacrol, thymol	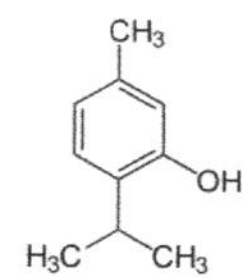 thymol 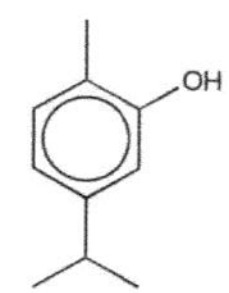 Carvacrol
Marjoram	e- and t-sabinene hydrates, terpinen-4-ol	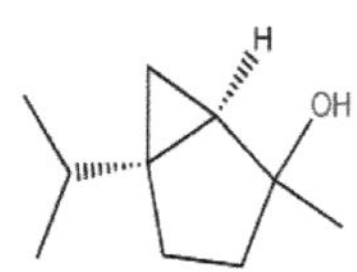 sabinene hydrate 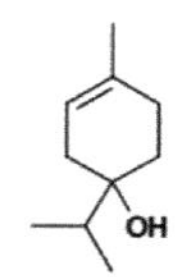 Terpinen-4-ol

TABLE 1.2 *(Continued)*

Spices	Important constituents	Structures
Savoury	Carvacrol	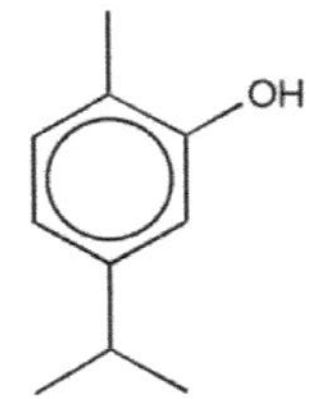 Carvacrol
Tarragon	Methyl chavicol, anethole	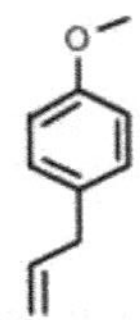 Methyl Chavicol 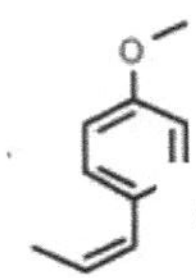 Cis-anethole 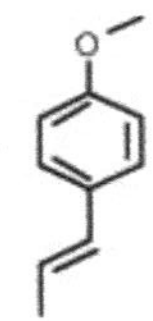 Trans-anethole
Allspice	Eugenol, β-caryophyllene	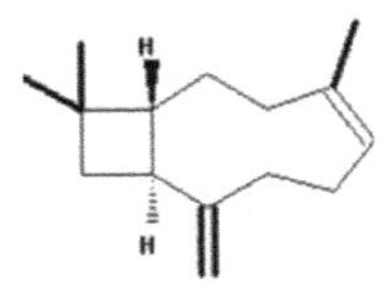 β-caryophyllene 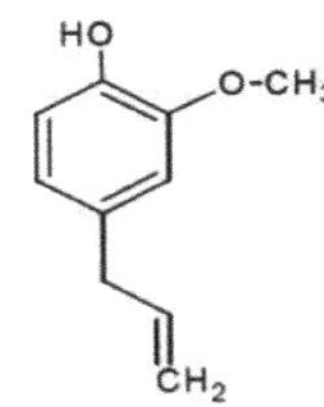 eugenol

TABLE 1.2 *(Continued)*

Spices	Important constituents	Structures
Cinnamon	Cinnamaldehyde, eugenol	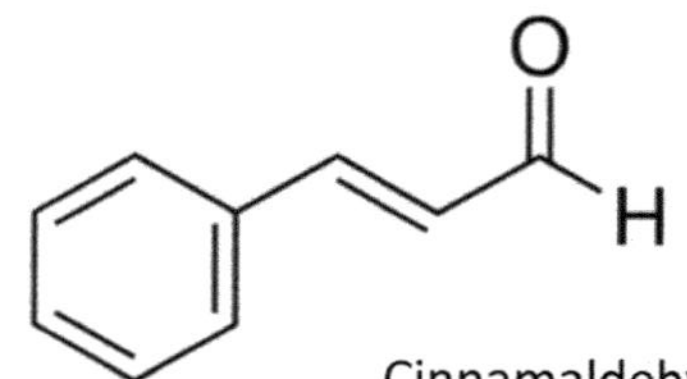 Cinnamaldehyde 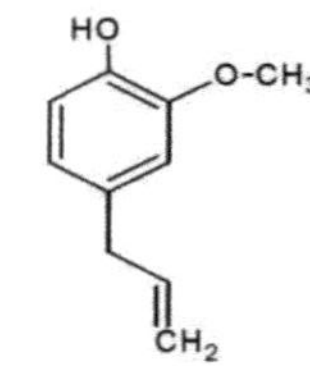 eugenol
Nutmeg	Sabinine, α-pinene, myristicin	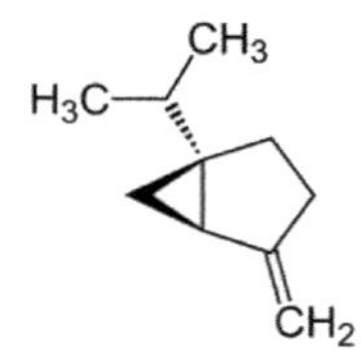 Sabinine 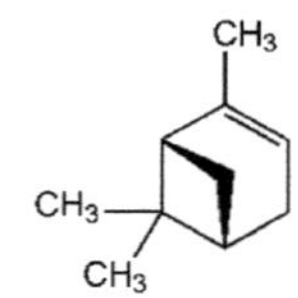 α- pinene 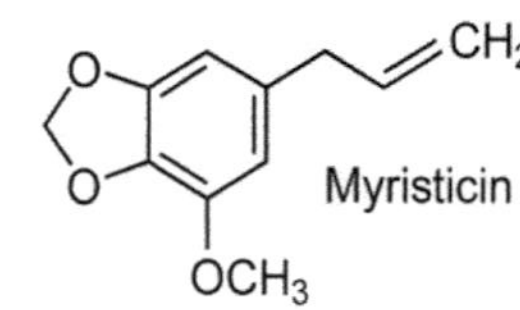 Myristicin
Saffron	Safranol	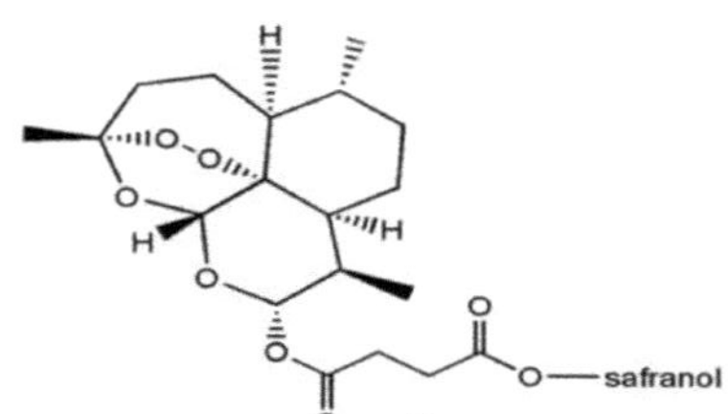

1.6 VALUE OF UNDERUTILIZED SPICES: NUTRITION AND BEYOND

1.6.1 CALORIFIC VALUE OF SPICES

All spices, like other food items, have their intrinsic energy and expressed in terms of calorific value (Table 1.3).

TABLE 1.3 Calorific Value of Spices.

Name of the spice	Calorific value (energy) (kilocalories; kcal)
Asafoetida	297
Cardamom	229
Chillies (green)	29
Cloves (dry)	286
Cloves (fresh)	159
Coriander seeds	288
Cumin seeds	356
Curry leaves	108
Fenugreek seeds	333
Mace	437
Mint	48
Nutmeg fruit	472
Nutmeg rind	52
Bishop's weed	363
Parsley	87
Pepper black (dry)	304
Pepper green	98
Poppy seeds	408
Tamarind pulp	283

Even as a food adjunct, the nutritive values of spices cannot be ignored. Some spices are rich in protein and fat (31.5 and 42.6%, respectively, as in mustard), some spices are rich in ash (16.7% as in basil leaves) and some other even rich in dietary fibres (43.3% as in red pepper). Chemical composition of volatiles on qualitative and quantitative aspects was evaluated from major, minor, seed and leafy spices using modern instrumental techniques such as gas chromatography–mass spectrometry (GC–MS). The major identified compounds included terpenoids, aromatic compounds, alcohols, hydrocarbons, aldehydes, ketones, esters, ethers, acids, and so forth (Rao, 2008). 'Charak Samhita', one of the greatest Indian medical treatises has many references on the use of spices as medicine that are mentioned in subsequent text.

1.6.1.1 DIGESTION AND ABSORPTION

Many spices such as mint, ajowan, cumin, fennel, coriander, horseradish and garlic are used as ingredients of commercial digestive stimulants as well as of home remedies for digestive problems such as flatulence, indigestion and intestinal disorders. Disregarding the previous empirical reports on the digestive stimulant action of spices, the beneficial attributes have been authenticated in recent years through exhaustive animal studies (Platel and Srinivasan, 2004). For its unique ability to overcome indigestion and colonic spasms by reducing the gastrocolic reflex, peppermint is taken at the end of a meal (Spirling and Daniels, 2001). The Ayurvedic recommendations of fennel (for infants and young children) and ginger (to be taken with salt) are also there to improve indigestion. Spices such as cloves and cardamom are known to be wrapped in betel leaves with betel nuts and chewed after meals to increase salivation and aid digestion.

1.6.1.2 REDUCING SUGAR LEVEL

A unique human disorder of carbohydrate metabolism, causing great concern in recent times, is known as diabetes. It is associated with high level of blood glucose and glycosuria resulting from dysfunction of pancreatic β-cells and insulin resistance. In advance stages of diabetes, even the metabolism of protein and lipid is altered. Drug, diet, change in lifestyles and spices therapies are the major approaches used for treatment and control of diabetes. Cinnamon is effective in reducing glucose level in type 2 diabetic individuals. The effective ingredient in cinnamon for such hypoglycaemic function has been identified as methyl hydroxychalcone polymers (Khan and Safdar, 2003).

A few spices were also found with great anti-diabetic potential (Srinivasan, 2005). Supplementation of fenugreek seed can improve the glycaemic status and decrease insulin resistance in mild type 2 diabetic patients (Gupta et al., 2001). The possible protective effects of *Murraya koenigii* leaves extract against β-cell damage and antioxidant defence systems of plasma and pancreas in streptozotocin-induced diabetes in rats was evaluated suggesting a therapeutic protective nature of *M. koenigii* (Arulselvan and Subramanian, 2007). Pimentol from allspice and biflorin from clove markedly inhibited the formation of malondialdehyde through inhibition of advanced glycation end products and pentosidine which are biomarkers of diabetes mellitus (Oya et al., 1997; Du Toit et al., 2001). Cinnamon contains compounds that aid in the fast, efficient metabolism of sugars, helping to maintain balanced blood sugar and insulin levels (Khan et al., 2003).

1.6.1.3 AS ANTI-CARCINOGENIC

Systematic research during the past few decades with experimental models, limited human studies, and some epidemiological data has indicated that cancer is preventable by dietary intervention. A few naturally occurring chemopreventive substances capable of inhibiting, retarding, or reversing the multistage carcinogenesis were identified. A wide range of phenolic substances available from many spices have been reported to possess substantial anti-carcinogenic and anti-mutagenic activities. Spices such as celery, parsley, cumin, dill, fennel, and coriander are having anti-carcinogenic potential basically due to the bioactive substances called phthalides which are believed to increase the glutathione S-transferase level (Wildman, 2000).

1.6.1.4 AS APHRODISIAC

The aroma of spices ensures a stimulating environment for the invigoration of romantic encounters. The Arabs had their 'Perfumed Garden' and the Hindus their 'Kama Sutra', each of which extolled flavoured spices such as nutmeg, clove, galangal, cardamom and ginger. The Romans came to favour cinnamon and pepper and the Chinese were most impressed with ginger.

1.7 COMPOSITION OF SPECIFIC UNDERUTILIZED SPICES

1.7.1 CARDAMOM (LARGE)

The large cardamom fruit comprises approximately 70% seeds and 30% skin (Pruthi, 1993). The following are the chemical composition which, however, varies with variety, growing location and age of the crop and produce (Table 1.4).

TABLE 1.4 Proximate Composition of Large Cardamom.

Composition	Percentage present in fruits
Moisture	8.49%
Volatile oil	2.8% v/w
Protein	6.0%
Total ether extract	5.31%
Non-volatile ether extract	2.32%
Volatile ether extract	3.0%
Crude fibre	22.0%
Starch	43.21%
Alcohol extract	7.02%
Total ash	4.01%
Water-soluble ash	0.90%
Ash insoluble in acid	0.42%

Large cardamom oil is depicted by flat cineol odour, harsh in aroma of low-quality flavour against the warm spicy, aromatic odour of cardamom. The large cardamom oil contains α-terpinyl acetate in low quantity (traces to 5%) whereas 1,8-cineole (65–80%) is very rich. The monoterpene hydrocarbon (5–17%) contains limonene, sabinene, terpinenes and pinenes. The oil comprises approximately 5–7% terpineols. The high cineole and low terpenyl acetate probably account for the very harsh aroma of this spice (Pruthi, 1993).

1.7.2 CELERY

The celery seeds have following chemical and biochemical constituents.

The seed on steam distillation yields oil-containing limonene (80%), as a major constituent. Other than that, it contains α-p-dimethyl styrene, N-pertyl benzene, caryophyllene, α-selinene, N-butyl phthalide and sedanenolide along with sabinene, β-elemene, trans-1,2-epoxy limonene, linalool, isovaleric acid, cis-dihydrocarvone, trans-dihydrocarvone, terpinene-4-ol, 1-cis-p-menth-2,8-diene-1-ol, trans-p-menth-2,8-diene-1-ol, α-terpineol, carvone, trans-8-diene-1-ol, perialdehyde and thymol. The seeds are also rich in vitamin B (Yan et al., 1998). Celery contains 1.5–3% volatile oil, primarily containing about 60–70% d-limonene and 10–20% b-selinene. The oleoresin extracted from celery contains 12–16% volatile oil. The seeds contain apiin, apigenin, caffeic acid and chlorogenic acid. Several other substances like rutaractin, apiumetin, and so forth are also present. The seeds are also rich in vitamin B. Fresh celery leaves and stalks contains moisture: 81.30 and 93.5%; protein: 6.0 and 0.8%; fat: 0.6 and 0.1%; fibre: 1.4 and 1.2%; carbohydrates: 8.6 and 3.5%; mineral matter: 2.1 and 0.9%; calcium: 0.23 and 0.3%; phosphorus: 0.14 and 0.4%; iron: 0.06 and 0.05%; vitamin A: 5800 and 7500 IU; vitamin B traces, vitamin C: 62 and 6 mg/100 g; calorific value; 64 and 18 cal/100 g, respectively. For celery seed, the average yields of concrete and essential oil were 0.71 and 0.16% (trichlorotrifluoroethane), 13.23 and 0.86% (ethanol), and 2.0 and 0.13% (dichloromethane), respectively. Twenty-seven compounds were identified. Isopulegone, α-ionone and epoxycaryophyllene may not have previously been reported in celery seed oil (Cu et al., 1990).

1.7.3 AJOWAN

Seeds contain medicinal values especially for curing indigestion, stomach pain and elements concerning digestive system. It is also used in cholera, diarrhoea, gastric and urinary trouble. Ajowan contains 2.5–4.0%volatile oil which is brownish-yellow in colour used in many Ayurvedic medicines and industries of which 'thymol' is the main constituent. Thymol (39.1%) was found as a major component along with p-cymene (30.8%), γ-terpinene (23.2%), β-pinene (1.7%)

and terpinene-4-ol (0.8%). In addition to volatile oil, it also contains moisture 8.9%, protein 15.4%, fat 18.1%, fibre 11.9%, carbohydrates 38.6%and minerals 7.1%.

1.7.4 ANISEED

Aniseed contains 1.5–6.0% of a volatile oil consisting primarily of trans-anethole and also as much as 8–11% of lipids rich in fatty acids, such as palmitic and oleic acids, as well as approximately 4% of carbohydrates and 18% of protein (Besharati-Seidani et al., 2005). Other studies have demonstrated the presence of eugenol trans-anethole, methyl chavicol, anisaldehyde, estragole, coumarins, scopoletin, umbelliferone, estrols, terpene hydrocarbons, polyenes and polyacetylenes as the major compounds of the essential oil of anise seed. The study of the essential oil of *Pimpinella anisum* L. fruits by GC and GC-MS showed the presence of trans-anethole (93.9%) and estragole (2.4%). Other compounds that were found with a concentration higher than 0.06% were (E)-methyl eugenol, α-cuparene, α-himachalene, β-bisabolene, p-anisaldehyde and cis-anethole (Ozcan and Chalchat, 2006). In another study for determination of the composition of essential oil of *P. anisum* L. fruits obtained from different geographical areas of Europe, in addition to the major components [trans-anethole (76.9–93.7%) and γ-himachalene (0.4–8.2%)], some other compounds such as trans-pseudoisoeugenyl 2-methyl butyrate, p-anisaldehyde and methyl chavicol were also identified in essential oil (Orav et al., 2008). The study of components of the whole plants and the seeds of *P. anisum* from Alberta showed that the major oil constituent (trans-anethole) was 57.4% of whole plant and 75.2% of seed oil. The other constituents of plant oil, present in amounts of 1–5%, were cis-anethole, carvone, β-caryophyllene, dihydrocarvyl acetate, estragole and limonene (Embong et al., 1997).

1.7.5 DILL

Dried ripe fruits contain volatile oil (3–4%) and fixed oil. The volatile oil is composed of anethole, phellandrene, d-limonene and carveol. Traditionally, it is used as an ingredient for flavouring curries. Seeds are stomachic, antipyretic, carminative and anthelmintic. It also cures cough and ulcers. Fruits are used as a condiment. It is an excellent remedy for children complaints such as flatulence and indigestion. It is mostly given in the form of dill water. Leaves are used as stimulating poultice. The main components of the vegetative herb essential oils were α-phellandrene (46.33%), limonene (13.72%), β-phellandrene (11.01%) and p-cymene (17.88%). p-cymene (33.42%), carvone (13.10%) and dill ether (19.63%) were the main components of the flowering herb, whereas carvone (62.48%), dillapiole (19.51%) and limonene (14.61%) were identified as the major compounds in seed essential oil (Hussein et al., 2015). Ruangamnart et al. (2015) reported that the main constituents of dill oils examined by GC–MS were dillapiole (19.98–48.9%), D-carvone (18.05–28.02%) and D-limonene (26.96–44.61%), and β-pinene (0–0.79%), β-myrcene (0.16–0.21%), decane (0.44–0.49%), 1,5,8-p-menthatriene (0.19–0.27%), undecane (0.34–0.38%), naphthalene (1.63–2.11%), cis-dihydrocarvone (0.38–0.95%), trans-dihydrocarvone (1.49–1.57%) and myristicin (0.67–1.41%) were the minor components.

Anethum graveolens contained essential oils, fatty oil, moisture (8.39%), proteins (15.68%), carbohydrates (36%), fibre (14.80%), ash (9.8%) and mineral elements such as calcium, potassium, magnesium, phosphorous, sodium, vitamin A and niacin. Fruits of *A. graveolens* contain 1–4% essential oil comprising of major compounds: carvone (30–60%), limonene (33%), α-phellandrene (20.61%), including pinene, diterpene, dihydrocarvone, cineole, myrcene, paramyrcene, isomyristicin, myristicin, myristin, apiol and dillapiole. *A. graveolens* essential oil also contained furanocoumarin, 5-(4-hydroxy-3-methyl-2-butenyloxy)-6,7-furocoumarin, oxypeucedanin, oxypeucedanin hydrate and falcarindiol (Al-Snafi, 2014; Ishikawa, 2002, Radulescu, 2010,; Stavri and Gibbons, 2005).

1.7.6 CINNAMON

Essential oils of cinnamon bark and leaf are widely used in food flavours, cosmetics and pharmaceuticals. It consists of a variety of resinous compounds, including cinnamaldehyde, cinnamate, cinnamic acid and numerous essential oils. However, the proximate composition is given below (Table 1.5):

TABLE 1.5 Proximate Composition of Cinnamon.

Moisture	9.9%
Protein	4.6%
Fat	2.2%
Fibre	20.3%
Carbohydrate	59.5%
Total ash	3.5%
Calcium	1.6%
Phosphorus	0.05%
Iron	0.004%
Sodium	0.01%
Potassium	0.4%
Vitamin B_1 (mg/100 g)	0.14
Vitamin B_2 (mg/100 g)	0.21
Niacin	1.9
Vitamin C	39.8
Vitamin A	175 IU/100 g
Calorific value (food energy)	355 cal/100 g

1.7.7 CASSIA CINNAMON

Cassia from China is less aromatic than that from Vietnam and Indonesia. Cassia from all the three countries possesses a sweet, aromatic and pungent flavour. Vietnamese or Saigon cassia is particularly highly esteemed. Volatile oil content of cassia varies from 1 to 2%, the principal component of which is cinnamic aldehyde (85–90%). The leaf oil from this species also contains a high per cent of cinnamic aldehyde. The oleoresin of cassia usually contains 25–40% volatile oil. Eugenol is found only in traces. Small amount of coumarin is also there. Trace components of cassia oil are benzoic acid, cinnamic acid, salicylic acid and the corresponding esters and aldehydes. Cassia bark contains significantly more slime (11%) than Ceylon cinnamon bark.

1.7.8 CURRY LEAF

Curry leaves are a popular leaf spice used in very small quantities for their distinct aroma due to the presence of volatile oil and their ability to improve digestion. Primary alkaloids found in the curry tree leaves, stems, and seeds are: koenimbine, mahanimbine, girinimbine, mahanine, undecalactone, isomahanine, 2-methoxy-3-methyl-carbazole, and so forth. Table 1.6 shows the proximate composition of curry leaf followed by comparative nutrient contents of fresh and dehydrated curry leaves.

TABLE 1.6 Proximate Composition of Curry Leaf.

Moisture	66.3%
Protein	6.1% fat (ether extract) 1.0%
Carbohydrate	18.7%
Fibre	6.4%
Mineral matter	4.2%
Calcium	810 mg/100 g of edible portion
Phosphorus	600 mg/100 g of edible portion
Iron	3.1 mg/100 g of edible portion
Carotene (as vitamin A)	12,600 IU/100 g
Nicotinic acid	2.3 mg/100 g
Vitamin C	4 mg/100 g
Thiamine and riboflavin	Absent

1.7.9 HORSERADISH

Horseradish is a root vegetable used as a spice. The intact horseradish root has hardly any aroma. When cut or grated, however, enzymes from the now-broken plant cells break down sinigrin (a glucosinolate) to produce allyl isothiocyanate, which irritates the mucous membranes of the sinuses and eyes. The proximate composition of horseradish is given below (Table 1.7):

TABLE 1.7 Proximate Composition of Horseradish.

Water	73.4%
Protein	3.2%
Fat	0.2%
Ash	1.8%
Total carbohydrates	21.4%
Fibre	2.4%

1.7.10 KOKUM

The principle antioxidant substance of *Garcinia indica* and other species is garcinol, also called as camboginol, which is a tri-isoprenylated chalcone. This compound is extracted from the dried fruit rind of the plant. Garcinia is a rich source of active compounds, including garcinol, xanthochymol, isoxanthochymol and hydroxycitric acid. These are flavonoids, benzophenones, xanthones, lactones and phenolic acids (Selvi et al., 2003). The fruits contain citric acid, acetic acid, malic acid, ascorbic acid, hydroxycitric acid and garcinol. The major constituent of kokum rind is garcinol, a polyisoprenylated benzophenone, isogarcinol and camboginol. Garcim-1, garcim-2 and cambogin are the chief oxidative products of garcinol, along with isogarcinol, gambogic acid, mangostin, clusianone, macurin, oblongifolin (A, B, C), and guttiferone (I, J, K, M, N). The rind of ripe kokum fruits consists of hydroxyacetic acid and hydroxycitric acid. It also contains 2.4% pigment as a mixture of two anthocyanins, namely cyanidin-3-sambubioside and cyanidin-3-glucoside in the ratio 4:1. Studies have shown that the fresh rind of kokum contains 80% moisture, 2% protein, 2.8% tannin, 5% pectin, 14% crude fibre, 4.1% total sugars, 1.4% fat, 2.4% pigment, 22% hydroxycitric acid and 0.06% ascorbic acid. Kokum leaves are reported to contain L-leucine, 75% moisture, protein 2.3, fat 0.5, fibre 1.24, carbohydrates 17.2 g, iron 15.14, calcium 250, ascorbic acid 10 and oxalic acid 18.10 mg/100 g (Krishnamurthy et al., 1982). Hydroxycitric acid lactone and citric acid are present in leaves and rinds in minor quantities (Jayaprakasha and Sakariah, 2002). Kokum seeds are rich in glycerides of stearic acid (55%), oleic acid (40%), palmitic acid (3%), linoleic acid (1.5%), hydroxyl capric acid (10%) and myristic acid (0.5%).

1.7.11 MINT

Leaf essential oil content of mint usually ranges from 0.5 to 4%. The major essential oil constituents are 1,8-cineole, carvone, limonene, linalool, linalyl acetate, menthol, menthone, menthyl acetate and piperitenone oxide. Some genotypes were found to have essential oil with a high content of a particular constituent such as piperitone oxide, carvone and linalool.

1.7.11.1 JAPANESE MINT (MENTHA ARVENSIS)

The Japanese mint forms a primary source of menthol. The fresh leaves contain 4–6.0% oil. The main constituents of the oil are menthol (65–75%), menthone (7–10%), menthyl acetate (12–15%) and terpenes (pipene, limonene and comphene).

1.7.11.2 PEPPERMINT (MENTHA PIPERITA)

The fresh herb contains essential oils ranging from 0.4 to 0.6%. However, the menthol content is lower in peppermint oil and varies between 35 and 50%. The other constituents are menthyl acetate (14–15%), menthone (925%), menthoufuran and terpenes like pinene and limonene.

1.7.11.3 BERGAMOT MINT (MENTHA CITRATE)

Linalool and linalyl acetate are the main constituents of bergamot mint oil. The oil is used in the preparations of scents, soaps, aftershave lotions and colognes.

1.7.11.4 SPEARMINT (MENTHA SPICATA)

Carvone (57.71%) is the main constituent of spearmint oil besides phellandrene, limonene, L-pinene and cinelole. The oil is used mostly as a flavouring in toothpastes and as a food flavouring in pickles and spices, chewing gum and confectionery, and soaps and sauces.

1.7.12 LONG PEPPER

The fruit contains a large number of alkaloids and related compounds, the most abundant of which is piperine, followed by methyl piperine, pipernonaline, piperettine, asarinine, pellitorine, piperundecalidine, piperlongumine, piperlonguminine, retrofractamide A, pergumidiene, brachystamide B, a dimer of desmethoxypiplartine, N-isobutyl decadienamide, brachyamide A, brachystine, pipercide, piperderidine, longamide, dehydropipernonalinepiperidine and tetrahydropiperine. Piperine, piperlongumine, tetrahydropiperlongumine, trimethoxycinnamoyl-piperidine, and piperlonguminine have been found in the root. Newly identified chemical constituents are 1-(3′,4′-methylenedioxyphenyl)-1E-tetradecene, 3-(3′,4′-methylenedioxophenyl)-propenal, piperoic acid, 3′,4′-di-hydroxy-biabola-1,10-diene, eudesm-4 (15)-ene-1-β,6-α-diol, 7-epi-eudesm-4(15)-ene-1-β,6-β-diol, guineesine, and 2E,4E-dienamide, (2E,4E,8E)-N-isobutylhenicosa-2,4,8-trienamide (Suresh Kumar et al., 2011; Kirtikar and Basu, 1980; Rastogi and Malhotra, 1993).

1.7.13 POPPY SEED

Poppy seeds contain many plant-derived chemical compounds that found to have antioxidant, disease-preventing and health-promoting properties. The proximate analysis of poppy seeds showed protein, moisture, ash, crude fibre and carbohydrates. It contains good levels of minerals such as iron, copper, calcium, potassium, manganese, zinc and magnesium. The outer husk is a good source of dietary fibre. Seeds are especially rich in oleic and linoleic acids. Linoleic acid is the major unsaturated fatty acid while palmitic acid is the main saturated one. Dried poppy seeds contain very small levels of opium alkaloids such as morphine, thebaine, codeine, papaverine, and so forth. Poppy seed oil also contains α-, β- and δ-tocopherols. Among the water-soluble vitamins, pantothenic acid is found at the highest level followed by niacin and thiamine. Volatile components of poppy seed includes 2-methylbutyraldehyde, hexanal, 2-heptanone, 2,5-dimethylpyrazine, 2-pentylfuran, 2-ethyl-3-methylpyrazine, 5-ethyl-1-acetaldehyde-1-cyclopenten-2-ethyl-3,6-dimethyl pyrazine. Volatile components of some other types also include 2-pentanone, toluene, hexanal, (Z)-2-heptenal, 1-octen-3-ol, 2-pentylfuran, 2-decenal, 2,4-decadienal, and so forth. Pyrazine compounds play an important role in the formation of the unique aroma of poppy seed oil (Guo et al., 2015).

1.7.14 PARSLEY

Seeds of *Petroselinum crispum* produced a high amount of essential oil. Root and leaf also possess the essential oil (Bruneton, 1999). Myristicin and apiol are the two main components of *P. crispum* essential oil which are responsible for its antioxidant activity (Zhang et al., 2006). α-pinene, sabinene, β-pinene, ρ-cymene, limonene, β-phellandrene, γ-terpinene, myristicin, elemicin, 1-allyl-2,3,4,5-tetramethoxy-benzene, carotol, eugenol and apiol were identified in *P. crispum* seed essential oil (Zhang et al., 2006; Wagner, 1996). Leaf essential oil contained β-elemene, β-caryophyllene, , γ-elemene, α-terpineol, α-pinene, α-thujene, toluene, camphene, hexanal, β-pinene, sabinene, 3-carene, m- and/or ρ-xylene, myrcene, α-phellandrene, β-phellandrene, α-terpinene, limonene, 2-pentylfuran, cis-β-ocimene, γ-terpinene, trans-β-ocimene, ρ-cymene, α-terpinolene, ρ-1,3,8-menthatriene, cis-hex-3-en-1-ol, 4-isopropenyl-1-methylbenzene, α-cubebene, benzaldehyde, α-copaene, cryptone, β-bisabolene, α-elemene, 2-(ρ-Tolyl) propan-2-ol, δ-cadinol and elemicin (Macleod et al., 1985). Analysis of volatile oil from *P. crispum* plant, callus and cell culture showed that monoterpenes were the main constituent. ρ-1,3,8-menthatriene was high abundant compound among monoterpenes followed by β-phellandrene and apiol. Moreover, aldehydes (nonanal and decanal) and also fatty acids (free and bound) were found in the volatile oil (Lopez et al., 1999). Evaluation of biochemical composition was done by Karklelienė et al. (2014) in different cultivars of parsley and reported that leaf showed a high amount of dry matter up to 19.4%, total sugar up to 3.34% and ascorbic acid up to 162.8 mg/100 g in parsley. Vokk et al. (2011) identified 34 essential oil components in parsley leaves. There were significant seasonal changes in the major constituents of parsley essential oils: myristicin and p-1,3,8-menthatriene content was higher in summer parsley (42.7 vs. 30.7% in winter parsley and 10.0 vs. 5.4%, respectively). In contrast, two other major oil components were less abundant in summer parsley: β-phellandrene (21.8 vs. 35.9% in winter parsley) and β-myrcene (4.5 vs. 8.7% in winter parsley).

1.7.15 SAFFRON

Crocus sativus L. contains four major bioactive compounds, namely crocin, crocetin, picrocrocin and safranal, all contributing to colour, taste and aroma, respectively (Melnyk et al., 2010). Crocin accounts for the yellow pigmentation from the stigmas; picrocrocin gives the rusty, bittersweet flavour; and safranal lends the earthy fragrance to the spice. Its major non-volatile components include crocin, α-crocin and carotenoids that include lycopene, zeaxanthin and both α- and β-carotenes, crocetin and picrocrocin. The major volatile components include terpene, terpene alcohol and terpene esters. Safranal is also a major volatile composite formed from picrocrocin as a result of the interaction of heat and enzymes during the drying process. Furthermore, antocyanins, flavonoids, vitamins (especially riboflavin and thiamine), amino acids, proteins, starch, mineral matter (Mn, Cu, K, Na, Fe and N), gums and other chemical compounds have been found in saffron.

The hydrophilic carotenoids of saffron includes crocins (about 6–16% of saffron's dry matter) depending upon the variety, growing conditions and processing methods (Gregory et al., 2005). Other minor components include 3-crocetin and y-crocetin, the mono- and dimethyl esters of crocetin, respectively, and mangi-crocin, an unusual xanthone-carotenoid glycosidic conjugate, have also been identified (Fernandez, 2004). The picrocrocin (1–13% of saffron's dry matter) is a colourless glycoside and is the main bitter principle of saffron, even though other components, such as flavonoids, are also responsible for saffron's bitterness (Alonso et al., 2001; Carmona and Alonoza, 2004). Picrocrocin is derived from the enzymatic degradation of zeaxanthine; in turn, the natural deglycosylation of picrocrocin gives safranal (Sampathu et al., 1984; Pfander and Schurteberger, 1982) responsible for the particular aroma of this spice. The safranal [30–37% of essential oil and 0.001–0.006% of dry matter (Carmona et al., 2007; Maggi et al., 2009)] has antioxidant potential (Kanakis et al., 2007) and cytotoxic effect on certain cancer cells (Escribano et al., 1996).

1.7.16 VANILLA

The active constituents of vanilla are responsible for its various biological and therapeutic activities. The flavour profile of vanilla contains more than 200 components, of which only 26 occur in concentrations greater than 1 mg/kg. The aroma and flavour of vanilla extract are attributed mainly to the presence of vanillin (4-hydroxy-3-methoxybenzaldehyde), which occurs in a concentration of 1.0–2.0% w/w in cured vanilla pods (Westcott et al., 1994; Bettazzi et al., 2006; Sharma et al., 2006). True vanilla pods possess a pure delicate spicy flavour that cannot be duplicated exactly by synthetic products. For this reason and because of limited supply, natural vanilla is able to command a premium price, leading to numerous efforts of its blending and adulteration.

The main constituent of *Vanilla planifolia* is vanillin, a methylprotocatechuic aldehyde (4-hydroxy-3-metoxy benzaldehyde) which constitutes 85%of the entire volatiles in vanilla beans. The extract of *V. planifolia* with vanillin contains a number of related phenyl propanoid (C6–C3) compounds. During curing, these compounds undergo a series of enzymatic reactions which bring about the characteristic aroma and flavour of vanilla. The other constituents reported are vanillic acid, anisaldehyde, hydroxybenzoic acid, anisic acid, anisyl alcohol, caproic acid, vitispiranes, eugenol, phenols, phenol ether, carbonyl compounds, acids, esters, benzyl ether, lactones, 25% carbohydrates, 15% fat, B complex, mineral salts such as magnesium, calcium, zinc manganese, potassium and iron which constitute 6%. The water content present in vanilla is around 35%.

1.7.17 STAR ANISE

The seeds of *Illicium verum* contain some volatile oil, resin and a large amount of fixed oil. The fruit (without the seeds) contains volatile oil, resin, fat, tannin, pectin and mucilage. The volatile oil amounts to about 4–5% and is almost identical with oil of anise. Chinese star anise contains anethol, phellandrene, safrole and hydroquinone ethyl ether. Japanese star anise, *Illicium anisatum*, is highly toxic. It contains a poisonous sesquiterpene lactone (called anisatin), shikimin, sikimitoxin, safrole, eugenol, and so forth. Volatile oil contains 94.2% of anethole (cis and trans). In the oil, estragole (1.4%), limonene (1.7%), linalool (0.3%), two terpineol isomers (0.3%) and linalyl acetate (0.3%) are also present. Caryophyllene (0.5%) and trans-bergamotene (0.7%) are the main compounds among sesquiterpenes.

1.7.18 ASAFOETIDA

As a food product, *Ferula asafoetida* usually contains 16% moisture, 297 kcal per 100 g, carbohydrates (67.8%), proteins (4%), dietary fats (1.1%), dietary minerals (7.0%) and dietary fibre (4.1%). As bioactive components of the oleoresin (40–60%) and gum (25%), it includes ferulsinaic acid (rearranged sequesterpene coumarin), ferulic acid at 1.3% of resin, 2-butyl propenyl disulphide (E and Z isomers) and diallyl disulphide as odorous compounds, umbelliferone, asaresinotannols, farnesiferols A–C, foetidin, kamolonol, luteolin, (2E)-3,4-dimethoxycinnamyl-3-(3,4 diacetoxyphenyl) acrylate, and so forth. The components of the essential oil (4–20%) are (E)-1-propenyl sec-butyl disulphide (20.6–40%), (Z)-1-propenyl sec-butyl disulphide (8–27%), 10-epi-γ-eudesmol (5.3–15.1%), (Z)-β-ocimene (2.4–7.8%), α-pinene (4.4–21.36%), β-pinene (4.2–47.1%), β-dihydroagarofuran (1.8–4.1%), γ-eudesmol (undetectable to 3.5%), guaiol (undetectable to 3.0%), agarospiral (undetectable to 3.0%), limonene (undetectable to 2.9%), α-phellandrene (undetectable to 2.9%), 1,2-dithiolane (undetectable to 18.64%), (E)-β-ocimene (1.4–2.9%), 5-epi-7-epi-α-eudesmol (undetectable to 2.1%), α-eudesmol (undetectable to 4.5%), β-eudesmol (undetectable to 1.1%), camphene (0.03–0.1%), and so forth.

1.7.19 BAY LEAF

The essential oil from the leaves (0.8–3%) contains mostly 1,8-cineol (up to 50%) and also eugenol, acetyl and methyl eugenol, α- and β-pinene, phellandrene, linalool, geraniol and terpineol. The dried laurel fruits contain 0.6–10% of essential oil. The aroma of this essential oil is mostly due to terpenes (cineol, terpineol, α- and β-pinene, citral) and also due to cinnamic acid and its methyl ester. The potential role of laurel essential oil as an antimicrobial agent was investigated, too (Atanda et al., 2007; Ozcan and Erkmen, 2001; Smith-Palmer, et al., 2001).

1.7.20 NUTMEG AND MACE

The principal constituents of the spices nutmeg and mace are steam volatile oil (essential oil), fixed (fatty) oil, proteins, cellulose, pentosans, starch, resin and mineral elements. The fixed oil content of sound nutmegs varies from 25 to 40% while that of mace is 20–30%. The major component of the fixed oil is trimyristin followed by others like essential oil (12.5%,) oleic acid (as glyceride), linolenic acid, unsaponifiable constituents, resinous material along with formic, acetate and cerotic acid (traces). The volatile oil constitutes 61–88% monoterpene hydrocarbons (viz. camphene, limonene, myrcene, α- and β-phellandrene, α- and β-pinene, sabinene, α- and γ-terpinene, terpinolene, α-thujene, etc.); simple oxygenated monoterpenes (viz. camphor, 1,8-cineole, etc.); monoterpene alcohols (viz. borneol, citronellol, para-cymen-8-ol, α-fenchol, geraniol, linalool, nerol, cis- and trans-piperitol, terpinen-4-ol, α- and β-terpineol, etc.); monoterpene esters (viz. borneol acetate, citronellol acetate, geraniol acetate, linalool acetate, nerol acetate, terpinen-4-ol, acetate, α-terpineol acetate, etc.); aromatic ethers (viz. elemicin, eugenol, methyl eugenol, ether, 5-methoxy eugenol, myristicin, safrole, etc.); sesquiterpenes (viz. α-bergamotene, bisabolene, cadinene, delta, β-caryophyllene, α-copaene, β-β-cubebene, α-farnesene, δ-germacrene, α-humulene, etc.); aromatic monoterpenes (viz. methyl-iso-propenyl p-benzene, p-cymene, etc.); alkenes; organic acids (viz. acetic, formic, lauric, myristic, octanoic, oleic, palmitic, pentadecanoic, stearic, etc.) and some miscellaneous (viz. trimyristin, α-styrene, yanillin, etc.).

1.7.21 BASIL

Basil contains a very low per cent (0.1–1.0%) of essential or volatile oil. An oleoresin containing 2–5% volatile oil is available. The principal components of the volatile oil are methyl chavicol (estragole), linalool and cineol. Basil is one of the spices which do have definite quality differences from one origin to another. The oil of *Ocimum basilicum* contained, as main components, methyl eugenol (78.02%), α-cubebene (6.17%), nerol (0.83%) and ε-muurolene (0.74%) (Ozcanl and Chalchat, 2002). The yield of essential oil from different plant parts varies between 0.15 and 1.59%, and it depends also on the seasonal factor and locality. Previously, as the characteristic compounds of basil essential oil, linalool, methyl chavicol, eugenol, estragol, thymol and *p*-cymene were found (Akgül, 1989; Khatri et al., 1995; Pino et al., 1996; Martins et al., 1999; Keita et al., 2000). Forty-nine constituents were identified in *O. basilicum*, representing that 88.1% of the oil methyl eugenol (78.02%), α-cubebene (6.17%), nerol (0.83%), α-muurolene (0.74%), 3,7-dimethyloct-1,5-dien-3,7-diol (0.33%) and α-cubebene (0.30%) were found as the major compounds. The essential oil from *O. basilicum* contained α-pinene, sabinene, α-pinene, myrcene, limonene and (Z)-α-ocimene as the most important monoterpene hydrocarbons.

1.7.22 ROSEMARY

Rao et al. (2010) reported that the oil was rich in monoterpene hydrocarbons (30.8%) and oxygenated monoterpenes (62.1%), constituting 92.9% of the oil. The major constituents of the oil were α-pinene (11.8%), camphene (5.3%), 1,8-cineole (20.1%), camphor (26.8%) and verbenone (6.2%), making up 70.2% of the oil. The oil is classified as 1,8-cineole-camphor type. This type of oil was reported from Algeria, France, India and Hungary (Boelens, 1985; Boutekedjiret et al., 1999; Boutekedjiret et al., 1998; Kumar et al., 2004; Lawrence, 1977, 1986, 1989, 1991, 1995, 1997, 2007; Mallavarapu et al., 2000; Mohan Rao, 2000; Rahman et al., 2007). Rosemary oils produced in high-altitude regions (Kumar et al., 2004; Mohan Rao, 2000; Rahman et al., 2007) and plains (Mallavarapu et al., 2000) of India exhibited similar composition with some variations in a few constituents.

1.7.23 SWEET FLAG

The dried rhizome of *Acorus calamus* contains the yellow aromatic volatile oils having asarone as a main constituent which contains the small quantity of sesquiterpenes and its alcohols; the rhizome also contains the choline, flavone, acoradin, galangin and acolamoneisocolamone and aerial parts of plant contains lutcolin-6,8-c-diglucoside; chemical constituents vary in ecotypes and polyploids. Asarone is a genotoxic substance causing genetic mutation and tumours. Phenylpropanoid promotes defence mechanism in herbivores and ultraviolet ray's protection. It yields around 1.5–3.5% volatile oil known as calamus oil. The important constituents of this oil are asarone and its b-isomer. It is volatile, yellowish-brown oil with pleasant and slightly sweet odour which is extracted by steam distillation. Calamus oil has expectorant action used as a remedy for asthma.

The essential oil composition of sweet flag rhizomes depends on ploidy. Diploid cytotypes are characterized by the absence of β-asarone; European and North American triploid cytotypes contain 3–19% of β-asarone, whereas the Indian, Indonesian and Taiwan tetraploid cytotypes contain up to 96% of β-asarone. Photochemical studies have reported the presence of glycosides, flavonoids, saponins, tannins, polyphenolic compounds, mucilage, volatile oil and bitter principle. The plant has been reported for the presence of glucoside, alkaloid and essential oil containing calamen, clamenol, calameon, asarone and sesquiterpenes. It also contains a bitter glycoside named acorine along with eugenol, pinene and camphene. The plant has been extensively investigated and a number of chemical constituents from the rhizomes, leave and roots of the plant have been previously reported which includes β-asarone, α-asarone, elemicine, cisisoelemicine, cis- and trans-isoeugenol and their methyl ethers, camphene, p-cymene, α-selinene, b-gurjunene, β-cadinene, camphor, terpinen-4-ol, aterpineol and a-calacorene, acorone, acrenone, acoragermacrone, 2-deca-4,7-dienol, shyobunones, linalool and preisocalamendiol are also present. Acoradin, galangin, 2,4,5-trimethoxy benzaldehyde, 2,5-dimethoxy benzoquinone, calamendiol, spathulenol and sitosterol have been isolated from *A. calamus*. Alcoholic extracts of the triploid *A. calamus* were characterized by a higher percentage of β-asarone (11%), which was the main compound, followed by higher percentages.

1.7.24 THYME

Thyme contains 0.8–2.0% volatile oil of which the main component is thymol. Other major components are p-cymene and d-linalool. Appreciable amounts of camphene, g-terpinene and carvacrol are also reported (Table 1.8).

TABLE 1.8 Chemical Composition of Thyme.

Compound Thyme	Volatile oil (%)	Thyme liposoluble fraction (%)
Monoterpene hydrocarbons		
α-pinene	1.23	–
Camphene	0.63	–
β-pinene	0.32	–
Myrcene	1.63	–
α-terpinene	0.8	–
p-cymene	30.53	0.81
Limonene	0.62	–
Sabinene	4.24	
Monoterpene esters		
1,8-cineole	1.24	

TABLE 1.8 *(Continued)*

Compound Thyme	Volatile oil (%)	Thyme liposoluble fraction (%)
Monoterpene alcohols		
Linalool	2.73	–
α-terpineol	1.24	–
Geraniol	0.64	–
Borneol	3.16	–
Monoterpene ketones		
Camphor	0.83	–
Monoterpenic esters		
Bornyl acetate	0.7	–
Terpenoidic phenols		
Thymol	30.86	1.01
Carvacrol	3.37	–
Sesquiterpene hydrocarbons		
Caryophyllene	2.48	–

1.8 USES OF SPECIFIC UNDERUTILIZED SPICES

1.8.1 LARGE CARDAMOM

Large cardamom is used as a spice and for aromatic purpose. It contains 2–3% essential oils, possesses carminative, stomachic, diuretic and cardiac stimulant properties, and it is also used as a preventive measure for congestion of lungs, inflammation of eyelids, digestive disorders and in the treatment of pulmonary tuberculosis gargle in infection of teeth and gums, throat and respiratory trouble. Large cardamom possesses a pleasant aromatic odour, because of which it is widely used in India for flavouring in vegetables and food preparations. It is also reported that seeds of large cardamom are considered as an antidote for either snake venom or scorpion venom. Planting of cardamom is usually done during the onset of the monsoon, that is months of May to July. Under favourable conditions, the crops will start giving fruits from 36 months or 3 years. Flowering starts in March to May and will be ready for harvesting from September to October, and may extend up to November in higher altitudes.

1.8.2 CELERY

Celery has been used as an aphrodisiac, anthelmintic, antispasmodic, carminative, diuretic, emmenagogue, laxative, sedative, stimulant and toxic. It is known as a mild diuretic and urinary antiseptic and has been in the relief of flatulence and griping pains. In the medicinal–herbal market, celery oil or oil extract, as well as ground seed or root are touted as herbal and dietary supplements that 'promote and regulate' healthy blood pressure, joint health and uric acid levels (Ahmed et al., 2002). Root tinctures have been used as a diuretic for hypertension and urinary disorder.

The dried ripe fruits are used as spice and condiment for flavouring food. Leaves and stalks are used as salads and in soups. The seed oil is used in the flavouring of different kinds of food-canned soups, meats, sausages and particularly in the flavouring of the popular celery salts and culinary sauces. Celery leaves and stalks are used as salad and sauces, vegetables juices, stew soups, and so forth. It is also widely used in meat seasonings, in flavouring beverages, confectionaries, ice creams and baked goods. It is figured as a natural medicine in different cultures. In modern medicine, it is used as a stimulant and for treating asthma and liver diseases. The seeds can be used in pickling fish and in salads, salad dressings, and other dishes where celery flavour is desired. Ground celery is used in a large variety of products like meat dishes, snack foods, gravies and sauces to provide a flavour-enhancing effect. The leaf stalks and roots give flavour as well as food value to soups and salads. Blanched celery leaves are eaten raw or cooked as a vegetable. Whole seeds can be added to bread dough or when making cheese biscuits. Celery salt and celery pepper are made by grinding the seeds with either salt or peppercorns in the required proportions. Celery seeds have stimulant and carminative properties. Since they have tranquilizing effect, they are prescribed as a decoction in psychiatric, epilepsy-like diseases. The fatty oil from seeds is antispasmodic and nerve stimulant. The roots possess diuretic property. The oleoresin is also used in a large variety of food items. Celery seed is used in ground form as dry seasoning powder and for extraction of oleoresin. The oil has been successfully employed in rheumatoid arthritis and probably acts as an intestinal antiseptic. The oil from the seeds is used medically to treat asthma, flatulence and bronchitis. The oil is also used as an important ingredient in novel perfumes.

1.8.3 AJOWAN

The plant is used traditionally as a stimulant, carminative, flatulence, atonic dyspepsia, diarrhoea, abdominal tumours, abdominal pains, piles, bronchial problems, lack of appetite, galactagogue, asthma and amenorrhoea. Medicinally, it has been proven to possess various pharmacological activities such as antifungal, antioxidant, antimicrobial, anti-nociceptive, cytotoxic activity, hypolipidaemic, antihypertensive, antispasmodic, broncho-dilating actions, anti-lithiasis, diuretic, abortifacient, antitussive, nematicidal, anthelmintic and anti-filarial activity. The roots are diuretic in nature and the seeds are brownish-grey, hot, pungent, stomachic and appetizer and possess excellent aphrodisiac as well as carminative, laxative and diuretic properties. It is used in pickles, biscuits, confectionery and beverages. An important use of seeds in medicine is as a remedy for indigestion. In India, ajowan seeds are distilled partly in primitive native stills and somewhat in more modern and large-scale distilleries located in Rao (Indore), Gwalior and Dhar. Its oil and thymol are used in pharmaceuticals. Oil of ajowan is an almost colourless to brownish liquid, possessing a characteristic odour and sharp burning taste. The oil exhibits anti-inflammatory, fungicidal, anthelmintic, antioxidant activity, analgesic antimicrobial, antispasmodic and anti-aggregatory effects on humans. A part of thymol may be separated from its oil in the form of crystal, which is sold in Indian market under the trade name of *Ajowan Ka Phool* or *Sat Ajowan* and is much valued in medicine as it has nearly all the properties ascribed to ajowan seed. This is used in surgery as an antiseptic and is found to be of great value in the treatment of hook and other worms. The aqueous solution of thymol is an excellent mouthwash, and hence, it is a constituent of most of the toothpastes. The essential oil and the thymol are used in India as medicine particularly for the cases of cholera. Thymol is a strong fungicide and germicide, and hence used against a range of microorganisms. Thymol is used in toothpaste and perfumery. It is also used in the preparation of lotions and ointment. It inhibits the bacterial-resistant microbial pathogens and is useful as a plant-based fourth-generation herbal antibiotic formulation.

1.8.4 ANISEED

It is widely distributed and mainly cultivated for the seeds. The seeds of anise contain an essential oil (1–4%) and active substances such anethole used in various pharmaceutical and food industries (Klaus et al., 2009). Today, anise seeds are an important natural raw material which is used for pharmaceutics, perfumery, food and cosmetic industries (Ross, 2001). Recently, this spice plant has drawn more consideration of consumers due to the antimicrobial, antifungal, insecticidal and antioxidative effect of this herb on human health (Ozcan and Chalchat, 2006, Tepe et al., 2006, Tirapelli et al., 2007). Fresh leaves may be used in salads, especially apple and seeds in cookies and candies. While the entire plant is fragrant, it is the fruit of anise, commercially called anise seed that has been highly valued since antiquity. The delicate fragrance is widely used for flavouring curries, breads, soups, cakes, candies, desserts, non-alcoholic beverages and liqueurs such as anisette and arak. Aniseed is widely used to flavour pastries; it is the characteristic ingredient of a German bread called Anisbrod. In the Mediterranean region and Asia, aniseed id is commonly used in meat and vegetable dishes. It is used in Italian sausage, pepperoni, pizza topping and other processed meat items. The volatile or essential oil, obtained by steam distillation of the crushed anise seed, is valuable in perfumery and soaps and has been used in toothpastes, mouthwashes and skin creams. The essential oil is used to flavour absinthe and Penod liqueurs. Anise oil is sometimes used as an adulterant in the essential oil of liquorice. The oil is sometimes used as sensitizer for bleaching colours in photography. The seeds are chewed after a meal in India to sweeten the breath. It makes a soothing herbal tea and has been used medicinally from prehistoric items. As a medicinal plant, anise has been used as a carminative, antiseptic, antispasmodic, expectorant, stimulant, and stomachic. In addition, it has been used to promote lactation in nursing mothers and as a medicine against bronchitis and indigestion. Oil of anise is used today as an ingredient in cough medicine and lozenges and is reported to have diuretic and diaphoretic properties. If ingested in sufficient quantities, anise oil may induce nausea, vomiting, seizures and pulmonary oedema. Contact of the concentrated oil with the skin can cause irritation.

1.8.5 DILL

Dill seeds are used, both whole and ground, as a condiment in soups, salads, processed meats and sausages. The green herb issued as a flavourings agent in various preparations. Dill stems and blossom heads are used for dill picking and for flavouring soups. Ground seed is an ingredient for seasoning various food materials. Dill fruits (seeds) is used, both whole and ground form, as a spice and condiment. It is widely used in soups, sauces, salads, processed meat and sausages and for other culinary purpose. Dill stems and blossom heads are used for dill pickles. Dill foliage, fruits and their volatile oil are used extensively for culinary and medicinal purposes. Dill oil and leaves are also used for flavouring purpose. The essential oil is used in the manufacture of soaps. Both the seeds and oil are used in medicine for treatment of flatulence and colic. The leaves are used as flavouring substance. The entire plant is aromatic, and the small stems and immature umbels are used for flavouring soups, salads, sauces, fish, sandwich fillings, and particularly pickles. The leaves freshly chopped may be used alone or in dill butter for broiled or fried meats and fish, in sandwiches, fish sauces, and creamed or fricasseed chicken. The major commercial use of dill is in the form of dill weed oil, used in the pickle industry. Dill has a warm, slightly sharp flavour somewhat reminiscent of caraway. The whole seeds and the seed oil have carminative properties and have been used in treating flatulent colic. They are often taken as 'dill water' to relieve digestive problems and flatulence. It is used widely to cure insomnia and hiccups. Occasionally, dill is used to perfume cosmetics. A medicinal oil is distilled from leaves, stems and seeds. It was used as a remedy for indigestion and flatulence and as milk secretion stimulant. Moreover, it is used as an anti-convulsion, anti-emetic, anti-cramp (in children), as a wound healer, to increase the appetite and strengthen the stomach (Ishikawa, 2002; Kaur and Arora, 2009).

1.8.6 CINNAMON

Bark is used for gastrointestinal upset, diarrhoea and gas. It is also used for stimulating appetite for infections caused by bacteria and parasitic worms and for menstrual cramps, the flu (influenza) and the common cold. Cinnamon is used as a spice and as a flavouring agent in beverages. Cinnamon bark has carminative, astringent, stimulative and antiseptic properties. It checks vomiting and relieves flatulence. Leaf oil and bark oil are used in the manufacture of perfumes, confectionary, soaps, toothpastes, hair oils and face creams.

1.8.7 CASSIA CINNAMON

Cassia bark is used as a flavouring agent in cooking, especially in savoury dishes and particularly in liqueurs and chocolate. It is an ingredient in mixed spice and pickling spices. It is good with stewed fruits. Southern Europeans prefer it to cinnamon. Cassia buds, the dried, unripe fruits of *Cinnamomum cassia* and *C. loureirii*, have a cinnamon-like aroma and a warm, sweet, pungent taste akin to that of cassia bark. The whole buds are added to foods for flavouring. The cinnamic aldehyde is a good antifungal agent. The volatile oil is used in some inhalants, in tonics and as a cure for flatulence, sickness and diarrhoea.

1.8.8 CURRY LEAF

Berry pulp is edible with a sweet but medicinal flavour in general; neither the pulp nor seed are used for culinary purposes. Leaves are used in many dishes and as an herb in Ayurvedic medicine in India and neighbouring countries. Curry leaves help to reduce diabetes if it is taken on empty stomach due to presence of glucose called koenigin which fight against diabetes. Due to its aromatic properties, it is used in soap-making ingredient, diffusers, body lotions, scent, air fresheners, potpourri, perfume, body fragrance, bath and massage oils, aromatherapy, health clinics, facial steams, hair treatments, and so forth.. Curry leaves are used for rituals in the absence of *tulsi* leaves, such as pujas. Being rich in calcium and vitamin A, it helps in building bones and good eye sight and delays cataract. Curry leaf cures osteoporosis and weak bones. Curry leaf helps avoiding mental disturbance by supplying calcium to brain, good for old-age people. Chewing fresh leaves on empty stomach helps in eliminating bad cholesterol and reducing excess weight. Curry leaves has anti-diabetic, antimicrobial and anti-carcinogenic properties.

1.8.9 HORSERADISH

Horseradish sauce made from grated horseradish root and vinegar is a popular condiment in the United Kingdom and Poland. In Russia, horseradish root is usually mixed with grated garlic and small amount of tomatoes for colour. In the United States, the term 'horseradish sauce' refers to grated horseradish combined with mayonnaise or salad dressing. Among the medicinal values, it is used to prevent diarrhoea and dysentery, influenza, rheumatism, cough and headache. Horseradish is considered a powerful effective diuretic, used by herbalists for centuries to treat kidney stones. It has high sulphur. Roots are used for indigestion and putrefaction in the digestive tract and clear lung problems as well. Narcotine is used to cure the whooping coughs and asthma. It stimulates the mucus surface throughout the entire body.

1.8.10 KOKUM

Kokum fruit is a popular condiment used in several states of India for making vegetarian and non-vegetarian 'curry' preparations, including the popular 'solkadhi' (Elumalai and Chinna Eswaraiah, 2011; Kirtikar and Basu, 1991). A healthy soft drink is made from kokum to relieve sunstroke, due to its heat neutralizing property. Kokum fruits are squeezed in sugar syrup to make a soft drink, named as 'Amrutkokam' (Devasagayam et al., 2004), which is quite popular during summer season. Kokum has a long history in Ayurvedic medicine as it was traditionally used to treat sores, dermatitis, diarrhoea, dysentery, ear infection and to facilitate digestion. The ripened rind and juice of kokum fruit are commonly used in cooking. The dried and salted rind is used as a condiment in curries. It is also used as a garnish to give an acid flavour to curries and for preparing attractive, red, pleasant flavoured cooling syrup. Kokum butter is used as an edible fat and is nutritive, demulcent and antiseptic. The rind has antioxidant property. The fruit is anthelmintic and cardiotonic and useful in piles, dysentery, tumours, pains and heart complaints. The fruit juice is given in bilious affections. The root is astringent. An edible fat, known as 'kokum butter' is considered nutritive, demulcent, astringent and emollient. It is suitable for ointments, suppositories and other pharmaceutical purposes. Kokum seeds are used for oil extraction. That oil is called kokum butter and used in curries, cosmetics, medicines and costly confectionery preparations in foreign countries. The kokum fruit acts as an antioxidant, acidulant and appetite stimulant and helps in fight cancer, paralysis and cholesterol. The kokum fruit is a good digestive tonic and used to improve skin health. Kokum has been reported for the treatment of dysentery, tumours, heart complaints, stomach acidity and liver disorders (Bhaskaran and Mehta,

2006; Krishnamurthy et al., 1982). Kokum seed is a good source of fat called as 'kokum butter'. Kokum is a minor oil seed crop and butter has food and non-food applications. The oil is traditionally extracted by boiling the kernel powder in water and the oil that is collected at the top is skimmed off. The yield of oil (fat) is about 25–30%. Kokum fat has been reported to be used in chocolate and confectionary preparations. It is also used is manufacture of soap, candle and ointments. An ointment made out of kokum fat, white dammar resin (resin exuded by *Vateria indica* tree) and wax is said to be effective in treating carbuncles. It is reported that Italy and some other foreign countries are importing kokum fat from India for confectionary preparations. Kokum fruit appears to be a promising industrial raw material for commercial exploitation in view of its interesting chemical constituents.

1.8.11 MINT

Mint has several benefits which include proper digestion and weight loss, relief from nausea, depression, fatigue and headache, treatment of asthma, memory loss, and skin care problems. Mint is a well-known mouth and breath freshener. Many products such as toothpaste, chewing gum, breath fresheners, candy and inhalers are prepared from mint as their base element. Extracts from mint leaves inhibit the release of certain chemicals, which aggravate severe nasal symptoms often found in cases of hay fever and seasonal allergies. Foods like ice cream and chocolates, as well as in alcoholic and non-alcoholic beverages are also prepared from mint.

1.8.12 LONG PEPPER

Long pepper, *Piper longum* (Linn.) used in the Indian System of Medicine, is the dried unripe fruit. It is used for the treatment of cold, cough, bronchitis, asthma, fever, muscular pains, insomnia, epilepsy, diarrhoea, dysentery, leprosy, and so forth. Roots and thicker basal stem portion of the plant are also presently used in many Ayurvedic preparations. It improves appetite and digestion, as well as treats stomach ache, heartburn, indigestion, intestinal gas, diarrhoea, and cholera. For women, it is normally used to treat menstrual cramps and infertility. Long pepper has been found to reduce blood glucose levels in diabetic patients. It is known to possess liver-protective functions and believed to prevent jaundice. It also promotes weight loss and has minimal or no side effects on the body. It is known to reduce body fat and discard the stagnant fatty toxins from the body, thereby preventing obesity.

1.8.13 POPPY SEED

Capsules of dried seeds are used for decorations, but they also contain codeine, morphine, and other alkaloids. Opium pods can be boiled in water to produce a bitter tea that induces a long-lasting intoxication. Mature poppy pods can be crushed and used to produce lower quantities of morphinans. Poppy seed oil is used for culinary purposes. It is free from narcotic properties and is used, along with olive oil, as a salad oil. The oil is used (for the presence of linoleic acid) to prepare soft soap, ointment emulsion and various compositions for skin care. It is used to cure diarrhoea, dysentery and scalds. Opium used with poison hemlock to put people quickly and painlessly to death, but it was also used in medicine. During the 18th century, opium was found to be a good remedy for nervous disorders. It was used to quiet the minds of those with psychosis, help with people who were considered insane, and also to help treat patients with insomnia due to its sedative and transquillizing properties.

Poppy seeds do not contain narcotic properties. They are used to flavour bread, cakes, rolls and cookies and often sprinkled on top of dishes to be used as a garnish or as a spice with cheese, eggs, salad, cookies, cakes, bread, pastries, salads, sauce, curries and noodles. Poppy seeds have antioxidant, disease-preventing and health-promoting properties. The seeds, being especially rich in oleic and linoleic acids, help to prevent coronary artery disease and strokes by favouring healthy blood lipid profile. Poppy seeds outer husk is a good source of dietary fibre, thereby eases constipation. Dried poppy seeds soothe nervous irritability and act as painkillers. Its seed extractions are found useful in the pharmacy and many traditional medicines in the preparations of cough mixtures, expectorants, and so forth.

1.8.14 PARSLEY

Parsley is commonly used for garnishing and seasoning of foods. They are eaten fresh, incorporated in salad and used as an ingredient in soups, stews and sauces. It is also used as a seasoning in meat and poultry. The roots are used as a vegetable in soups. The dried leaves and roots are used as condiments. The herb is possessing diuretic, carminative, antipyretic properties. The juice of the fresh leaves is used as an insecticide. Parsley herb oil and parsley seed oil are obtained by steam distillation of seeds. Dried leaves and dried roots of parsley are used as spices. A tea made from leaves or roots is used to treat jaundice, coughs and menstrual problems, rheumatism, kidney stones and urinary infections. The juice expressed from them soothes conjunctivitis and eye inflammations. Parsley has been used as carminative, gastro tonic, diuretic, antiseptic of the urinary tract, anti-urolithiasis, antidote and anti-inflammatory and for the treatment of amenorrhoea, dysmenorrhoea, gastrointestinal disorder, hypertension, cardiac disease, urinary disease, otitis, sniffle, diabetes and also a various dermal disease in traditional and folklore medicines. A wide range of pharmacological activity including antioxidant, hepatoprotective, brain-protective, anti-diabetic, analgesic, spasmolytic, immunosuppressant, anti-platelet, gastroprotective, cytoprotective, laxative, estrogenic, diuretic, hypotensive, antibacterial and antifungal activities have been exhibited for this plant in modern medicine. In folk medicine, the aerial part of *P. crispum* is used to treat haemorrhoids, the stem for urethral inflammation, and the root is used to pass kidney stones and improve brain function and memory. *P. crispum* is used as a carminative, stomachic, emmenagogic, abortifacient and nutritive agent. Seven studies have shown that *P. crispum* has hypoglycemic, diuretic, hypolipidemic, antimicrobial, anticoagulant and hepatoprotective activities.

1.8.15 SAFFRON

In foods, saffron is used as a spice, yellow food colouring, and as a flavouring agent. Medicinally, it is used for asthma, cough, whooping cough and as an expectorant. It is also used to cure insomnia, cancer, atherosclerosis, intestinal gas (flatulence), depression, Alzheimer's disease, fright, shock, spitting up blood (haemoptysis), pain, heartburn and dry skin. Women use saffron for menstrual cramps and premenstrual syndrome. Men use it to prevent early orgasm and infertility. Saffron extracts are also used as a fragrance in perfumes and as a dye for cloth. The stigmas are used for making medicine. . Saffron is also used for to increase interest in sex and to induce sweating. Saffron is applied directly to the scalp for baldness by some people (alopecia).

1.8.16 VANILLA

In recent years, there has been growing interest in natural and healthy foods, especially with regards to the ingredients such as flavouring agents and preservatives. Among the variety of natural flavours in use today, vanilla occupies a prominent marketplace and has been in use for the preparation of ice creams, chocolates, cakes, soft drinks, pharmaceuticals, liquors, perfumery and in nutraceuticals. The extract of this plant is useful in treating hysteria, rheumatism and other low forms of fever. The principal constituent of *V. planifolia* is vanillin, chemically known as methylprotocatechuic aldehyde. It is used to flavour medicinal syrups. In addition, this orchid contains alkaloids, flavonoids, glycosides, carbohydrates and other phytochemicals. All the parts of this plant, namely leaves and stem possess some measurable inhibitory action against the pathogens. Vanilla is used mainly as a flavouring material; a critical intermediary in a host of pharmaceutical products and as a subtle component of perfumes. As a flavouring agent, it is used in the preparation of ice creams, milk, beverages, candies, confectionaries and various bakery items. Vanilla has stomachic, digestive and choleretic agents that increase the bile secretion, mildly invigorating and aphrodisiac properties. The seeds of the vanilla pod are used to flavour ice creams, liquor, soft drinks and candies. Vanilla is also used in the production of pharmaceuticals, cosmetics, tobacco and handicrafts.

Vanilla is world's most popular flavourant for numerous sweetened foods, several commercial food products, liquors, perfumes, and so forth. Vanilla extracts or essence is extracted with alcohol and contains the aroma and flavour principles and sweetening/thickening agents. They are widely used as flavouring par excellence for ice creams, soft drinks, chocolates, confectionary, candy, tobacco, baked foods, puddings, cakes, cookies, liquors and in perfumery. Vanilla sugar is a mixture of vanilla extracts and sugar. Vanilla tincture is used for pharmaceutical uses. Of late, technical grade vanillin is used as a chemical intermediate in the production of a number of pharmaceutical products.

1.8.17 STAR ANISE

Star anise is used in cooking as a spice or drunk as a tea and has numerous important health benefits. It is a good source of two antioxidants: the essential oil linalool and vitamin C that helps protect the body against cellular damage caused by free radicals and environmental toxins. It can fight influenza because of its high shikimic acid content. It can treat sleep disorders, boost the immune system, help combat illnesses and regulate hormonal function. Shikimic acid of star anise is used as an ingredient in Tamiflu—the influenza medication also been used against swine flu.

The stars are available whole or ground to a red-brown powder. The bulk of the oil in commerce is obtained from the star anise fruit in China. Apart from its use in sweetmeats and confectionery, it contributes to meat and poultry dishes, combining especially well with pork and duck. It is also one of the ingredients used to make the broth for the Vietnamese noodle soup called *pho*. Star anise is an ingredient of the traditional five-spice powder of Chinese cooking. Chinese stocks and soups very often contain the spice. In the West, star anise is added in fruit compotes and jams and in the manufacture of anise-flavoured liqueurs, the best known being anisette. It is an ingredient of the mixture known as 'Chinese Five Spices' (Morton, 2004). The water-soluble extract of *I. anisatum* promotes hair growth and may be a useful additive in hair growth products (Sakaguchi et al., 2004). Star anise is also used in different Indian curry powders for meat preparations.

1.8.18 ASAFOETIDA

Asafoetida has been a popular spice in Europe since the Roman times and has been used much in the Middle Ages (for example, to flavour barbecued mutton), but has fallen in dishonour thereafter. It is still an important ingredient in Persia, and is popular with Brahmins and Jains in India who refuse to eat onions and garlic. In Indian cuisine, it is normally not combined with garlic or onion, but is seen as an alternative or substitute for them; it is nearly always used for vegetable dishes. The Tamil (South Indian) spice mixture 'sambar podi' frequently contains asafoetida. Asafoetida is extensively used for flavouring curries, sauces, and pickles.

Usage differs a little bit from the powdered form and the pure resin. The resin is very strongly scented and must be used with care; furthermore, it is absolutely necessary to fry the resin shortly in hot oil. This has two reasons: first, the resin dissolves in the hot fat and gets better dispersed in the food, and second, the high temperature changes the taste to a more pleasant impression. A pea-sized amount is considered as a large amount, sufficient to flavour a large pot of food. Powdered asafoetida, on the other hand, is less intense and may be added without frying, although then the aroma develops less deeply. Lastly, powdered asafoetida loses its aroma after some years, but the resin seems to be unperishable.

Asafoetida is also used in medicines because of its antibiotic properties. It is an interesting alternative to onion and garlic, even for Western dishes. In ancient Rome, asafoetida was stored in jars together with pine nuts, which were alone used to flavour delicate dishes. Another method is dissolving asafoetida in hot oil and adding the oil drop by drop to the food. If used with sufficient moderation, asafoetida enhances mushroom and vegetable dishes, but can also be used to give fried or barbecued meat a unique flavour. Asafoetida is a useful antidote for flatulence. There are claims for it being used to cure bronchitis and even hysteria.

The whole plant is used as a fresh vegetable, and the inner portion of the full-grown stem being regarded as a delicacy. The horrible smell of fresh asafoetida does not seem to qualify as a valuable food enhancement, but after frying (and in small dosage) the resin, the taste becomes rather pleasant, even for Western taste buds. The so-called powdered asafoetida is the resin mixed with rice flour and therefore much less strong in taste, but easier in the application.

1.8.19 BAY LEAF

The smooth and lustrous dried bay leaves are usually used whole and then removed from the dish after cooking; they are sometimes marketed in powdered form. The crushed form is a major component in pickling spices in processed meats and pickle industry. Ground bay is utilized in many seasoning blends and products. Oil of bay and bay oleoresin is used in soluble pickling spices.

Bay leaves are used as a flavouring in soups, stews, meat, fish, sauces and in confectionaries. Both leaves and fruits possess aromatic, stimulant and narcotic properties. The dried leaves and essential oils are used extensively in the food industry for seasoning of meat products, soups and fishes. The fruits contain both fixed and volatile oils, which are mainly used in soap making (Bozan et al., 2007). Traditionally, it is used in rheumatism and dermatitis (Kilic et al., 2004), gastrointestinal problems, such as epigastric bloating, impaired digestion, eructation, and flatulence, the aqueous extract is used in Turkish folk medicine as an anti-haemorrhoidal, anti-rheumatic, diuretic, as an antidote in snakebites and for the treatment of stomach ache (Gulcin, 2006; Baytop, 1984). The essential oil obtained from the leaves of this plant has been used for relieving haemorrhoid and rheumatic pains. It also has diuretic, antifungal and antibacterial activities (Zargari, 1995). The essential oil from the leaves is also used as spice and food-flavouring agent and has wider application in traditional medicines of different countries. The major functional properties are antimicrobial, antifungal, hypoglycaemic, anti-ulcerogenic, and so forth.

1.8.20 NUTMEG AND MACE

In foods, nutmeg and mace are used as spices and flavourings. Nutmeg and mace are used for diarrhoea, nausea, stomach spasms and pain, and intestinal gas. They are also used for treating cancer, kidney disease, and insomnia. It increases menstrual flow, acts as a hallucinogen, and as a general tonic. Nutmeg and mace are applied to the skin to treat rheumatic pain, mouth sores and toothache. Nutmeg oil is used as a fragrance in soaps and cosmetics.

1.8.21 BASIL

The dried leaves and tender four-sided stems are used as a spice for flavouring and for extraction of essential oil. Apart from flavouring numerous foods, it is used for seasoning in tomato paste products. The sweet basil oil is widely used in perfumery compounds. It has application in areas of medicine and also used as an insecticide and bactericide. The leaves of basil are used in folk medicine as tonic and vermifuge (Heath 1981; Lawrence 1985). Basil tea taken hot is good for treating nausea, flatulence and dysentery (Baytop, 1984). It is used in pharmacy for diuretic and stimulating properties, and in perfumes and cosmetics for its smell; in fact, it is a part of many fragrance compositions (Baritaux et al., 1992; Khatri et al., 1995). Its oil has been found to be beneficial for the alleviation of mental fatigue, colds, spasms, rhinitis, and as a first-aid treatment for wasp stings and snake bites. The essential oil has antifungal, physicochemical and insect-repelling properties (Lahariya and Rao, 1979; Dube et al., 1989; Özcan, 1998; Martins et al., 1999). It is also regarded as highly antiseptic and has been applied in both to prevent post-partum infections. One can inhale the vapours of the infusion of the leaves of *Ocimum* or take a bath to improve the general conditions and to ameliorate the respiratory function (Martins et al., 1999).

The essential oil showed antibacterial (Khorana, 1950), antifungal (Kaul and Nigam 1977), insecticidal and larvicidal activity (Chopra et al., 1941). Methyl chavicol and methyl cinnamate obtained from the essential oil of *Ocimum basilicum* were found to be mainly responsible for the insecticidal activity of the oil against *Tribolium castaneum, Sitophilus oryzae, stagobium paniceum* and *Bruchus chinensis* (Deshpande and Tipnis, 1977). The leaves, fresh or dry, may be used to improve the flavour of tomato dishes, cucumbers, green salads, eggs, ricotta cheese mixes and shrimp. It is a popular culinary flavouring, typical of Mediterranean cuisines. Oil of basil is used in perfumery, soaps, cosmetics and liqueurs. It is a good insect repellent. Medicinally, it is used to sooth pain and treat vomiting, nervous stress and headaches.

1.8.22 ROSEMARY

Rosemary has a wide range of uses in food processing. Fresh tender tops are used for garnishing and flavouring of cold drinks, pickles, soups, and so forth. Dried and powdered leaves are used as a condiment. Rosemary is a popular culinary flavouring added to meat dishes, baked foods and Mediterranean recipes. The fresh or dried leaves may be used sparingly for special accent with cream soups made of leafy greens, poultry, stews and sauces. Rosemary oil is used in processed meats for flavouring. Rosemary extract has antioxidant properties in food products.

Unique compounds and oils include rosmarinic acid and essential oils such as cineol, camphene, borneol, bornyl acetate and α-pinene, providing carminative, antidepressant, antispasmodic, antimicrobial, carcinogen blocker, liver-detoxifier and anti-rheumatic, anti-inflammatory, antifungal, antibacterial, and antiseptic properties. Moreover, research provides ample evidence that rosemary not only improves memory but also helps fight cancer. Antioxidant activity of rosemary extracts is shown to inhibit carcinogenesis and it has got antitumour and anti-inflammatory activity too. Carnosol and carnosic acid are the antioxidants present in rosemary oil (Chen et al., 1992; Bauman et al., 1999). It is also widely used for making medicinal tea and in aromatherapy.

1.8.23 SWEET FLAG

Sweet flag is mainly used in medicine. The oil is used to cure gastritis. In the form of infusion, it is carminative and possesses emetic and anti-spasmodic properties. It is used in the perfumery industry. It also has insecticidal properties. The water–ethanolic extract of sweet flag exhibit antioxidant property. Dried root (rhizome) of sweet flag is used in medicinal preparation and flavouring liquors. The dried rhizomes constitute the drug calamus of commerce. In Ayurveda, the rhizomes are considered to possess antispasmodic, carminative and anthelmintic properties. Extracts of whole plants of sweet flag have been employed for medical purposes since the time of ancient Greece, and its leaves have provided floor covering as well. American Indians had so many medicinal used for the rhizomes and roots. Physicians used it for stomach cramps and gas and as a tonic and stimulant. Its root is used in India to treat toothache, fever and menstrual problems.

Traditionally, *A. calamus* is used to treat appetite loss, bronchitis, chest pain, colic, cramps, diarrhoea, digestive disorders, flatulence, gas, indigestion, nervous disorders, rheumatism, sedative and vascular disorders. The plant has a long history from various countries and has been in use for at least around 2000 years in China and India. Many Native American tribes used it as an anaesthetic for toothache and headaches. The ancient Chinese used it to lessen swelling and for constipation. In India, it is used to cure fevers, asthma and bronchitis, and as a sedative. The root was also believed to be used by the ancient Greeks and included in the traditional remedies of many other European cultures. *A. calamus* was also known to many early American settlers and used for a number of folk remedies. *A. calamus* was also widely used by Canadian trappers working for the Hudson Bay Company, using it as a stimulant, chewing a small piece whenever they feel tired. The unpeeled, dried rhizome was listed in the US Pharmacopoeia until 1916 and in the National Formulary until 1950 for medicinal use of humans. Both the leaves and rhizome are apparently psychoactive, due to the presence of asarones, which have mescaline-like hallucinogenic properties if taken in sufficient quantities. In lesser amounts, it has stimulating and tonic effects. According to Arabic, Roman and later European folk botany, the plant is also an aphrodisiac. The plant is mentioned by many of the great classical writers on medicine, from Hippocrates (460–377 B.C.) and Theophrastus (371–287 B.C.) onwards. According to Dioscorides, the smoke of *A. calamus*, if taken orally through a funnel, relieves cough. Celsus records that the plant was readily available in the markets of India almost 2000 years ago.

1.8.24 THYME

Thyme is a widely used herb. The leaves, usually blended with other herbs, may be used in meats, poultry stuffings, gravies, soups, egg dishes, cheese, and so forth. It is used to season tomato soups, fish and meat dishes, liver and pork sausages, headcheese, cottage and cream cheese. Thyme oil is used in the treatment of bronchitis. It has antispasmodic and carminative properties. It possesses antioxidant and antimicrobial properties (Nickavar et al., 2005; Pirbalouti et al., 2011). Thyme oils are used in some processed meats and some sauce and prepared foods applications. Leaves make a tonic and stimulating tea, used to treat digestive complaints and respiratory disorders. Antiseptic and vermifuge essential oil (thymol) is added to disinfectants, toothpaste, perfumes, toiletries and liqueurs.

KEYWORDS

- **aroma**
- **astringent**
- **thymol**
- **limonene**
- **terpinene**

CHAPTER 2

UNDEREXPLOITED SPICE CROPS: SCENARIO AND PRESENT STATUS

CONTENTS

2.1 INTRODUCTION

Dioscorides, in the 17th century B.C., refers to plenty of spices and medicinal herbs in his work *Materia Medica.* Later, a Roman doctor Galenus dealt with the medical effects of many spices in 137 A.D. After that, scientific descriptions or reviews appeared about spices and herbs in books of grasses or in herbaria. In this way, with the passage of time and keeping pace with global technological advancements, the positive sides of herbs and spices for human well-being are revealed. Many spices are selectively chosen and facilitated by the growers, end users and researchers for their promotion and use. For example, sage (*Salvia officinalis*) and thyme (*Thymus vulgaris*) were sacred ceremonial herbs of the Romans. They were associated with immortality and were thought to increase mental capacity in ancient times. But many others remain unexplored, unutilized or underutilized.

India has a vast compendium of knowledge in traditional medicine wherein spices are extensively used. The non-traditional use of spices and spice products is an emerging field and holds great export potential. The Indian form of medicine using herbals including spices is 'Ayurvedic medicine'. Indians describe this as a system of 'science of life' which looks at a holistic treatment. The fact that these so called underutilized spices can also greatly contribute in human health and well-being is being slowly understood by the people around the world.

2.2 ROLE OF UNDERUTILIZED SPICES IN GENERAL

Underutilized spices usually occupy key role in the following areas:

1. Food security: Growth and marketing of underutilized spice crops may be part of a focused effort to help the poor for subsistence and income, thereby, reducing the risk of overreliance on limited major spices. In fact, these crops precisely preserve the cultural and dietary diversity.
2. Better nutrition: The quality of food, as such, may be enhanced through underutilized spices rich in specific bioactive chemicals, nutraceuticals, pharmaceuticals, etc.
3. Sustainable use of inputs: It increases sustainability of agriculture through a reduction of inputs, especially the nitrogen fertilizers,

fossil fuel-derived materials, water and other chemicals. Smallholder farmers and housewives can use traditional processing means.

4. Buffer against the effects of climate change: The potential of these crops is being increasingly understood in areas within the pursuit of risks arising out of carbon footprint in agriculture and the impending threats of climate change.
5. Increased resilience:
   - *Economic resilience* – People can increase their household income by using crop diversity not only for subsistence needs but also for sale of diverse products,
   - *Natural (agro-ecosystem) resilience* – To withstand extreme weather events; natural or nature-like diverse ecosystems favour distribution of resource capture in time and space which results in variability of resource conversion and products.
   - *Social resilience* – To address adverse conditions through collective action. Local knowledge systems are based on the surrounding plant diversity and by working together to manage and protect this resource, social cohesion within a community is increased.

2.3 GLOBAL TRADE IN SPICES: HISTORICAL PERSPECTIVE

- The spices found their way into the Middle East during the pre-Christian era. They were awfully regarded as wonders behind the curtains of mystical tales and folklore. Prehistoric records mentioned India's southwest coast path, especially Kerala as a major spice trade centre since 3000 B.C., which marks the beginning of the spice trade.
- The Greco-Roman world was followed by trading along a new route called the Incense route. The sea routes to India and Sri Lanka were governed by the Indians and Ethiopians.
- With the rise of Islamic powers, the land caravan routes were closed down making this route the paramount route of importance in the old world. Maritime trade routes led to tremendous growth in trade and commercial activities. Soon the European powers sensed great profit from the above trade and began inquiring into these sources.
- The trade was transformed by the European Age of Discovery where the route from Europe to Indian Ocean through the Cape of

Good Hope in South Africa was found by the Portuguese who soon landed in India and set up trading hubs in the country.
- This trade became an interlinkage for the global economy and became an important driver of the same.

2.4 ROLE OF UNDERUTILIZED SPICES IN GLOBAL ECONOMY

Traditionally, spices have been produced and consumed very extensively in the Asian countries, such as India. However, the changing demographics in the western countries have also given a boost to the spices market. Uncertain economic conditions coupled with inefficient logistics, especially for the underutilized spice crops, act as major setback for the producing countries such as India. Government policies resulting in rejection of imported minor spices also hamper the growth in international markets. Increasing demand from the organic segment and greater therapeutic potential of the apparently underutilized spices offers great opportunity for the world market.

According to a new market report published by Persistence Market Research, 'Global Market Study on Seasoning and Spices: Salt and Salt Substitute to Witness Highest Growth by 2019', the global market for seasonings and spices in value term is expected to grow from US$12,530.5 million in 2013 to US$16,628.6 million in 2019 growing at a compound annual growth rate (CAGR) of 4.8%.

Globally, Asia-Pacific and Europe were the largest markets for seasonings and spices in 2012, in terms of volume and value, respectively. These regions are expected to continue their dominance during the forecast period. Seasonings and spices market in Asia-Pacific increased from 409.4 million kg in 2008 to 511.1 million kg in 2012. The European seasonings and spices market is expected to grow to US$5723.3 million in 2019. Certain countries, such as the United Kingdom, the United States and Spain, have low per capita consumption and low expected CAGR, which indicate that the seasonings and spices market is yet to pick up in these countries as the consumers have started building interest in spices only in the last few years. Countries such as India and China have high consumption (although low per capita consumption owing to their large population base) and high growth rate, which will serve as growth drivers for the global market as consumers in these countries are demanding traditional as well as new products (Table 2.1 and Table 2.2).

TABLE 2.1 Area, Production, Export and Import of a Few Underutilized Spices (India).

	2015–2016 (Estimate)		2015–2016 (Export estimate)		2015–2016 (Import estimate)	
	Area (ha)	**Production (t)**	**Quantity (t)**	**Value (Rs., lakhs)**	**Quantity (t)**	**Value (Rs., lakhs)**
Cardamom (large)	26,387	5300	600	7332.50	3410	30,795.48
Coriander	585,710	557,000	40,100	42,680.50	25,305	17,467.10
Cumin	701,560	372,290	98,700	156,699.00	2000	3440.90
Fennel	46,760	78,570	15,320	17,239.60	-	-
Fenugreek	124,710	134,100	33,300	23,380.00	-	-
Celery	4070	5510	5800	5776.50	-	-
Ajowan	24,010	17,180	-	-	-	-
Clove	2310	1200	-	-	20,235	104,542.35
Nutmeg	21,110	14,400	4050	20,928.25	490	1955.20

Source: State Agri/Hort. Departments/DASD Kozhikkode Cardamom: Estimate by Spices Board (http://indianspices.com/2016.pdf).

TABLE 2.2 Main Exporters and Major Export Markets of a Few Underutilized Spices in the World.

Sl. No.	Underutilized spices	Main exporters	Major export markets
1	Cinnamon	Sri Lanka (Ceylon cinnamon), Indonesia, China and Vietnam (for cassia cinnamon)	The United States, Mexico, Canada, Nicaragua, India, Pakistan, Saudi Arabia and United Arab Emirates (UAE)
2	Nutmeg	Guatemala, Indonesia, India, Nepal, Sri Lanka and Vietnam	Saudi Arabia, UAE, Germany, Netherlands, the United Kingdom and the United States
3.	Cloves	Madagascar, Tanzania, Comoros, Brazil, Indonesia and Sri Lanka	Belgium, Germany, Netherlands, UAE, the United Kingdom and France
4	Vanilla	Madagascar, Comoros, Mauritius, Uganda, Indonesia and Mexico	The United States, Germany, France and Canada

India exported 665 t of large cardamom in 2014–2015. Sikkim, which grows large cardamom in 17,000 ha of land, produces 4000 Mt (90% of the country's production) of the spice annually (The Economic Times, January 25, 2016).

In 2015 alone, Sri Lanka's spice exports amounted to US$377 million from US$264 million the previous year. According to data compiled by the Massachusetts Institute of Technology's Observatory of Economic Complexity, Sri Lanka exported US$128 million worth of cinnamon in 2014, which accounted for 28% of global cinnamon exports for that year.

2.5 PRESENT SCENARIO OF A FEW UNDERUTILIZED SPICES

2.5.1 LARGE CARDAMOM

Cardamom is among the world's oldest spices and is the third most expensive spice following saffron and vanilla. Large cardamom (*Amomum subulatum*) is one of the world's extremely ancient spices. It grows in the wild form and is also domesticated in the sub-Himalayan region at altitudes ranging from 1000 to 2000 m mean sea level. It is one of the main cash crops cultivated in Sikkim, Nagaland, Uttaranchal, Darjeeling and some other parts of the north-eastern region (NER) and covering an area of about 23,500 ha. The Middle East, South Asia, Southeast Asia and Europe are the main markets for cardamom consumption. Coffee consumption appears to be a strong driver of demand for cardamom in Saudi Arabia. Ready-ground cardamom coffee in a retail store in Arabia will typically amount to 5 or 10 g of ground spice per 250 g of coffee. However, for special occasions or to honour a guest with a particular display of generosity, large quantities of cardamom may be used. India is the world's largest consumer market for cardamom, but trails Saudi Arabia in imports. The United States remains a minor importer of cardamom. Guatemala is the leading supplier of cardamom in the world, producing approximately 23,000 Mt annually. The large cardamom grades are Badadana, Chottadana, Kanchicut and non-Kanchicut. Large cardamom prices reached a peak in May 2011, but have been steadily declining since then. Cardamom is graded on the basis of colour, clipping (i.e. pods with the tips trimmed), size, whether bleached or unbleached, the proportion of extraneous matter present and product origin. Because of the high value of cardamom, it is generally packaged in double-layered bags (42–50 kg) and is seldom transported in boxes without bags. Consumption of cardamom in India and Saudi Arabia strongly correlates with income trends.

2.5.2 CELERY

The native habitat of celery is the lowland of Italy from where it spread to Sweden, Egypt, Algeria, Ethiopia and India. Initially, it was cultivated as a food plant in France, in 1623. In India, it is cultivated in north-western Himalayas, Punjab, Haryana and western Uttar Pradesh. India, China, Southern France, Italy, Pakistan and Egypt are the major producers, and Southern France and China are major competitors for India. In India, Punjab, Himachal Pradesh, Haryana and Uttar Pradesh are the important celery-growing states for seed purpose, whereas in the rest parts of the country, it is cultivated as a salad crop. Punjab contributes about 90% of the total Indian production. In India, it is cultivated in the area of 4070 ha and the production was 5510 t during 2013–2014. The United States, South Africa and Netherlands, Canada, Germany, Japan, the United Kingdom and Australia are the important countries that import celery seed from India. The United States is the largest importer. Export of celery from India during 2014–2015 was 5650 t and 4302.10 (in lakhs Rupees) and it was 5600 t and 3661 (in lakhs Rupees) during 2013–2014. During 2012, the US consumers used an average of 6.0 lb of fresh celery per person per year. The US imported 93.9 million pounds of celery in 2012 valued at $16.9 million. Most of the US fresh celery imports are supplied by Mexico.

2.5.3 AJOWAN

It originated in the Eastern Mediterranean region and in Southwest Asia, probably in Egypt, and came to India with the Greeks who were called *Yavanas* by South Indians. It is very widely cultivated in black soil, particularly along the riverbank in Egypt and many other countries like India, Iran and Afghanistan. It is sometimes used as a spice mixture favoured in Eritrea and Ethiopia. It is cultivated in Iraq, Iran, Afghanistan, Pakistan and India. Ajowan is cultivated in the Mediterranean region, southwest Asian countries, Iran, Iraq, Afghanistan and Egypt; however, the major producer and exporter is India. In India, the major ajowan producing states are Rajasthan and Gujarat, where Rajasthan produces about 90% of India's total production and it grows on small-scale in Uttar Pradesh, Bihar, Madhya Pradesh, Punjab, Tamil Nadu, West Bengal and Karnataka. India exports ajowan seed to around 46 countries. The major importing countries are Pakistan, Saudi Arabia, the United States, the United Arab Emirates (UAE), Malaysia, Indonesia, Nepal, South Africa, Kenya, Bangladesh, Canada and the United Kingdom. A large quantity of ajowan seed from India, Egypt, Persia and Afghanistan has been reported to be exported to Germany for distillation of the essential oil and extraction of thymol (NHB, 2010). Iran and Egypt are major competitors for India in the international market.

2.5.4 DILL

Indian dill (*Anethum sowa*) is native to Northern India, is stronger than the European dill and is grown in Rajasthan, Gujarat, Maharashtra, Andhra Pradesh, Madhya Pradesh and Uttar Pradesh besides other states where it is grown in small pockets. In India, it is widely cultivated with the name of sowa in the states of Rajasthan, Gujarat, Jammu and Kashmir, Uttar Pradesh, Orissa, Madhya Pradesh and Punjab. European dill (*Anethum graveolens*), also called true dill, is native to Asia, Mediterranean region of Europe and Africa, and is cultivated in England, Germany, Romania, Turkey, the United States and Russia. It is cultivated commercially in Mediterranean region, Europe, central southern Asia, Netherlands, India, North America, Hungary, Germany, Romania, South Russia, Bulgaria and on a smaller scale in France, Sweden, Belgium, Poland, Greece, Spain, the United Kingdom, Turkey and the United States. Major producers are India, Egypt, Mexico, Netherlands and Pakistan. Germany and Hungary are major competitors for India in the production. In India, it is cultivated mostly in Rajasthan, Gujarat, Jammu & Kashmir, Orissa, Maharashtra, Andhra Pradesh, Madhya Pradesh, Uttar Pradesh and Punjab. The largest importer of celery seeds from India is the United States, followed by South Africa and Netherlands. India also exports celery to China, Canada, Germany, the United Kingdom, Australia and Japan.

2.5.5 CINNAMON

Cinnamon played a major role in world history by motivating Christopher Columbus to discover the new world and Vasco Da Gama to South

India and Sri Lanka. The traditionally known cinnamon was the peeled cinnamon bark rolled into the quill form, which facilitates storage and transportation. Cinnamon is native to Sri Lanka and extensively grown there with a significant share in the global trade. To extend brand promotion on 'pure Ceylon cinnamon (PCC)', the Export Development Board issued licences to qualified exporters to use the PCC logo on their value-added cinnamon exports. In 2014, Sri Lanka produced 16,230 Mt of cinnamon from 29,512 ha area and exported 13,949 Mt with revenue to the tune of Rs. 18,255 billion. During 2015, export earnings from cinnamon declined marginally by 0.2% due to the decline recorded in export volumes in spite of the increase recorded in price levels. In India, it is cultivated in the lower elevations of Western Ghats in Kerala and Tamil Nadu (Central Bank of Sri Lanka, Annual Report, 2015).

2.5.6 CASSIA

Both wild and cultivated Cassia is found in Southeast Asia, South China (Guangxi and Kwangtong provinces), Burma (Myanmar), Laos and Vietnam. It was introduced into Indonesia, Sri Lanka, South America and Hawaii. In Vietnam, it is found in many provinces from the north to the south, but is concentrated in the provinces of Quang Ninh, Yen Bai, Nghia Lo, Tuyen Quang, Ninh Binh, Thanh Hoa, Nghe An, Hue, Quang Namand Quang Ngai and in Tay Nguyen plateau. Indian imports of Cassia from Vietnam in 2015–2016 stood at 19,405 t valued at 242.22 crore (The Hindu Business Line, 2017). Presently, international cassia and cinnamon prices remain high due to lower output and increasing demand. Increase of around US$1000/t on Indonesian cassia prices has been observed since August 2017. This was mainly due to lower production in this origin (www.agranet.com, 2017). China's cassia production in 2015–2016 was around 28,000 Mt, excluding 2000 Mt carry forward stocks from the last year crop of 2014–2015, thus bringing the total availability of cassia from China up to 30,000 Mt for the year 2015–2016 (www.cassiafromchina.com, 2017).

2.5.7 CURRY LEAF

Green curry leaves are picked as soon as they are mature and have maximum flavour. The flavour of fresh curry leaves is used to enhance the taste of various cuisines. Both fresh as well as dry curry leaves and dehydrated curry leaf powder are popular marketable items. Fresh curry leaves are provided in suitable packaging to maintain the quality in handling and delivery. India exports almost 1.2 lakh kg of curry leaves to the European Union (EU). In 2013, the EU had banned importing curry leaves from India, citing concerns related to high pesticide residue. It had later stipulated that exports must be accompanied by results of sampling and analysis and by a health certificate verified by authorized representatives.

2.5.8 KOKUM

Kokum is a tropical evergreen tree of moderate to large size. It is found at an altitude of about 800 m from sea level. It originates and is grown in the Western Ghats of India, South Konkan region of Maharashtra, Coorg, Wayanad and Goa and is found in evergreen and semi-evergreen forests and as a home garden tree (Chandran, 2005). Its cultivation is confined to the coastal hilly regions of Maharashtra and Goa states and is popularly known as 'Ratamba'. Some *kokum* trees are also found in Tamil Nadu, the Western Ghats of Karnataka and Kerala as well as parts of West Bengal, Assam and Gujarat, evergreen forests of Assam, Khasi and Jaintia hills. It also flourishes well on the lower slopes of the Nilgiri Hills.

2.4.9 MINT

The main centre of origin for mint is believed to be in the Mediterranean basin and from there, it spreads to the rest of the world by both natural and artificial means. The Japanese mint is cultivated on a large scale in Brazil, Paraguay, China, Argentina, Japan, Thailand, Angola and India. Peppermint is grown in the United States, Morocco, Argentina, Australia, France, Union of Soviet Socialist Republics (USSR), Bulgaria, Czechoslovakia, Hungary, Italy, Switzerland and on a small scale, in many European countries. The United States is the major producer of peppermint and spearmint. In India, the mint cultivation is mainly confined to Uttar Pradesh and Punjab with a total area of around 10,000 ha.

2.5.10 PARSLEY

Parsley is native to southern Europe and western Asia, and it also grows in the United States, Mexico, Dominican Republic, Canada, West Germany, Haiti, France, Hungary, Belgium, Italy, Spain and Yugoslavia. Parsley is a cold weather crop, growing best in rich, moist soil. The United States is the largest producer. In India, it grows better at higher altitudes. Parsley is believed to have originated in Sardinia and has been grown in England since the middle of the 16th century (Prakash, 1990). Over the centuries, parsley has been moved from the old world to the new and is now naturalized in India, the Americas (Prakash, 1990), Australia and New Zealand (Low et al., 1994).

2.5.11 SAFFRON

It is native to Southern Europe and cultivated in Mediterranean countries, particularly in Spain, Austria, France, Greece, England, Turkey, India and Iran. Saffron cultivation has long centred on a broad belt of Eurasia bounded by the Mediterranean Sea in the southwest to India and China in the northeast. The major producers of antiquity—Iran, Spain, India and Greece—continue to dominate the world trade. In recent decades, cultivation has spread to New Zealand, Tasmania and California. Saffron is cultivated largely and harvested by hand. Saffron is considered as one of the world's most expensive spices as a large amount of labour is involved in harvesting. Saffron is a traditional production of Iran and the production share of this product is 93.7% of the total world production, 82% of it being exported. In fact, Iran is the largest producer and exporter of saffron on the international scale (Table 2.3). Major importer countries of Iran's saffron are: The UAE, Spain, Turkmenistan, France and Italy. The UAE is the largest customer with a total purchase of 78.8 t of saffron per year. Spain is the second largest with 56.4 t (Ghorbani, 2008). Consumers use saffron for medicine, in foodstuff and cooking and as a remedy for many ailments. Currently, the price of a kilogram of Afghanistan's saffron in the global markets is around $2000 (The World Bank, 2015).

TABLE 2.3 Area and Production of Saffron Cultivation in the World.

Countries	2014		2015	
	Area (ha)	**Production (t/year)**	**Area (ha)**	**Production (t/year)**
Iran	85,000	280	86,000	300
Greece	800	3	900	7
Morocco	800	3	800	4
India	950	2	950	3
Spain	120	1	50	2
Others	3000	3	2000	2

2.5.12 VANILLA

It is native to the Atlantic Coast from Mexico to Brazil. The important vanilla producing countries are Madagascar, Mexico, Tahiti, Malagasy Republic, Comoro, Reunion, Indonesia, Seychelles and India. Vanilla cultivation on a systematic basis began with the introduction of it into Java, Seychelles, Tahiti, Comoro Islands, Martinique, Madagascar, Uganda, etc. in the 19th century and early part of the 20th century. At present, Malagasy Republic is the major producer of vanilla. It was introduced in India in 1835. It is now cultivated in very limited areas in Kerala, Karnataka and Tamil Nadu.

The major vanilla producing countries are Madagascar, Indonesia, Mexico, Comoros and Reunion. In India, vanilla cultivation is increasing in Kerala, Karnataka and Tamil Nadu since the early 1990s. The area under cultivation at present is about 6470 ha with production of 1050 Mt (NHB, 2014–2015).

2.5.13 LONG PEPPER

Long pepper is native to the Indo-Malayan region. It is found growing wild in the tropical rainforests of India, Nepal, Indonesia, Malaysia, Sri Lanka, Timor and Philippines. Indian long pepper is mostly derived from the wild plants but also occurs in the hotter parts of India, from central Himalayas to Assam, Khasi and Mikir hills, lower hills of West Bengal and evergreen forests of Western Ghats from Konkan to Kerala, and also recorded from Car Nicobar Islands (Satyavati, 1987). The plant grows in evergreen forests of India and is cultivated in Assam, Tamil Nadu and Andhra Pradesh. Long pepper is cultivated on a large scale in limestone soil and in heavy rainfall areas where relative humidity is high. A large quantity of long pepper is imported by India from Malaysia, Indonesia, Singapore and Sri Lanka.

2.5.14 SWEET FLAG

Sweet flag is native to most northern latitude countries around the world, widely dispersed around the United States. It is found wild or cultivated in India and Sri Lanka up to 1800 m. Sweet flag thrives best in marshy and moist places under variable climates. The plant is grown in clayey loams and light alluvial soils of riverbanks. It is distributed throughout the tropics and subtropics, especially in India and Sri Lanka. It is found in marshes, wild or cultivated, ascending the Himalayas up to 1800 m in Sikkim. It is plentiful in the marshy tracts of Kashmir and Sirmoor, and in Manipur and Naga Hills. It is regularly cultivated in Koratagere taluk in Karnataka. It is now found widely on the margins of ponds and rivers in most English countries.

2.5.15 HORSERADISH

It is native to Eastern Europe and also grown in the United States, to a small extent in gardens both in North India and hill stations of South India. In the United States, about 7 million kg of horseradish are processed annually for consumption as food. Consumption of freshly grated horseradish is growing fast and is the main engine of growth in the horseradish segment of Austria. Around 20% of the total prepared horseradish volume of Austria goes abroad. One of the main export countries is Germany. Emerging markets are Russia, Hungary and Switzerland (http://www.steirerkren.at/en/strong-demand/facts-and-figures.html).

2.5.16 ASAFOETIDA

The species are distributed from the Mediterranean region to Central Asia. In India, it is grown in Kashmir and in some parts of Punjab. The major supply of asafoetida to India is from Afghanistan and Iran. There are two main varieties of asafoetida: *Hing Kabuli Sufaid* (milky-white asafoetida) and *Hing Lal* (red asafoetida). Asafoetida is acrid and bitter in taste and emits a strong disagreeable pungent odour due to the presence of sulphur compounds therein. The white or pale variety is water soluble, whereas the dark or black variety is oil soluble. Since pure asafoetida is not preferred due to its strong flavour, it is mixed with starch and gum and sold as compounded asafoetida mostly in bricket form. It is also available in free-flowing (powder form) or in tablet forms.

2.5.17 BAY LEAF

This ornamental tree is indigenous to the Mediterranean area and the southeast part of Europe (Stace, 2010). It is widely cultivated in Europe, Iran, California, America and Arabian countries. It is not cultivated as a commercial crop in India.

2.5.18 NUTMEG AND MACE

It is believed to have originated from Moluccas Islands (Indonesia). Indonesia is the worlds' exporter (50%) of nutmeg and mace is the world. Grenada is the second largest exporter of nutmeg and mace in the world. In India, nutmeg is mainly cultivated in Thrissur, Ernakulam and Kottayam districts of Kerala and parts of Kanyakumari and Tirunelveli districts in Tamil Nadu.

Indonesia produces more than 75% of the world's total nutmeg and mace. However, an export regulation introduced by the Indonesian government in 1986 facilitated cartehsation of the market leading to a dramatic rise in world prices. Later in 1990, accumulation of unsold stocks in Indonesia and smuggling of nutmeg and mace through Singapore led the government to abolish the regulations. As a result, world prices fell steeply and since then have remained low. With Grenada still struggling to recover the output after being hit by Hurricane Ivan in 2004, Indonesia holds position as the main global producer. Up to the end of 2016, the Indonesian nutmeg and mace market remains mostly unchanged from its state of play back (Gale, 2016).

2.4.19 BASIL

It is indigenous to the lower hills of Punjab and Himachal Pradesh and is cultivated throughout India. It is widely cultivated in Iran, Japan, China and Turkey and also in Southern France, Egypt, Belgium, Hungary and other Mediterranean countries as well as in the United States.

2.4.20 ROSEMARY

Rosemary is native to the Mediterranean region and is cultivated in Europe and California in the United States. It is also grown in Algeria, China,

Middle East, Morocco, Russia, Romania, Serbia, Tunisia, Turkey and to a limited extent in India. Temperate climate is suitable for the cultivation of rosemary. The soil properties influence the yield and the composition of rosemary oil.

2.4.21 THYME

It originated in the Mediterranean region, northern Africa and parts of Asia (Montvale, 1998). It is native to Southern Europe. Thyme is produced from cultivated and wild harvested plants in most European countries, including France, Switzerland, Spain, Italy, Bulgaria, Portugal and Greece. Apart from Europe, it is grown in Australia, North Asia, North Africa, Canada and northern United States. In India, it is cultivated in the western temperate Himalayas and the Nilgiris. Most of the thyme produced is for the fresh and dried market. Yields of *T. vulgaris* for fresh herb production can be 5–6 t/ha and for dry herb production can be 2 t/ha. Under irrigation, thyme will yield about 15 t of plant material per hectare per year, at an oil recovery rate of 0.5–1% or 75–150 kg/ha/year. Thyme oil displays anti-ageing properties in mammals. Therefore, on the world market, thyme oil is among the top 10 essential oils and the demand continues to rise. Some 90% of the thyme oil of world trade is produced in Spain. The recent report on thyme essential oil market throws light on the various factors governing the market across the globe. It forecasts the market size and revenue to be generated by each of the segments. The present slow growth of the economy and the impact of the government's latest initiatives have been taken into account while forecasting the growth of the thyme essential oil market.

2.4.22 ANISEED

The importance of anise throughout history dates back to the Roman Empire, where it was used as a spice and fragrance. It began being cultivated in England as a food crop in the 1500s and has since spread throughout the world, becoming a staple spice and popular herbal remedy in many different cultures. Western cuisines have long used anise to flavour dishes, drinks and candies. Anise seed has a wonderful sweetness and hints of liquorice flavours. It provides a perfect balance to the acidity. The plant is native to Asia Minor and Egypt, and is cultivated in central, southern and Eastern Europe, erstwhile southern USSR, Macedonia, Syria, Tunis, Morocco, India, China, Chile and Mexico. Production is primarily from Turkey and China.

2.4.23 STAR ANISE

Star anise is indigenous to south-eastern China. Commercial production is limited to China and Vietnam. In India, it is produced to a small extent in Arunachal Pradesh. This star-shaped spice has an intensely sweet liquorice taste and smell, and has many uses both alone and in combination. It is a key ingredient in the Chinese five-spice powder. A chemical compound vital for Tamiflu synthesis can be extracted from star anise in relatively high quantities. During the 2006s flu scare, about 90% of the world's pharmaceutical star anise was coming from Guangxi province in China (Garner-Wizard, 2005). Prevailing wisdom among developed nations and the mega pharmaceuticals they harbour states that the developing world ought to be able to manage handsome profits from their bioresources like star anise.

2.6 SWOT ANALYSIS

It would be wise to see underutilized spices as part of a whole in harnessing the role of spices for a particular region, rather than as 'stand-alone' crops, thereby to achieve a holistic step forward towards a desired goal. The strengths, weaknesses, opportunities and threats on underutilized spice crops are discussed in subsequent text.

2.6.1 STRENGTHS

- Wide genetic diversity
- Vibrant industry and trade
- Availability of location-specific technological inputs—integrated pest management, integrated disease management and Integrated nutrient management

2.6.2 WEAKNESSES

- Stagnant productivity
- Absence of high degree of resistance to biotic and abiotic stresses
- Lack of quality planting materials
- Low adoptability of technologies, mechanization and hygienic practices
- High contaminants
- Unable to meet global safety standards
- Price instability
- Lack of effective transfer of technology

2.6.3 OPPORTUNITIES

- As these crops possess very high therapeutic values, growing awareness and health consciousness among people will escalate the industrial demand and their products in the future years
- Newer applications in functional food, bioactive constituents and pharmaceuticals
- Increasing demand in non-traditional areas, having intercropping potential with many plantation crops
- Strengthening of existing germ plasm stock by exploiting maximum indigenous and exotic variability available for broadening the genetic base. Molecular profiling of the available biodiversity in these spices
- Development of a network of certified nurseries and biocontrol units in private–public partnership mode
- Development of processing technology for value-added products
- Protected cultivation, urban cultivation and off-season cultivation
- Development of standard package of practice on organic farming, good agricultural practices and contract farming
- Intellectual property rights and farmers varieties—participatory plant breeding

2.6.4 THREATS

- International competition, biopiracy and patenting
- Sanitary, food safety and biosafety issues

- Market fluctuations
- Declining labour force, soil fertility and water resources
- Climate change

KEYWORDS

- **pharmaceuticals**
- **food security**
- **export markets**
- **global economy**
- **SWOT analysis**

CHAPTER 3

UNDEREXPLOITED SPICE CROPS: AGROTECHNIQUES

CONTENTS

3.1 LARGE CARDAMOM

3.1.1 SYSTEMATIC POSITION

Kingdom: Plantae

Division: Zingiberales

Class: Liliopsida

Order: Zingiberales

Family: Zingiberaceae

Genus: *Amomum*

Species: *subulatum*

Large cardamom (*Amomum subulatum* Roxb.), a member of Zingiberaceae family under the order Scitaminae is a perennial soft-stemmed plant, commonly known as large cardamom. It is one of the ancient spices in the world. It grows in the vicinity of mountain streams in swampy, cool and humid areas in the shade of forest trees, of which nitrogen-fixing trees are the more suitable shade trees. It is domestically grown in the sub-Himalayan region, at altitudes ranging from 1000 to 2000 m mean sea level (MSL). It is mainly cultivated as cash crop in the eastern Himalayan region including Sikkim and the Darjeeling hills in India, the eastern part of Nepal and southern Bhutan (Sharma et al., 2000). Sikkim is the largest producer of large cardamom in India and second largest in the world, after Nepal. Nepal and Bhutan are the other countries where large cardamom is cultivated. It is used as a spice and in several Ayurvedic preparations. It contains 2–3% of essential oil and possesses medicinal properties such as carminative, stomachic, diuretic, cardiac stimulant, antiemetic, etc. The characteristic pleasant aromatic odour of large cardamom makes it popular to be extensively used for flavouring many food preparations in India.

3.1.2 BOTANY

The plant is perennial bush growing in shady areas with sheathed stem, growing to a height of 2–5 m and the root is a tuberous rhizome. Leaves are 30–60 cm in length and 5–15 cm in width. Soft pods are born on the leafy stalk that grows from the plant base at ground level. The inflorescence is spike, short in nature and produces dense flowers when directly arise from the base of the plant. The flowers are green in colour with a white-purple vein tip. Capsules are harvested before they are ripe to avoid them from splitting during the drying process.

3.1.3 CLIMATE AND SOIL

Large cardamom falls under the category of pseophyte (shade-loving plant). Humid, subtropical and semi-evergreen forests on the steep hills of the sub-Himalayan region are the natural habitat of large cardamom. It grows at an altitude of 600–2350 m. It is cultivated at higher altitudes in warmer areas and lower altitudes in cooler areas. By being dormant, it can withstand up to a temperature of 20°C during winter. Hailstorm and frost are dangerous. Continuous rainfall during flowering is detrimental as it hampers the activity of pollination by bees. This affects the flowering and results in poor capsule formation.

Deep, well-drained loamy soils containing medium availability of phosphorous and potassium having a pH of 4.5–6 are suitable for cultivation of large cardamom. Soil testing is not done in areas of north-eastern regions where large cardamom is cultivated. These soils are rich in organic matter and nitrogen as the plants are cultivated under local varieties of trees.

3.1.4 CULTIVATION AND INTERCULTURE

Ramsey, Sawney and Golsey are among the few varieties of large cardamom suitable for cultivation at an altitude of 1500 m. The foliage varies in colour from green to light green with stem appear in maroon colour. Sawney grows best at an altitude of 1000–1500 m. Plants are tall and robust with dark green leaves and greenish to purple stem. Plants flower during May and the flowers are yellow in colour with maroon colour capsules. Golsey grows best at an altitude of below 1000 m. The leaves are dark green in colour with greenish stem. The fruits are oval in shape. Other popular cultivated varieties are Bebo, Bharlangey and Ramla. Some popular cultivars of large cardamom cultivated in north-east regions are Ramsey, Sawney, Golsey, Varlangey and Seremna. Bebo, Boklok Tali, Jaker and Belak, etc.

Large cardamom propagation is done through seeds and suckers. Propagating materials such as seeds enable to produce large number of seedling. Planting material should be free from viral diseases. Protection of the nursery should be given for fresh infection by giving the adequate care. Plants raised from seeds need not necessarily be high yielders even if they are collected from very productive plants due to cross-pollination. The pollination is done by wild bees and honey bees. Planting suckers should be true to type and high productivity if they are collected from high yielding plants.

Virgin forest soil and loamy soils with gentle to medium slope are well suitable for cultivation of large cardamom. Plants grow well in shade, whereas waterlogged conditions are detrimental. Utis (*Alnus nepalensis*) is the most common and preferred shade tree for large cardamom, whereas the other shade trees for growing cardamom are siris (*Albizzia lebbeck*), panisai (*Terminalia myriocarpa*), malito (*Macaranga denticulate*), asare (*Viburnus eruberens*), pipli (*Bucklandia* spp.), argeli (*Edgeworthia gardneri*), kharane (*Symplocos* spp.), bilaune (Maesa cheria), dhurpis and Khasi cherry, katuse, Faledo (*Erythrina indica*), jhingani (*Euria tapanica*) and chillowne (*Schima wallichii*).

June to July months, when the soil contains enough moisture, are suitable for cultivation of cardamom. The land selected should be free of weeds for new planting. Old plants should be removed if it is replanting. Pits of size 30 × 30 × 30 cm are dug for the preparation of contour of the hill. For normal cultivation, the spacing of 1.5 × 1.5 m is sufficient but for robust cultivars such as Sawney, Varlangey, Ramsey, a spacing of 1.8 × 1.8 m is required. The pits are filled with top soil mixed with farm yard manure (FYM) at the rate of 1–3 kg/pit and left for weathering for a fortnight. The suckers are planted in the middle of the pit. Planting should not be very deep in the pit. Seedlings/suckers are staked after planting and the base of the plant is mulched with dry leaves. Optimum soil fertility should be maintained, well-decomposed FYM and oil cakes may be applied at the rate of 2 kg per plant at least once in 2 years during April to May. Application of inorganic fertilizers may not be necessary if all the crop residues are recycled in the plantation.

Weed control in the plantations is necessary for the maximum utilization of the available soil moisture and nutrients by the plants. Weeds can be effectively controlled by either hand weeding or sickle weeding depending upon the intensity of weed growth. Weeding around the plant base can be pulled out by hand and the weeds in the interspace need only be slashed with a sickle. Dried shoots and other trashed materials can be used as mulch around the plant base to conserve soil moisture and to prevent weed growth.

Water sources should be available in large cardamom growing areas for irrigating the crop by gravity flow, either through pipes, sprinklers or flood irrigation through open channels. It is observed that where irrigation is provided, the productivity will be higher. During the dry month, watering is to be done for better plant yield. Depending on the availability of water sources, sprinkler or flood irrigation channels can be adopted. Hose irrigation can be done at the rate of 40–50 l per plant at fortnightly intervals.

3.1.5 PLANT PROTECTION

Large cardamom is free from the attack of major pests. However, the sporadic incidence of leaf-eating caterpillars (*Artona chorista*) is mainly observed during May to July and October to March. The caterpillar feeds on the lamina from under the surface of the leaf and finally defoliates the leaf completely leaving the midribs. These insects are controlled by their natural enemies. If insecticides are used to control them, the natural enemies will also disappear leading to the outbreak of these pests in epidemic form. The best management is to inspect the plantations during the month of May to July and October to March, to handpick the infected leaves along with the caterpillars and destroy them by burning. However, cardamom is seldom attacked by fungal and bacterial diseases. Only minor diseases such as leaf streak or rot diseases are found in isolated areas. The major diseases found in large cardamom are viral diseases, namely chirke and foorkey. They are mainly observed in the large cardamom growing tracts of Sikkim and Darjeeling and cause considerable crop loss. These diseases spread due to a drastic change in the ecosystem, inadequate rain in dry months, failure in planting varieties suitable to their altitude and absence of good agricultural practices by the farmers. The chirke disease is characterized by the appearance of pale streaks slowly turning to brown on the tender leaves as well as withering and drying of the plants. This leads to reduction in growth and yield of the affected plants and ultimately they perish. It is transmitted by aphids and also by using infected plant suckers. It is also transmitted mechanically through the garden tools which are used for harvesting. In foorkey, the number of small tillers appear at the base of the affected plants which become stunted and fail to give any yield. Even the inflorescence is noticed to produce unproductive spikes. Plants affected by the viral diseases cannot be cured but the losses can be reduced by adopting appropriate management practices, keeping constant vigilance for detection of disease-infected plants, uprooting and destroying the affected plants immediately after the occurrence of symptoms, repeating detection and uprooting at regular intervals. Moreover, seedlings produced in certified nurseries should be used. Propagation by suckers is recommended only through certified multiplication nurseries.

3.1.6 HARVESTING AND POSTHARVEST TECHNOLOGY

First harvesting can be done in 2–3 years after planting of sucker or seedlings. However, stabilized yields are obtained from the fourth year up to 10–12 years. Yield depends on the proper management practices followed by farmers such as regular rouging coupled with replanting, weeding, mulching plant bases, winter/summer irrigation, shade regulation, etc. harvesting of large cardamom starts in August/September in areas of low altitude and continues up to December in high-altitude areas. Harvesting is done only in one round. Hence, the harvested produce contains capsules of various maturity stages. Harvesting should be done when the seed of the top capsules in the spike attain dark grey colour. *Elaichi chhuri*, a special type of knife is used for harvesting of large cardamom. The stalk of the spike is cut very close to the leafy shoot. Individual capsules are separated manually after harvesting and cured to induce moisture level to 10–12%. Curing is done in two ways: direct heat drying *which is Bhatti* curing system and another is indirect heat drying. Large cardamom is also cured by flue pipe curing system.

Cardamom dried in kilns, locally called *Bhatti*. It consists of a platform made of bamboo mats/wire mesh, laid over a four-walled structure made of stone pieces with a V-shaped opening in the front for feeding firewood. Capsules are spread over the platform and are directly heated by the heat generated from the firewood. For curing 100 kg of green cardamom in this traditional kiln, 70–80 kg of firewood is required. Both green and dry woods are used so as to generate huge volume of smoke for curing the cardamom. It takes 60–72 h for curing depending on the thickness of the cardamom spread. Cardamom cured under this system turns dark brown or black in colour. If smoke percolates through cardamom, it loses its original colour, gets smoky smell and fetches low price.

A construction with corrugated tin sheets as roof and fitted with furnace, flue pipes made of galvanized iron sheets of 20-gauge thickness and 25-cm radii and chimney ventilation system, etc. are required for flue pipe curing system. The flue pipes run from an end to other along the centre of the room below the wire gauge fittings. The pipe is connected to the furnace at one end and the other end to the chimney to lead the smoke up through the roof in the air. Water vapour is let out through ventilation provided in the room during the process of drying. Ventilation and charging of fuel are regulated based on the temperature inside the room which is known through the thermometer kept in the room.

When the temperature inside the chamber reaches 40°C, cardamom is spread over the wire mesh floor and shelves in one or two layers. After spreading of cardamom, the curing room is kept closed and the room temperature is brought to 45–50°C by passing the hot air through the furnace into the flue pipes. This temperature is maintained for about 3–4 h. Sweating of capsules takes place and moisture is released at this stage. Ventilators are then opened for letting out the accumulated moisture completely from the chamber. Ventilators are then closed and temperature is again maintained at 40–45°C. For uniform drying, capsules are stirred one or two times. It is important to note that curing once started should be a continuous process till the drying is over. The whole process of curing takes about 28–29 h. The cured cardamom capsules are collected in trays and rubbed or processed in cardamom polishing machine for removing the tail. The cleaned product is then packed in polythene lined gunny bags and stored in wooden boxes. Cured cardamom on an average gives 25% by weight of fresh cardamom.

Drying of cardamom in flue pipe curing system has many advantages over the Bhatti system. In flue pipe curing system, the original colour (pink) and flavour (sweet camphor aroma) of cardamom is retained, better market price is realized, curing expenses are low, firewood consumption is less, total curing takes only 28–29 h and uniform drying is ensured.

Machines for removing capsule tails in cardamom have not been developed. The tails are partially detached when the capsules are rubbed against the wire mesh after curing. During this process, the bristly outer layer of capsules may also be removed. Usually, the tails of capsules are cut manually with scissors by local traders. As revealed by local dealers in Sikkim, extra labour needs to be employed for this labour-intensive step at the cost of about US$0.41/kg of capsules. College of Agricultural Engineering and Post Harvest Technology in Ranipool, Sikkim, and the Indian Cardamom Research Institute (Spices Board) in Tadong, Sikkim, jointly invented a cardamom polisher for possible use as a tail-cutting machine (Yurembam, 2010). The Indian Cardamom Research Institute is also in search of alternative capsule–tail-cutting devices. Such a device will have a strong potential application in the large cardamom value chain, provided it is cost-effective.

Cardamom capsules after drying are usually packed in polythene-lined jute bags. Polypropylene and ethylene terephthalate/polyethylene have been reported to considerably reduce moisture and volatile oil exchange under normal storage conditions (Sulochanamma et al., 2008). The hot capsules taken out of the curing chamber are allowed to cool and then are placed into the bags, which are sealed and stored on wooden platforms to avoid moisture absorption, which could lead to mould growth. Storage stability has been maintained for large cardamom capsules with up to 11% moisture content (Naik et al., 2000). Loss of capsule weight and insect damage can also occur during storage. Moisture content of 13–15% is conducive to insect breeding; therefore, the Central Food Technological Research Institute in Mysore has recommended use of fumigants such as methyl bromide (0.016 kg/m^3), phosphine (0.0015 kg/m^3) and ethyl formate (0.30 kg/m^3) to control insect infestation without affecting quality (Naik et al., 2005). No report addresses the need for consumer packs (smaller packages) suitable for local markets, which may also have benefits in this regard.

Cardamom capsules after finishing are commercially graded in local markets in the following names such as badadana (big capsules) or chotadana (small capsules) and, as discussed previously, as kainchi-cut (capsule tail removed) or non-kainchi-cut (capsule tail intact) (Sharma et al., 2009). The difference in capsule size may be due to the difference in cultivar or preharvest conditions. For example, Golsey cultivar yields bigger capsules. Grading of capsules based on size is done using manual screens. Use of mechanical grading machines is so far not reported, except for manually operated sieves in Nepal. Quality grading is only done by local dealers and wholesalers, who employ large numbers of labourers for this purpose. The Bureau of Indian Standards (1999) has established quality standards for large cardamom capsules based on the Prevention of

Food Adulteration Act of 1954. Some importing countries allow only the product that conforms to these standards.

3.2 CELERY

3.2.1 SYSTEMATICS

Division: Spermatophytes

Subdivision: Angiospermae

Class: Mangnolisisa

Order: Apicedes

Family: Apiaceae

Genus: *Apium*

Species: *Graveolens*

Celery (*Apium graveolens* L.) also known as 'Karnauli' or 'Ajmod' is an important minor spice that grows in India. It is an aromatic, herbaceous and hardy biennial that belongs to the family Umbelliferae such as fennel, parsley, coriander and dill. It is mainly grown for its seeds, leaves, oleoresin and essential oil and also for the long fleshy leaf stalk used as salad but the dried fruits or seeds (produced in the second year) which are used as spice are of economic importance.

3.2.2 BOTANY

This herbaceous plant has a well-developed succulent tap root system and a branched stem with ridged angular jointed branching. The leaves are pinnate or trifoliate, 7–18-cm long with an oblong shape. The inflorescence is a sessile compound umbel with a flower head producing mass of fruits. The flowers are very small and white or creamy white or greenish white in colour with five petals that are ovate and cute with inflexed tips. The carpels are semi-terete, sub-pentagonal and the primary ridges are distinct and filiform. The fruit consists of two united carpels each producing a small single seed which is about 1.5–2 mm in length, ovoid to globose and greenish brown. Celery is naturally cross-pollinated but not self-incompatible.

3.2.3 CLIMATE AND SOIL

Celery is a cool season biennial and moisture-loving plant that is grown as a winter crop in colder climates and on the hills, whereas in plains it is annual. It requires cool and warm dry climate during growth period but prolonged cold temperatures are not congenial for crop growth. Optimum temperature is between 16 and 18°C. It can be grown in wide range of soils; however, well-drained loamy soils with optimum pH of 6–7 are better. Saline, alkaline and waterlogged soils are not suitable for celery. A combination of 12–15 and 22–25°C day and night temperature, respectively, gives 80% seed generation within 2-week period.

3.2.4 CULTIVATION AND INTERCULTURE

Ajmer Cellery-1 is a new variety developed through Punjab local selection at National Research Centre on Seed Spices (NRCSS), Ajmer, Rajasthan in 2006. It is suitable for cultivation in semiarid regions under irrigated conditions. It produces an average yield of 8.01 q/ha with essential oil content of 2.4% from seeds. The other high yielding varieties include exotic collection (EC)-99,249–1 and Regional Research Laboratory (RRL) 85–1.

Celery is propagated by seeds that can be sown directly or transplanted. The seeds are small, irregular in shape and difficult to sow. Usually, these are sown in the nursery and then transplanted after 2 months by raising the seedlings in the nursery beds. A seed rate of 1.5 kg/ha is sufficient. They germinate slowly and young seedlings require special attention.

To raise nursery, the land should be brought to fine tilth by two to three ploughing and manures and fertilizers should be mixed in soil properly at this time. The bed should be of 3 × 1 m size in about 200 m^2 area for transplanting 1 ha land. The first fortnight of November is the most suitable time for sowing of seeds in a nursery. About 2 kg of seeds are sufficient for 200 m^2 nursery area. At the time of land preparation, 500–600 kg of FYM should be mixed in soil along with 1.6 kg nitrogen, 0.8 kg phosphorous and 0.4 kg potash as basal dose before seed sowing. As the seeds are minute, required quantity of seeds is mixed with dry sand in the ratio of 1:5. Beds are mulched with thin layers of soil and FYM mixture till germination. The germination percentage in seeds is about 50%. Hand weeding for two to three times should be done depending on weed growth. First, light irrigation is given just after sowing and second after 5–8 days after sowing

(DAS) and afterwards irrigations are given as and when required to keep the soil moist up to full germination. The irrigations are then given at an interval of 10–15 days. The seedlings of 50–60 days old are ready for transplanting when they are about 10-cm tall.

In the main field, land should be brought to fine tilth by one to two ploughing followed by two or three harrowing and planking. Celery has a small root system and is a poor nutrient remover, so there needs to be a good supply of nutrients in the soil. Before planting, required amount of well-rotten organic matter or compost, and a complete fertilizer is to be incorporated. Transplanting of seedling in main field should be done during evening time and it should be ensured that optimum moisture is available in the soil. Suitable time for taking up transplanting of celery is the first fortnight of January. The seedlings should be transplanted at a spacing of 45 × 20 cm. Balyan et al. (1990) obtained a mean seed yield per plant much lower (12.65 g) at 30-cm spacing than at 40–70-cm spacing (18.80–19.95 g), whereas mean seed yield and oil per hectare were maximum (1503 and 32.33 kg/ha, respectively) at 40-cm spacing justifying it to be as the most economic one.

Once the crop is established, it requires three to four hoeings to keep the weeds under check. Care should be taken so that the roots are not injured by implements, as they are shallow. First intercultural operation/ hoeing should be done 20–25 days after transplanting. Intercultural operations should be repeated thrice based on need at an interval of 20–25 days. If necessary, one or two hand weedings can also be done.

Celery is a shallow rooted crop with high moisture requirements and hence it requires irrigation. Water requirement of celery sharply increased at the start of heart development at 8 weeks after planting, reached maximum at 16 weeks and declined during ripening. Irrigation at the time of seed development is critical. First irrigation should be applied on the day of transplanting and subsequent irrigations should be given at fortnightly intervals and at weekly intervals when nearing seed maturity. The increase in seed yield under more frequent irrigation (50 or 25/50 mm cumulative pan evaporation [CPE]) may be attributed to improvement in yield attributes (number of umbels/plant, seed weight/umbel and test weight) compared with irrigation at 75 and 100 mm CPE (Saini et al., 1989). The crop requires about 10–12 irrigations during the crop period.

Celery being shallow rooted crop draws majority of nutrients from the upper surface of the soil as it requires quite high nutrients. The general recommended dose is 25–30 Mt FYM which is mixed in soil at the time of land preparation and 80 kg nitrogen applied in splits of 40 kg as basal dose and 40 kg as top dressing, 30 days after transplanting and the entire dose of 40 kg phosphorous and 20 kg potash per hectare are applied as basal dose in rows. Saini et al., in the year 1989, reported that an increase in N rate from 50–150 kg/ha increased the seed yield. Similar findings were obtained from Shukla and Sharma (1996) with 30–50 ha of FYM before planting followed by 80–200 kg nitrogen in two splits, half as basal and remaining half after 1 month as top dressing and full dose of 30–40 kg phosphorus, 20 kg potassium per hectare at the time of planting. Magnesium deficiency can be controlled by spraying magnesium sulphate at the rate of 12 kg/ha.

3.2.5 PLANT PROTECTION

Moderate temperatures with adequate water suits best for celery growth. This cool and cloudy situation makes free water to remain on foliage for prolonged periods which promotes to cause several leaf diseases and it might be a limiting factor for successful celery production. The diseases include early blight caused by *Cercospora apii* which shows symptoms of ash-grey necrotic leaf spots that are about 10 mm in diameter. In the beginning, they appear as yellow circular spots on leaves which on petiole infection turn as elongated lesions. Two species of *Septoria* cause late blight and leaf blotch, respectively. The former is more prevalent, and incidence is favoured by mild temperatures and free moisture. Chlorotic spots occur on mature leaves and stalks, later turning grey with black pinpoint-sized pycnidia (spore-forming structures) embedded in the spots. Under severe conditions, these leaves will blacken and the plant withers. Copper oxychloride can be fairly effective as a protective spray against late blight infection. Prophylactic sprays of several fungicides can effectively control Cercospora leaf spot which are not currently registered but hoped for registration for use on celery. The causal fungi of the above diseases can be transmitted on celery seed. If seed transmission occurs then infection of young seedlings will be initiated in the seedbed. To avoid this, seed should be heat treated where fresh seeds are held in water preheated to a temperature of 48°C, for 30 min (as a comparison, the fresh cruciferous seed is treated at 50°C).

Similar to the diseases, there are insects such as cutworms that are problematic immediately after transplanting. A number of insecticides are registered for cutworm control. Some may be used as bait formulations, but most would be sprayed onto the damp soil in the row, or applied next to the seedling transplants and gently watered into the soil. Of other insect pests, chewing insects such as armyworm and Plusia looper might attack leaves. Aphids might also become a problem. The red spider mite is another pest which is often not noticed. No specific insecticides or miticides are registered for use on celery.

3.2.6 HARVESTING AND POSTHARVEST TECHNOLOGY

Harvesting is done during May in plains and November, March to April in hills. It should be done when about 80% umbels turn to light brown colour. After harvesting, the crop should be dried on a clean surface and threshed to separate the seeds. Grading is done with the help of sieve or vibrator. Seed crop is harvested during March in plain and June to July in hills. In cooler climates and on the hills, celery is a perennial plant and produces seeds only in the second year. It takes about 4–5 months from the time of sowing to seeding. In the plains, the crop matures within about 3 months of transplanting.

The average yield of celery is about 1000–1200 kg/ha of dried seed. The seeds yield 2–3% of pale yellow volatile oil with a persistent odour. The volatile or essential oil seed is isolated by steam distillation. The seeds should immediately be sent for distillation to avoid the loss of oil by evaporation. The wastes from the distillation are usually redistilled.

3.3 AJOWAN

3.3.1 SYSTEMATIC POSITION

Kingdom: Plantae

Division: Magnoliophyta

Class: Magnoliopsida

Order: Apiales

Family: Apiaceae

Genus: *Trachyspermum*

Species: *ammi*

Ajowan (*Trachyspermum ammi* L.), also known as Bishop's weed, Carom seed and ajowan or omum, belongs to the family Apiaceae and is a popular seed spice in India. *Trachyspermum ammi* (Linn.) is a Greek word where *Trachy* means rough and *spermum* means seeded, whereas *ammi* is the name of the plant in Latin. The name ajowan is originated from the Sanskrit words *yavanaka* or *ajomoda*. It is an aromatic herb (2n=18) having a cross-pollinating breeding behaviour. It is the native to India and has a huge genetic variability for different traits. The economic part is seed which is used as spice. It is grown mainly in plains, but flourishes equally well at higher altitudes, in plateaus and on hills. The fruit pods are sometimes called seeds as they have seed-like appearance.

3.3.2 BOTANY

Ajowan is an annual herb which is small and erect growing to a height of 60–90 cm. The stems are straight, branched with leafy stems and small feathery leaves containing 4–12 rays of flower heads each bearing 6–16 flowers. The fruits are minute, aromatic, ovoid, greyish browned, cordate and cremocarp with a persistent stylopodium. It has two mericarps compressed and is about 2-mm long and 1.7-mm wide with five ridges. The inflorescence is a compound umbel with 16 umbellets each containing up to 16 flowers that are actinomorphic, white, male and bisexual. The corolla and stamens are five alternating with bilobed petals with inferior ovary and a knob-like stigma. Leaves are pinnate with a terminal and seven pairs of lateral leaflets. It is a cross-pollinated crop with self-fertile flowers, but cross-pollination occurs through insects.

3.3.3 CLIMATE AND SOIL

Cool weather and cloudiness for about a week after sowing and occasional drizzling during active growth are conducive to the successful cultivation of ajowan It can be grown in all kinds of soils but does well in loams or clay loams rich in humus and well drained. In India, Afghanistan and Egypt it is grown under both rainfed and irrigated conditions. Cool and dry climate favours good growth but the crop is affected by frost. High atmospheric humidity invites diseases and insects hence avoidance of high humidity especially after flowering is beneficial.

3.3.4 CULTIVATION AND INTERCULTURE

Several varieties are popular and grown for the quality production of this seed spice. Among them, there are few Indian varieties that are discussed below.

Gujarat ajowan-1 is a selection from germ plasm developed at Sardarkrushinagar Dantiwada Agricultural University, Jagudan. It is a late maturing variety that takes 176 days and yields about 2269 kg/ha. It is non-shattering and is mildly susceptible to powdery mildew and resistant to insects.

Pant Ruchika is a pure line selection for local collection is another late maturing variety that takes 170–175 days is developed from G. B Pant University of Agriculture and Technology, Pantnagar, Uttaranchal in 2001. Average yield is 600–800 kg/ha. The plant is erect and bushy and the seed is light brown and attractive.

RFA-68 is a selection from local germ plasm that grows in Pratapgarh area is a medium maturity variety that flowers in 90 days and takes about 150 days to mature and is developed at A. R. S. Substation, Udaipur Agricultural University, Rajasthan. It yields on an average 900 kg/ha.

Ajmer ajowan-1 (AA-1) is a variety developed at NRCSS that yields about 500–1400 kg/ha and is a selection from Pratapgrah local NRCSS AA-61. It is tall, late maturity group (160 days) and is suitable for early and rabi sowing under irrigated and limited available water conditions. The seeds are medium size and contain 3.4% volatile oil.

Ajmer ajowan-2 (AA-2) is a selection from Gujarat local NRCSS AA-19 developed at NRCSS. It is a bushy plant, early maturing in 147 days and is moderately tolerant to drought. Both rainfed and irrigated conditions suit best for its growth and yields on an average 500–1200 kg/ha. The seeds are medium with 3.0% volatile oil.

AA 93 is an early maturing variety developed at NRCSS, Ajmer. It starts flowering in 46 days after sowing and matures in 123 days only. It matures about 30–40 days early in comparison to normal existing improved cultivars which take 150–170 days. The plant is erect and can resist the lodging caused due to dew. Average yield is 900–1400 kg/ha.

Lamsel-1 and Lamsel-2 are developed at PRS, ANGRAU, Guntur, Andhra Pradesh through mass selection. The former is a tall early maturity (140 days) variety yielding 1000–1400 kg/ha and the latter is a spreading bushy type with more branches, requiring more spacing yielding 1000–1200 kg/ha.

Rajendra Mani is developed from Department of Horticulture, Tirhut College of Agriculture RAU, Dholi, Bihar.

Azad ajowan-1 is a new high yielding disease resistant variety developed at Department of Vegetable Science, C. S. Azad University of Agriculture and Technology, Kanpur through thorough screening of the available pure genetic stocks for various yield contributing character and other desirable traits including major diseases. Then based on the earliness, yield potential and disease reaction, the accession 9401 was found most promising by giving 15 days earlier yield of 11.74 q/ha, 56.74% higher over the check. It was also found resistant to *Sclerotinia sclerotiorum* infection under artificial epiphytotic conditions. However, that accession 9401 was released as a variety with the name of Azad ajowan-1 in the year 2001 by the U. P. State Variety Release Committee.

Ajowan is propagated by seed. The soil is reduced to fine tilth and prepared for sowing in October to November in most parts of India and May to June in some regions. Choosing an appropriate planting date is one of the most important cropping techniques by which environmental factors such as moisture, temperature and light can be optimally controlled. Early planting increases the probability of frost injury but seed yield and essential oil per cent are high (Mohammadjavad et al., 2014), whereas late planting has negative effects on the biological and seed yield due to shortened growth period. According to Naruka et al. (2012), line sowing at a spacing of 45 × 30 cm resulting in high yield potential is in practice but in some areas a spacing of 30 × 20 cm is also followed. The experimental results of Muvel et al. (2015) with a spacing of 45 × 30 cm significantly increased the plant height, fresh weight per plant, dry weight per plant, number of umbels per plant, number of umbellets per umbel, 1000 seed weight, yield per plant, seed yield, straw yield, biological yield, chlorophyll content of leaves, carotenoids content of leaves and essential oil content in seed and crop geometry of 50 × 25 cm resulted 10% higher seed yield (Mehta et al., 2015) and higher seed yield, biological yield were obtained at 30 × 15 cm row spacing (Tripathi and Dwivedi, 2009).

About 3–4 kg of seed is required for sowing in 1 ha area. Premnath et al. (2008) experimented with sowing on October 30 which significantly increased plant height, number of primary and secondary branches per plant, plant spread, number of umbels per plant, number of umbellets per umbel, number of seeds per plant and yield of seeds per plant and per ha. The probable reason for these results might be due to the suitability of climatic factors. Similar results have also been reported by Malhotra (2002). Germination of seed is completed in 15 days depending upon the prevailing temperature. Light irrigation is beneficial during sowing. At least two weedings are required and thinning is carried out at the time of first weeding. Irrigations are given at 20–30 days intervals depending on the prevailing weather conditions and soil moisture content. But in no case field should be kept deficient of moisture after flowering. Even Tripathi and Dwivedi (2009) reported similar results of higher plant height, more primary branches and maximum number of umbels at the wider spacing of 40 × 25 cm and significantly higher seed yield, biological yield and harvest index were obtained at 30 cm × 15 cm row spacing. The required seed rate is around 4 kg/ha and the plant to plant spacing is 20–30 cm. The ripe fruits germinate quickly within 12 days. Since it is a slow growing crop during the initial it is necessary to keep the field free from weeds. It is not greatly affected by diseases and pests.

Ajowan is a cool season plant. After 95 days, it faces critical crop–weed competition which can be controlled by Oxadiargyl (6% EC). Yield loss occurs during water stressed conditions and the probable reasons for that might be its sensitivity in the seed filling stage, change in assimilation allocation favouring the roots, photosynthesis reduction and inaccessibility to required nutrients as a result of drought stress. It is a resistant plant which can withstand medium drought stress in reproductive stage. Mohammadjavad et al. (2014) reported that water stress reduces the grain yield but it had no significant effect on the morphological characteristics and essential oil per cent and irrigation at 18 days interval gave higher yield attributes namely umbel/plant (184.32), seeds/umbellates (16.97), umbellates/umbel (17.64), test weight (2.01 g) seed yield (11.69 q/ha), net return (Rs. 21,885/ha) and benefit cost ratio (BCR) (1.15) over irrigation at 15 and 12 days interval (Mehta et al., 2015).

Moussavi-Nik et al. (2011) suggested that an irrigation interval of 7 and 14 days had an insignificant effect on the number of umbels per ajowan plant and umbels per square meter but an irrigation interval of 21 days caused significant decrease. The results of Yogita et al. (2013) revealed that foliar application of plant growth regulators significantly enhanced the vegetative characters, that is, plant height, no. of branches and leaf area of ajowan in addition to photosynthetic pigments, total carbohydrate, essential oil percentage, essential oil yield/plant and thymol content. Gibberellic acid application showed significant improvement in all yield attributes, yield and seed quality parameters over control (Shetty and Rana, 2012).

Fertilizer application at 75 kg N and 30 kg P_2O_5/ha and application of sheep manure at 10.0 t/ha with seed inoculation by *Azotobacter* are better for realizing higher yield and net return. According to Premnath et al. (2008) application of higher doses of nitrogen enhanced protein and chlorophyll synthesis leading to marked improvement in vegetative growth of the plant as well as yield and yield attributes of the crop under eastern Uttar Pradesh conditions. Similarly, higher doses of N, P and K increased plant height, stem diameter, number of branches per plant, number of umbels per plant, plant spread, seed yield and dry matter yield. Highest seed yield (14.67 q/ha) and dry matter yield (5851.83 kg/ha) were obtained with application of 80 kg N + 40 kg P_2O_5 and 40 kg K_2O/ha (Sathyanarayana et al., 2015a). and increasing application of N, P and K increased fertility status of soil (available nutrients, namely N, P and K after harvest of crop) and highest essential oil yield (44.70 l/ha) was obtained from T5 treatment, which was 80 kg N + 40 kg P_2O_5 and 40 kg K_2O/ha (Sathyanarayana et al., 2015b) and among the various levels of fertilizer tried, the 60:30:30–1 kg ha level of fertilizer significantly increased the growth, yield and quality attributes of ajowan. However, the levels of fertilizers non-significantly affected the fresh weight per plant (g) at 90 DAS and harvest index (%). The maximum benefit of cost ratio (2.05:1) was found with 45×30 cm + 60:30:30 kg NPK per hectare (Muvel et al., 2015).

3.3.5 PLANT PROTECTION

This crop is not greatly affected by diseases and pests. However, pests such as Tobacco and Gram caterpillars appear which can be controlled by spraying chlorpyrifos 2.5 ml or Quinalphos 2 ml/l of water. Even mites and aphid affect this crop where spraying with wettable sulphur at 3 g/l of water can control mites and 1.6 ml/l spray of monocrotophos for aphids, respectively. Diseases such as powdery mildew in ajowan can be kept under control by spraying with Karathane 1 ml or Carbendazim 1 g/l of water.

3.3.6 HARVESTING AND POSTHARVEST TECHNOLOGY

The small white flowers of the ajowan bloom in November and December in plains and mid-summer in hills in India and the harvesting takes place from February to May. The harvesting index is when the flower production ceases and the seed starts maturing and becomes greyish brown in colour. The crop matures in around 160–180 days after sowing. When flower heads turn brown it is harvested by cutting with sickle or pulled out with roots and allowed to dry completely. After drying, fruits are separated by careful rubbing followed by winnowing. It yields about 400–600 kg/ha under a rainfed farming and 1200–2000 kg/ha under irrigated conditions.

The harvested crop is first transported to a clean threshing floor, where it is dried hygienically in a thin layer for 1 or 2 days before threshing to separate the seeds. The seed is then dried in the shade where it produces more oil content than the sun-dried ones. The drying of ajowan seed in India is often carried out in zero energy solar drier tunnels to avoid entry of dust and foreign matter. Then it is cleaned in a screening mill and processed through a gravity separator. Seeds that have been well dried (8–9% moisture), cleaned and graded by sieving arc stored in polyethylene lined gunny hags in a cool dry place.

3.4 DILL

3.4.1 SYSTEMATIC POSITION

Kingdom: Plantae

Division: Magnoliophyta

Class: Magnoliopsida

Order: Apiales

Family: Apiaceae

Genus: *Anethum*

Species: *graveloens*

Dill (*Anethum graveloens* Linn.) is a minor spice crop that belongs to the family Umbelliferae and is mainly grown during rabi season. Its seed is used commercially as a spice and condiment and has got medicinal value. It originally comes from eastern Mediterranean region (Sokhangoy et al., 2011). It is considered to be native to the Mediterranean West Asia and is one of the oldest cultivated seed spices from ancient times. Initially, dill was used as herbs for flavouring in dynastic Egypt and for flavouring and medicine by the Greeks and Romans. The term *Anethum* comes from the Greek word *anethon*, meaning 'dill'. It is now widely cultivated in Europe, India and North America and has been cultivated since ancient times (Bailer et al., 2001). It is used as a vegetable, a carminative, an aromatic and an antispasmodic (Hornok, 1992; Sharma, 2004), and as an inhibitor of sprouting in stored potatoes (Score et al., 1997). In the Middle Ages it was thought to have magical properties and was used in witchcraft, love potions and as an aphrodisiac. The whole plant is aromatic with young leaves and the fully developed green fruit serves for flavouring purposes. Both seeds and leaves are valued as spice. Two types of dill exist in cultivation, namely European dill and Indian dill. A variant called East Indian dill or Sowa (*Anethum var. sowa* Roxb. ex, Flem.) occurs in India and is cultivated for its foliage as a cold weather crop throughout the Indian subcontinent, Malaysian archipelago and Japan.

3.4.2 BOTANY

It is a herbaceous annual and a long-day plant that grows to a height of 1–1.2 m with finely feathered blue green fern-like compound leaves which are light green in colour and a stem which is hollow smooth and shiny. It has long and fusiform tap root which is 10–15-cm length with a few secondary rootlets. It produces small terminal compound umbels with creamy yellow flowers and dark brown seeds. The fruit or seed is broad, oval in shape and is about 3.5-mm long with three longitudinal dorsal ridges and two wing-like lateral ridges. The seeds and leaves emit a pleasant aromatic odour and warm taste.

3.4.3 CLIMATE AND SOIL

Dill is grown as an irrigated annual crop both in temperate and tropical regions up to 1000 m above MSL. It can resist frost to a limited scale during vegetative stage. A dry and relatively high temperature is desirable during seed production. It is grown as winter season crop for commercial seed production in India and extreme heat during the critical periods of maturity is unfavourable to herbs, causing the reduction in the yield of their fruits. Dill is fairly tolerable moisture stress among the seed spices and also to mild soil acidity and alkalinity with optimum pH range of 5–7 and it can be grown in eroded soil.

3.4.4 CULTIVATION AND INTERCULTURE

There are two important varieties of dill that has been discussed hereunder.

NRCSS-AD-1 (AD-1–43) is a European type of dill which gives an average yield of 1470 kg/ha and contains 3.5% essential oil. It has been developed at NRCSS, Ajmer in 2005 through selection from an exotic source. It is 134-cm tall with dark green leaves and is late in flowering and takes about 142 days to maturity. It is suitable for cultivation under irrigated conditions, however, susceptible to powdery mildew. It has export value.

NRCSS AD-2 (AD-1–6) is an Indian type which gives an average yield of 1460 kg/ha. It has been developed at NRCSS, Ajmer (2005) through selection from local germ plasm. It is suitable for cultivation under both irrigated and rainfed conditions and has bold and dark brown seeds that are compact and require pressure to split.

A. graveolens can be reproduced by seeds which are not actually true seeds but small dry fruits called schizocarps. These fruits split upon maturity, with each half containing one seed which remains viable for 2–3 years without special storage measures and the germination rate is about 75% and 1000 seed weight of 4–5 g. The suitable time of sowing for dill was mid-October on a well prepared land that is ploughed three to four times with a country plough and is pulverized and levelled before sowing for better germination. Application of about 10.0 t ha of sheep manure with seed inoculation by *Azotobacter* is better for realizing higher yield and net return. Seed rate is around 3–4 kg for 1 ha. The method of sowing may be by broadcasting or drilling in rows with the spacing of 45 cm × 20 cm and depth of 1.5–2.0 cm. Deep sowing should be avoided. The seeds should be planted in rows at the rate of 15–20 ft and when it attains 7–10-cm height, it is thinned to 3 or 4 plants/ft 3 weeks after sowing. Germination commences after a week in tropical regions and may take 2 weeks in warm temperate conditions and its growth and development need proper spacing and a particular set of climatic condition which is influenced by time of sowing. Dill sown on October 15 at 40 × 10 cm spacing was found most suitable for realizing higher growth, seed yield, net returns and BCR. However, branching and flower initiation time were not influenced significantly by different crop geometry (Meena et al., 2013; Meena et al., 2015). According to the results of Saeid Zehtab et al. (2006) in Iran, dill has successful seed production when it is sown in early spring (3–18 April), since there was no significant difference in seed yield between the sowing dates of 3 April and 18 April.

Fertilizer dose of 60 kg N and 30 kg P_2O_5 per hectare is used for better seed yields (13.7 q/ha) in dill. The European dill required high nitrogen and moderate phosphorus for higher seed oil. Meena et al. (2012) revealed that the application of nitrogen up to 90 kg/ha significantly increased its uptake at different growth stages (45, 75 and 105 DAS) and at harvest where it was found increase up to 120 kg/ha, whereas phosphorus uptake at various crop growth stages was up to 40 kg P_2O_5 per hectare but in seed significantly improved up to 60 kg P_2O_5 per hectare. According to Wandera and Bouwmeester (1998), an increase in the nitrogen rate accelerated the ripening of the crop and the seed and carvone yields were largest in the 30–60 kg/ha nitrogen application. Fatemeh et al. (2013) observed an increase in the number of seeds per plant and seed yield nearly 32 and 32.5% with mycorrhizal application. A dose of 105 kg N per hectare was found best for higher seed and oil production while carvone content increased significantly up to 70 kg N per hectare and dillapiole content in oil decreased with increase in nitrogen levels (Kewalanand et al., 2001) as well as Zahedi et al. (2014) revealed that an application of 32 t/ha vermicompost is recommended to gain the highest seed and essence yield.

No irrigation is required when dill is grown in black cotton soils but usually two to three irrigations are given in light soils. If the seed is sown by broadcasting, first irrigation may be given immediately after sowing for better germination and the subsequent irrigations are given according to requirements of the crop and prevailing climate. At the time of flower initiation and seed development, stage sufficient soil moisture should be present in the soil. Water deficit during seed filling reduced seed vigour significantly, for producing high-quality dill seeds and it is necessary to provide sufficient water for plants during seed filling, especially for delayed sowings (Saeid Zehtab et al., 2006). The plot should be kept clean by proper weeding and hoeing. First weeding and hoeing are done in about 3–4 weeks after sowing and the next whenever needed. After germination, the crop remains 40–67 days in vegetative stage.

3.4.5 PLANT PROTECTION

Dill crop is not affected by any serious pests due to acrid odour. However, crop is occasionally infested with leaf-eating caterpillars and powdery mildew which damage the leaves and can be controlled by spraying 0.01% of monocrotophos once or twice depending upon the incidence and aphids (*Myzus* spp.) suck flowering axils causing loss in growth vigour which on weekly sprays of malathion (0.2%) in water controls the infestation. The incidence of powdery mildew is seen on all green parts of the plant that can be controlled by spraying sulphur combined with 3 g/l of water twice. First spraying is done as soon as the disease appears in the field and then after 15 days. Sometimes, it attacks the crop at flowering stage causing severe damage for which spraying of Bordeaux mixture three to four times at weekly intervals is recommended.

3.4.6 HARVESTING AND POSTHARVEST TECHNOLOGY

Crop duration varies from 130–160 days where flowering starts by the end of December and continues till the second week of January and the crop will be ready for harvest by the middle of March. It is pollinated mostly by bees and other insects. Higher quantity and quality of seeds were obtained when umbels were harvested at seed turning grey or maturity stages. Kewalanand et al. (2001) shows that carvone content increases while apiol content decreases with maturity. The seed is the ripe fruit of the plant, actually formed by two united carpels and is light to dark brown in colour. It is harvested and shade dried for 7–10 days. Delay in harvest results in shattering of grains. The leaves are used in the fresh state, whereas the fruiting tops in either fresh or dried state. Essential oil content increased gradually till the seed ripening after which it starts declining.

Threshing is done by hand or with a small stick. It has been reported that the yield of oil was 2.88% when seeds were dried in shade, but only 1.03% when dried in the sun. The seed is cleaned and stored in gunny bags for marketing. On an average, the yield of seed under rainfed conditions varies from 7–9 q/ha and contains 3–4% oil; however, in east Indian dill (sowa), the seed yield is higher (1 Mt/ha).

3.5 CINNAMON

3.5.1 *SYSTEMATIC POSITION*

Division: Tracheophyta
Subdivision: Spermatophytina
Class: Magnoliopsida
Order: Laurales
Family: Lauraceae
Genus: *Cinnamomum*
Species: *verum*

Cinnamon (*Cinnamomum verum*) is one of the earliest and most important tree spices crop which is mainly cultivated in India mostly for the dried inner bark of the tree. The fragrance of bark is warm sweet agreeable taste and used as a spice in culinary preparations. It is precious not only as a flavouring agent for food but as a medicine and perfume (Yousef and Tawil, 1980). Cinnamon is a native of Sri Lanka and is cultivated in lower elevations of the Western Ghats in Kerala and Tamil Nadu. It is mainly used as a culinary herb such as other cinnamons in the traditional eastern and western medicine (Sudan et al., 2013). The 'true' cinnamon or spice cinnamon is the dried inner bark of *C. verum*.

3.5.2 *BOTANY*

Cinnamon trees are grown up to the height of 10–15 m. Leaves are ovate-oblong in shape, 7–18-cm long. The flowers are arranged in panicles, which are greenish in colour, and have a distinct odour. The fruit is a purple berry (1 cm) containing a single seed.

3.5.3 CLIMATE AND SOIL

Cinnamon is a hardy plant and adopted a wide range of soil and climatic conditions. Cinnamon trees are grown in lateritic and sandy patches with poor nutrient status in the west coast of India. Mixture of humus or vegetative mould with sandy loam soil is the best for sweet and fragrant bark. It is grown at an altitude of 1000 m from the mean sea level. It is raised as a rainfed crop, an annual rainfall of 200–250 cm is ideal. For the cultivation of cinnamon, a hot and moist climate with an average temperature of 27°C is suited. Prolonged dry weather is not conducive for the successful growth of the crop.

3.5.4 CULTIVATION AND INTERCULTURE

Cinnamon's two high-yielding varieties released from Indian Institute of Spices Research (IISR) are suitable for cultivation in various regions of India. These varieties are Navashree and Nithyashree. In the initial year, yield potential of 56 and 54 kg dry quills/ha/year, respectively, when one seedling or cutting is planted in a hill.

Cinnamon is propagated through seeds. Cinnamon start flowering in January and the fruits start ripening in June to August under west coast conditions. The fully ripened fruits are either picked from the tree or the fallen ones are collected from the ground. The seeds are removed from the fruits, washed to make free of pulp and sown without much delay as the seeds have a low viability. The seeds are sown in sand beds or polythene bags are mainly used for seeds sowing which contain a mixture of sand, soil and well rotten cattle manure (3:3:1). The seeds start to germinate within 15–20 days. Irrigation should be given in frequent interval for maintaining adequate moisture. The artificial shading is to be given for seedling till they are about 6 months old. Cinnamon is also propagated vegetatively through rooted cuttings, air layering, etc.

The pits are dug 50 × 50 × 50 cm size at a spacing of 3 × 3 m. Before planting pits are filled with compost and top soil. Planting of cinnamon is done in June to July to take advantage of the monsoon for the establishment of seedlings. Transplanting should be done with 10–12-month-old seedlings or well-rooted cuttings. Three or four seedlings or rooted cuttings or air layers can be planted in each pit. Sometimes, the seeds are directly dibbled in the pits that are filled with compost and soil.

Sr. No.	Name of variety	Institution	Special characters
1	IISR Navashree	IISR, Calicut, Kerala	200 kg dry quills per hectare. Bark oil 2.7%, leaf oil 2.8%, bark oleoresin 8.0%, bark recovery 40.6%
2	IISR Nithyashree	IISR, Calicut, Kerala	200 kg dry quills per hectare. Bark oil 2.7%, leaf oil 3.0%, bark oleoresin 10.0%, bark recovery 30.7%
3	YCD-1	Horticultural Research Station, Yercaud, Salem District	360 kg dry bark/ha, bark oil 2.8%, leaf oil 3.0%, bark recovery 35.3%
4	PPI (C) 1	Horticultural Research Station, Yercaud, Salem District	973 kg fresh bark per hectare. Bark oil 2.9%, leaf oil 3.3%. Bark recovery 34.22%
5	Konkan Tej	Dr. B. S. K. K. V, Dapoli	378.3 g fresh bark per plant. Bark oil 3.20%, leaf oil 2.28%, bark recovery 29.16%
6	RRL (B) C-6	Regional Research Laboratory (RRL), Bhubaneswar, Orissa	Leaf oil 0.8%.
7	Sugandhini (ODC-130)	Aromatic and Medicinal Plants Research Station, Odakkali, Kerala	1.2 kg fresh bark tree per year. Bark oil 0.94%, leaf oil 1.6%, bark recovery 51%
8	Konkan Tejpatta	Dr. B. S. K. K. V., Dapoli	Higher yield of leaves (1.94 kg/plant/year) and 80 g dried bark per plant per year

Partial shade should be provided in the initial years is advantageous for healthy, rapid growth of plants. Weeding is to be done two times in a year during June to July and October to November and digging of the soil once in a year around the bushes during August to September is recommended.

In first year, fertilizer dose of 20 g N, 18 g P_2O_5 and 25 g K_2O per seedling is recommended. Fertilizers dose is increased gradually to 200 g N, 180 g P_2O_5 and 200 g K_2O for plants of 10 years and above. Two equal splits of fertilizers are applied in May to June and September to October. During summer, mulching is to be done with green leaves (25 kg) and during May to June, application of FYM (25 kg) is also recommended. Cinnamon is grown as rainfed crop and annual rainfall ranging from 200 to 250 cm is ideal. In the initial years, irrigation is given during summer months twice a week. The requirement of water depends upon the soil moisture level and growth of plants.

3.5.5 PLANT PROTECTION

Leaf spot and die back, seedling blight and grey blight are the major diseases of cinnamon, whereas cinnamon butterfly and leaf miner are the notable insect pests. The causal organism of leaf spot and dieback is *Colletotrichum gloeosporioides*. Brown small deep specks appear on the leaf lamina, which later coalesce to form irregular patches. In some cases, the leaves shed leaving shot holes. Later, the entire lamina is affected and the infection spreads to the stem causing die back. Pruning of the affected branches and spraying with suitable chemicals are found effective. Application of Bordeaux mixture (1%) is recommended to control the disease. The causal organism of seedling blight is *Diplodia* sp. This disease occurs on seedlings in the nursery. The fungus causes light brown patches which girdle the stem resulting in mortality. The disease is controlled by the application of 1% Bordeaux mixture. Grey blight is caused by *Pestalotia palmarum.* Small brown spots appear which later turn grey with a brown border. The disease can be controlled by spraying 1% Bordeaux mixture. Cinnamon butterfly (*Chilasa clytia*) is especially found in nursery or in younger plantations during the post-monsoon period. The larvae feed on tender and slightly mature leaves. Defoliation of the entire plant leaving only midribs with portions of veins is the typical symptom. The adult butterflies are larger in size and occur in two forms. One has blackish brown wings and outer margin with white spots. Other has black wings with bluish white markings. Fully grown larvae are pale yellow with dark stripes on the sides and measure about 2.5 cm in length. Quinalphos (0.05%) are sprayed on tender and partly mature leaves for the control of pest. Leaf miner (*Conopomorpha civica*) seriously affects seedling in nursery during the monsoon period. The adult is a minute silvery grey moth. The larvae are pale grey initially and become pink later. They feed on the tissues between the upper and lower epidermis of tender leaves resulting in linear mines that end in 'blister' like patches. Leaves become crinkled and the mined areas dry up leaving large holes on the leaves after infested. Spraying 0.05% quinalphos during emergence of new flushes for the pest infestation controlled (Rajapakse and Kumara Wasantha, 2007). Leaf feeding caterpillars and beetles also occur sporadically on cinnamon feeding on tender flushes.

3.5.5 HARVESTING AND POSTHARVEST TECHNOLOGY

The cinnamon tree is coppiced or cut back periodically when it attain height of 10–15 m. When the plants are 2 years old, they are coppiced at a height of about 12 cm from the ground during June to July. Stump after coppiced is covered with earthing up. Then it encourages the development of side shoots from the stump. This is repeated for every side shoot developing from the main stem during the succeeding season, so that the plant will assume the shape of a low bush of about 2-m height and shoots suitable for peeling would develop in a period of about 4 years. After fourth or fifth year of planting, the first coppicing should be done (Anon, 1973).

In Kerala conditions, the shoots are harvested from September to November. Coppicing is done in alternate years and shoots having 1.5–2.0-cm thickness and uniform brown colour are ideal for bark extraction. A 'test cut' can be made on the stem to judge its suitability for peeling with the help of sharp knife. Coppicing can be commenced immediately, if the bark separates readily, when they are about 2 years old, stems are cut close to the ground. After removing the leaves and terminal shoots then the shoots are bundled.

The harvested shoots are cut into pieces of 1.00–1.25-m length. Cutting is done by scraping and peeling operations. Peeling is a specialized operation, requiring skill and experience which is done with specially made knife, which has a small round end with a projection on one side to facilitate ripping of the bark. First, the rough outer bark is scraped off. Then the scrapped portion is polished with an aluminium or brass rod to facilitate easy peeling. A longitudinal slit is prepared from one end to the other. The bark can be easily removed by the knife between the bark and the wood. The shoots which are cut in the morning are peeled on the same day. The peels shoot are gathered and kept overnight under shade. They are dried in shade for the first day and then 4 days in sunlight. During drying, the bark contracts and assumes the shape of a quill. The smaller quills are inserted into larger ones to form compound quills.

The quills are graded from 0, being the finest quality, to 0 the coarsest quality. The small pieces of the bark, left after preparing the quills are graded as 'quillings'. The very thin inner pieces of bark are dried as 'featherings'. From the coarser canes, the bark is scraped off, instead of peeling, and this grade is known as 'scraped chips'. The bark is also scraped off without removing the outer bark and is known as 'unscraped chips'. The different grades of bark are powdered to get 'cinnamon powder'.

Dried cinnamon leaves and bark are distilled to obtain leaf and bark oils, respectively. The dried cinnamon leaves are steam distilled in special distiller. About 4 kg of bark oil could be obtained from a hectare of cinnamon plantation. Perfumes, soaps, toothpastes, hair oils and face creams are manufacture from leaf and bark oil and also as an agent for flavouring liquor and in dentifrices. Coppicing is done by cutting back the 2-year-old plant during June to July to a height of about 15 cm from the stump of cinnamon plant for encourage of side shoots and subsequently the plants assume the shape of a low bush of about 2-m height and a bunch of canes suitable for peeling crop up in a period of about 4 years. In case of seedling bushes, regular peeling operation could be commenced, from the fourth or fifth year, depending upon the extent of development of peeler shoots. Usually, coppicing is done in alternate years.

When the plant reaches a height of 1–2 m under favourable condition or after 2 or 3 years of transplanting, the first cutting is made. About 100 g of dried bark can be obtained per bush such as 200–300-kg dried bark can be obtained per hectare. The 35 kg of leaf oil can be obtained per hectare per year. The value-added products of whole and ground cinnamon are leaf and bark oil, seed oil, root bark oil and oleoresin.

3.6 CASSIA

3.6.1 SYSTEMATIC POSITION

Kingdom: Plantae

Division: Magnoliophyta

Class: Magnoliopsida

Family: Lauraceae

Genus: *Cinnamomum*

Species: *cassia*

Cassia (*Cinnamomum cassia*) known as Chinese cinnamon belongs to the family Lauraceae is similar to true cinnamon but it is more pungent, has less delicate flavour and is thicker than cinnamon bark. It provides dried edible aromatic bark and was known to the Chinese as early as 3000 B.C. It is the oldest spice which is also mentioned in the Bible where it was popularly used by the Pharaohs and came to Europe in the 17th century B.C., but it seems different botanical origin. Cinnamon and cassia are used together by Greek and Romans but the dominated Arab traders in spices trade shrouded the sources of mystery.

3.6.2 BOTANY

It is a small bushy evergreen tree that grows up to 18–20-m high and 40–60-cm diameter with a straight and cylindrical trunk with greyish brown bark that is 13–15-mm thick when mature. The previous year branchlets are dark brown, longitudinally straight and slightly pubescent but those of the current year are more or less tetragonal, yellow brown and densely tomentose with greyish yellow hairs. Terminal buds are small about 3-mm long with broadly oval scales. Leaves are simple, alternate, oblong-oval or narrowly elliptic to sub-lanceolate, 8–16 cm × 4–7 cm, coriaceous, thick, shining green, glabrous upper, greenish opaque and sparsely hairy under and the leaf base is acute with cartilaginous margins and involute. Petioles are 1.5–2.5-cm long and the triplicate branched panicles are axillary, 8–16-cm long, with three flowered branch and peduncle which is as long as half of the inflorescence. Flowers are white, 4.5-mm long with

3–6-mm long pedicels that are yellowish brown and tomentellate. Perianth is densely yellow-brown tomentellate outside and inside with 2-mm long obconical tube which is ovate-oblong and has subequal lobes. Stamens are nine in number, fertile with first two extrose whorls and the third four-celled introse whorl that is ovate and oblong. The innermost whorl consists of three staminodes. It possesses 1.7-mm-long ovary that is ovoid and glabrous with slender style and a small inconspicuous stigma. Fruits are ellipsoid, 7–10 mm in size, pink violet when mature, glabrous, shallow perianth, that is 4-mm long, truncate or slightly dentate and up to 7-mm broad at the apex.

3.6.3 CLIMATE AND SOIL

This crop requires a mean daily temperature of about 22°C and an annual rainfall of 1250 mm and grows at an altitude up to 300 m. The absolute temperature is in the range of 0–38°C. It needs acidic soil with pH 4.5–5.5 and prefers undulating hills.

3.6.4 CULTIVATION AND INTERCULTURE

Seeds and cuttings serve as the propagating material for cassia but the plants raised through cuttings marks very thin bark low essential oil contents in both leaves and barks. Therefore, ripe fruits from strong mother trees producing thick bark of good aroma should be selected for propagation.

Seeds should be collected in July to August and put in water and only those seeds that sink are used for sowing. Before sowing, seeds are to be kept in warm water (30–35°C) with 1% permanganate for 6–8 h and then taken out, kept in shade in humid sand in multilayers: one layer of 8–10-cm thick sand intercalated with one layer of seeds, and three such sand layers are usually made. Seeds are left for 7–10 days with occasional sprinklings of water. When seeds germinate the sprouts are transplanted to nursery beds or polybags. Alternatively, direct seeding in polybags can also be employed.

The nursery must be covered as seedlings prefer shady places. The covering frames should be 1.5–2-m high for easy watering and tending. About 80% shade is provided before transplanting. The land is ploughed 30–40 days before seed sowing adding powdered lime at the rate of

0.2 kg/m^2. Beds of 10-m length and 1–1.2-m width are made after second ploughing. Seeds are transplanted at the spacing of 10 cm × 10 cm. Watering regime varies with humidity in each area. It is especially necessary only on dry days and in general, the watering is adjusted based on rainfall.

Weeding is needed once in every month. In case of pests/disease, spraying at fortnights with appropriate insecticides such as Basudin or Wofatox 666: 1–2%, 4 l/m^2 or fungicide such as Bordeaux solution: 1–1.5%, 4 l/m^2 to prevent soilborne diseases is essential. Basal fertilizer dose of muck (30%) and phosphorus (2 kg/m^2) is given along with the additional dose of potassium 0.1 kg/m^2; nitrogenous or muck 0.3% through irrigation water or spraying.

The pits of dimension 40 cm × 40 cm × 40 cm are made at least 30-days prior to planting with an interval of 2 m between rows and 1.5 m between trees. After hardening of the seedling in the nursery it is removed together with a ball of earth and planted in the pit. Care is needed to prevent bud and root breaking. After planting in the pit, the piling up of soil at the base is important to prevent the stagnation of water during heavy rains. The best time for planting is October to November.

In the first year, in September and October cutting of the vegetation cover, weeding, light turning over the soil at the base, piling up of soil 5–10-cm high and 0.6–0.8-cm wide, adjusting the shade at 50%, and replacement of dried up seedlings are done while in second to fourth years, trimming the lower and diseased branches, piling up of soil at the foot of the plant and replacing dead trees are the important activities. The shade adjustment as per requirement is also done. During fifth to tenth years, clearing all the vegetation cover and cutting of diseased and stunted branches are done. From the sixth year, shade is completely removed. In a 10-year-old plantation, thinning is done for adjusting the density of trees to 1000–1250/ha.

3.6.5 PLANT PROTECTION

Cinnamon butterfly (*C. clytia*) is the most serious pest of cassia that effects both in nurseries and in plantations and is widely prevalent in India and Sri Lanka. Its larvae feed voraciously on tender and partly mature leaves, leaving only the midrib and portions of veins. Young plants are often completely defoliated and the plant growth is adversely affected under higher infestations. It is more severe during December to June when numerous tender flushes are present on the plants and coming to leaf miner (*C. civica*), it infests especially during the monsoon period in both cinnamon and cassia in nursery and plantation. The larvae mine into tender leaves and feed on the tissues between the upper and lower epidermis making linear and tortuous mines, which end in patches and the infested portions appear as blisters with larvae inside them. Leaves become crinkled and malformed, and the affected portions dry up resulting in large holes on the leaf lamina while the shoot and leaf webber (*Sorolopha archimedias*) webs tender shoots and leaves and feeds from within. Infestation is generally serious in the field during the post-monsoon period. Various other insects such as leafhoppers, psyllids, whiteflies, aphids, scale insects, mirid bugs, leaf-feeding caterpillars, leaf miners, leaf webbers, root grubs, leaf beetles and weevils have also been recorded on cinnamon in many regions. These species mostly infest foliage and can sometimes cause heavy damage in certain localities.

Diseases such as stripe canker, foot rot and witches broom are prevalent in cassia. Of them, the stripe canker caused by *Phytophthora cinnamomi* makes severe damage to forest trees and avocado and was first recorded on *C. verum* (Rands, 1922). Symptoms start as a vertical stripe on the stem with amber coloured exudate at the advancing margins, which hardens later. This disease is reported to be severe in ill-drained soils and causes up to 42% damage. The fungus *Fusarium oxysporum* is responsible for the foot rot which attacks and destroys the vascular system by infecting through the roots and is prevalent during the post-rainy season when the weather is humid and warm. Application of *Trichoderma* sp. is effective in examining the disease incidence, especially in nurseries while witches broom caused by phytoplasma is the more common one in cassia seedlings in nurseries and young trees in fields. Soaking the seeds in warm water (70°C) containing an antibiotic before sowing is the preventive measures for this disease.

3.6.6 HARVESTING AND POSTHARVEST TECHNOLOGY

Crops are ready to harvest when it attains the age of 20 years. The harvesting varies in north and south. There are two harvests: in south, one is from February to March and the other from July to September. Similarly, second harvest is from April to May in north and September to October in the south. The bark extracted from the first crop has more scales than that of the second one but the quality is better in second. Yield varies so much from tree to tree, and from year to year, that it is practically impossible to give the normal values and is generally higher during the dry season than during the wet season and it also depends on the age of the plant. In the south, the yield is around 6.5 Mt and in the north, it is about 3–5 Mt.

3.7 CURRY LEAF

3.7.1 SYSTEMATIC POSITION

Kingdom: Plantae

Division: Magnoliophyta

Class: Magnoliopsida

Order: Sapindales

Family: Rutaceae

Genus: *Murraya*

Species: *koenigii*

Curry leaf is used in very small quantities but a popular leafy spice for their distinct aroma contributed by the volatile oil present in it and their ability to improve digestion (Singh et al., 2014). *Murraya koenigii* is native to India and Sri Lanka which is grown in tropical to subtropical condition. Small white flowers are produced which are self-pollinated to produce small shiny black berries containing a single, large viable seed. Leaves are used in many dishes and as herb in Ayurvedic medicine in India and neighbouring countries and also possess antidiabetic properties The juice of the root is taken to relieve pain associated with the kidneys (Anonymous, 1988). Owing to its aromatic characteristic properties, it is used for soap making ingredient, diffusers, body lotions, scent, air fresheners, potpourri, perfume, body fragrance, bath and massage oils, aromatherapy, and health clinics, facial steams, hair treatments, etc. Curry leaves are used for rituals in the absence of tulsi leaves, such as *pujas*. Primary alkaloids present in the curry tree leaves, stems and seeds are koenimbine, mahanimbine, mahanine, isomahanine, girinimbine, undecalactone and 2-methoxy-3-methyl-carbazole.

3.7.2 BOTANY

Curry leaf is a less deciduous shrub (0.9 m), and is an attractive and aromatic crop. It is grown up to 6-m height (Himalayan regions) and 15–40-cm diameter. The bark is dark green to brownish, with numerous dots on it; its bark can be peeled off longitudinally, exposing the white wood underneath; the girth of the main stem is 16 cm. Leaves are exstipulate, bipinnately compound, 30-cm long, each bearing 24 leaflets and having reticulate venation; its leaflets are lanceolate, 4.9-cm long, 1.8-cm broad and having 0.5-cm long petiole. The flowers are bisexual, white, funnel-shaped, sweetly scented, stalked, complete, ebracteate, regular, actinomorphic, pentamerous, hypogynous, the average diameter of a fully opened flower is 1.12 cm; inflorescence, a terminal cyme, each bearing 60–90 flowers; fruits, round to oblong, 1.4–1.6-cm long, 1–1.2 cm in diameter; weight. The fully ripe fruits are black with a very shining surface and pulp. The roots are woody widely spread and produce many suckers. The tree bears flowers from February to May and the flowers are self-pollinated, and variability is very limited.

3.7.3 CLIMATE AND SOIL

Curry leaf can be grown in a wide range of soils. Red sandy loam with good drainage will be ideal for its normal and fleshy growth, which will result in better leaf yield. The optimum temperature requirement is 26–37°C. It can tolerate temperature up to 37°C and if the temperature goes below 16°C, it affects the growth of plants. This plant does not tolerate waterlogging; as soon as it gets too much water, it starts exhibiting yellow leaves. Therefore, good draining is important as part of curry leaves farming, keep the soil a bit loose and sandy type. It can be grown an elevation of 1500 m from MSL.

3.7.4 CULTIVATION AND INTERCULTURE

Dharwad (DWD) 1 and DWD 2 are two improved varieties of curry leaf. Both of these have a good aroma. They have an oil content of 5.22 and 4.09%, respectively. They are mainly cultivated in Karnataka. DWD 1 is sensitive to the winter season. During winter its growth is poor, whereas

DWD 2 is winter insensitive. It gives higher yield than DWD 1. 'Senkaampu' is a local cultivar which is grown in Tamil Nadu. *Suwasini* and DW-2 are promising varieties of curry leaf from Kerala Agricultural University. The leaves of *Suwasini* are dark green, shiny and highly aromatic. The leaves have an oil content of 5.2%, and can be dehydrated at 50°C and made into powder. Though the leaves of DW-2 are pale green and slightly less aromatic, the rate of shoot growth is very high in this variety. Aside from these, many promising local types are also being cultivated.

Seeds are the main propagating material of curry leaf. The well ripe fruits are collected from high-yielding plants for seedling raising. The seeds are sown in polybags and nursery filled with a mixture of 1:1:1 soil, sand and FYM. Seeds are germinated within 3 weeks. After 1 year, seedlings are planted in the main field and are also used as propagating materials. A number of root suckers are found near its plants. They are separated from the main plant during the rainy season and planted immediately in the main field.

Curry leaves are planted in the main field during the monsoon season. The main field is ploughed thoroughly, so that field should be weed free and fine tilth. Pits are dug 30 cm × 30 cm at a spacing of 4 m × 4 m. About 10 kg/pit FYM should be applied. Usually, inorganic fertilizers are not given in curry leaf. The plant may be given 10-kg FYM and NPK of 60:80:40 g/plant/year for higher yield. The fertilizers may be applied at the onset of the monsoon.

Irrigation should be applied in pits if there is no rain. The irrigation should be applied at 3-day intervals, if there is no rain. The field should be kept weed free. For the maintenance of 1-m height, the plant should be trained and pruned. Terminal buds of curry leaf are removed to encourage lateral branching. Minimum five to six branches are kept per plant.

3.7.5 PLANT PROTECTION

The most prominent pest in curry leaves are citrus butterfly and citrus psylla. Among them, early instar larvae of citrus butterfly (*Papilio demoleus)* are dark with white patches resembling the droppings of birds. When grown up, they turn deep green in colour, stout and cylindrical in shape. Hand-picking and destruction of the larvae can be done wherever possible. Pest can be controlled by hand-picking and destruction of the larvae eggs and by malathian spray (1 ml/l) and *Diaphorina citri*, the citrus psylla is known to infest branches and their aphids and scale insect (*Anoidiella orientalis*) have to be removed immediately. These pests can be controlled by spraying dimethoate (1 ml/l). The stem (bark) borer (*Indorbela tetraonis*) is reported to cause drying of the branches occasionally and mites are also found to infest the leaves, for which no control measures have been suggested. Curry leaf is affected mainly by leaf spot which is caused by *Phyllosticta murrayae* and *Cyilindrosporium* sp. It causes severe defoliation and can be controlled by spraying carbendazim at 1 g/lit of water. Spraying of sulphur compounds should be avoided. The other diseases like sap rot (*Fomes pectinatus*) and collar rot of seedlings (*Rhizoctonia solani*) are also noticed but economic loss due to these diseases are minimal.

3.7.6 HARVESTING AND POSTHARVEST TECHNOLOGY

Curry leaves should be harvested 15 months after planting. Commercial harvest can be started from 3-year-old plants. The normal yield can be obtained up to the age of 20–25 years with good management. Two and a half to three months intervals leaves can be harvested, the average yield is 20–25 Mt/ha. The young shoots of tender leaves are harvested, packed in clean gunny bags or tied in bundles and are transported. Sprinkling of water should be done on the bags per bundles. The leaves are dried ground into the powder and use as curry leaf powder. The curry leaf value-added products are volatile oil and dehydrated curry leaves.

3.8 KOKUM

3.8.1 SYSTEMATIC POSITION

Kingdom: Plantae

Division: Mangoliophyta

Class: Mangoliopsida

Order: Malpighiales

Family: Clusiaceae

Genus: *Garcinia*

Species: *indica*

Kokum *(Garcinia indica)* is an important underutilized tree spice belonging to the family Guttiferaceae. The dried rind is of economical use and is known as kokum. It is an ancient fruit tree of culinary, pharmaceutical, nutraceuticals and industrial uses and is used for making several vegetarian and nonvegetarian dishes including the popular *solkadhi* (Khare, 2007; Sies, 1996; Yoshikawa et al., 2000). This crop is found growing in the riversides, forests and wastelands and its genus includes 200 species out of which 30 different species are reported to be found in India. Among those 30, *G. indica* is confined to India (Konkan region of Maharashtra and Goa) and Sri Lanka only and is commonly known as Brindonia Tallow tree or Kokum Butter tree in English.

3.8.2 BOTANY

Kokum is a slender pyramid shaped, dioecious evergreen tree with drooping branches and grows up to a height of 18 m. Its leaves are 6–10-cm long, 2.5–3.5-cm broad, dark green coloured above and pale beneath, ovate or oblong and lanceolate. It is androdioecious type producing male and bisexual flowers on separate plants (Karnil et. al., 2001; Milind and Isha, 2013) and consists of two to three female and three to four male flowers that are pale yellow in colour and born either singly or in clusters. The male buds are short and roundish and the female buds are oval in shape and vary in size and their weight also varies from 50 to 180 g. The calyx and corolla have four sepals and petals where the petals are larger compared to sepals and are yellow to pink dorsally and dark pink ventrally. Male flower has long pedicel and numerous stamens forming short capitates column with oblong anthers. The fruits are of lemon size, dark purple in colour, globose or spherical, round, oblong, oval with pointed tips and contain five to eight large seeds. It has 7–10 ridges, an agreeable flavour and a sweetish taste.

3.8.3 CLIMATE AND SOIL

It is a tropical crop that prefers warm and moderately humid climates with mean annual temperature of 20–30°C and total rainfall range of 200–300 cm and flourishes well at 60–80% humidity and an elevation of about 800 m from MSL. It is a hardy crop that grows best in sandy, clay, loamy lateritic, alluvial soils of 1.2-m depth and pH of 6.7 having good water holding capacity and can tolerate waterlogged and drought conditions. Kokum prefers warm moderate humid zones and valley situations and grows well along river banks and seashores. It can be grown as rainfed crop in locations where coconut and areca nut can be cultivated but its cultivation on hilltops is not possible like that of mango or cashew nut.

3.8.4 CULTIVATION AND INTERCULTURE

Konkan Amruta (S-8), the first variety released in India in 1998 from Dr. B. S. Konkan Krishi Vidyapeeth, Maharashtra is an early bearer which matures in 78 days. It has apple shaped, attractive fruits that has a maximum shelf life of 15 days and can be harvested well before the onset of the monsoon with an average fruit yield of 138.28 kg/tree/year.

Konkan Hat is another variety that is released from Dr. B. S. Konkan Krishi Vidyapeeth, Dapoli, Maharashtra. Under Konkan conditions, it shows flowering in December and the fruits ripe in April to May. The fruits are big sized weighing 90–95 g and apple shaped with five to six seeds per fruit. The rind weight is around 48–50 g and the yield of 250 kg fruits per year can be expected from a 7-year-old tree. The total soluble solid (TSS) is noted as 9.2°Bx and acidity is 5.10%.

Propagation on large scale is done by seeds. As the crop is cross-pollinated, the seedling progeny shows heterogeneity and thereby variability and due to its dioecious nature, it shows 50% males whose sex can be determined only when it starts flowering and the remaining are female trees that produce fruits. Vegetatively grown plants are early in nature than the others which are slow in growth and takes around 7–8 years for first flowering. Various methods of vegetative propagation are tried in kokum to get female trees that obtain uniformity in yield and quality enhancement of fruits. Inarch grafting is done on 10–18 months old seedlings in December to January was found to be 90% successful in kokum and for this purpose about 3–4-month-old upright growing shoots from scion plant should be selected. If horizontal growing scion shoot is used, the graft will have straggling habit with top heaviness on one side. Soft wood grafting done during June gives maximum success of about 80% and found to be easy than inarching. The best season for grafting is October and it is done on 1-year-old rootstocks with 9 months old scion sticks that are of 6–8-cm length and the grafted plants could be successfully maintained either in the glasshouse or outdoors without shade (Haldankar et al., 1991). The graft comes up well in 2 months. Even though the orthotropic shoots are limited in number on mother plants, for normal growth habit, they should be used as scion material and if plagiotropic shoots are used, the resulting grafts will be dwarf, bushy and drooping with reduced fruiting area. Therefore, seedlings continue to be the major propagating material.

Before planting, the area should be cleared by removing bushes and trees in the month of April to May and the pits of 60 cm × 60 cm ×60 cm are dug at 6 m × 6 m and filled up with good soil, about 10-kg FYM and 1.0-kg single super phosphate or bone meal at the end of May. Kokum plants can be cultivated as monocrop or as a mixed crop in coconut and areca nut plantation and also in kitchen garden. Considering its growth habit and conical canopy, a spacing of 6 m × 6 m for sole plantation has been recommended by Dr. B. S. Konkan Krishi Vidyapeeth. Square and Hedge row system can be adopted where hedge row system provides scope for intercropping and better intercultural operations. In an established coconut plantation of 7.5–8 m spacing, kokum can be planted as a mixed crop in the centre of two coconut palms accommodating 300 plants/ha which increases the coconut yield by 34%. Similarly, in an areca nut plantation of 2.7 m × 2.7 m, it can be planted at the alternate centre of areca nut palm, whereas when grafts are used for planting the spacing can be reduced to 5 m. While planting in kitchen garden, kokum should be planted at least 4–5 m away from other tall plants and the planting is done at the onset of monsoon.

After planting, seedlings/grafts are required to be protected from stray cattle. In the first year, overhead shade is provided for protecting the young seedlings from the scorching heat. Weeds should be removed from time to time for good growth of seedlings/grafts. For the initial 2 years, about 10 l of water is given per week per plant during winter and summer months. Mulching of dry grass may be done around the basin of the plant to conserve the soil moisture.

Irrigation helps for better establishment of kokum plant and it is essential in the first 3 years after planting. Initially, 15 l of water per week in winter and twice a week in summer is advised in Konkan region of Maharashtra. The modern methods such as drip irrigation are beneficial than the conventional methods. The weed near kokum plant should be removed and used for mulching that helps in retaining soil moisture.

In the Konkan region of Maharashtra, for 1-year-old kokum plant an application of 2 kg FYM, 50 g N, 25 g P_2O_5 and 25 g K_2O is recommended. This dose is increased in same proportion every year up to 10 years and there onwards 20 kg FYM, 500 g N, 250 g P_2O_5 and 250 g K_2O is recommended and they are applied in the month of August after the heavy rain, in a circular trench around plant of about 30-cm deep and 30.5-cm wide and covered with the soil. Inorganic fertilizer application is not in practice in the Konkan region but FYM and other available organic manures are used and mostly they are said to be organic. However, the plantation is scattered and very small the certification becomes difficult. Foliar application of urea (0.5% twice) influences the length (3.98 cm), breadth (4.24 cm), circumference (13.22 cm) and fruit weight (32.67 g) as compared with others while in chemical composition highest TSS 15.93°B was recorded by urea (0.5% twice) spray and lowest acidity (3.73%) was noticed by monopotassium phosphate (0.5 twice) spray where it resulted highest reducing sugar (6.05%) nonreducing sugar (5.54%) and total sugar (11.59%).

The kokum plant is dioecious in nature and if the orchard is established by using seedlings, it is very difficult to identify the male and female plants at an early stage. Only after flowering (7–8 years from planting), the male and female plants can be identified. It is advisable to remove excess number of male plants and maintain only 10% to ensure adequate pollination, better fruit set and higher yield. Rest of the male plants can be converted into female trees by side grafting in the month of August to September where 80% success can be seen.

Kokum is an evergreen plant with attractive conical shaped canopy that is obtained without pinching the central stem when seedlings are planted. A full-grown plant can attain a height of about 10 m but as the height increases the lower portion of plant comes under shade and becomes less productive and furthermore, the fruits at the top of tall tree remain small in size and becomes unmarketable. So, the harvesting from tall plants of kokum is an important constraint. At Dapoli, maintaining the height of kokum tree at about 4–5 m by decapitating the apex has shown promise preliminary. When the grafts are planted, it is observed that only one branch grows in certain direction and this growth should be prevented by regular pinching. Growth in all directions should be tried to induce on a graft and the suckers from rootstock below graft union should be removed regularly.

The seedlings start flowering after 7–8 years of planting generally from December to January, whereas flowering in grafts is noticed after 3–4 years. Flowers are borne singly or as fascicular cymes on leaf axils and are tetramerous and the period from flower bud appearance to initiation of flowering is about 30 days, where pollination is through wind. The fruits are harvested after about 120 days of fruit set in the month of April to May and mostly in the month of May and June which is the start of the rainy season where 40–70% fruits are trapped in rains and hence lost. Presently in Konkan region alone, this loss is estimated to be of Rs. 157 lakhs. Not only farmers suffer seriously because of this loss but the processing industry is also adversely affected as large quantity of kokum fruits is required for value addition. Post-flowering foliar spray of potassium nitrate and monopotassium phosphate helps to prepone harvesting by about 10–34 days.

3.8.5 PLANT PROTECTION

There are no serious pests and diseases found in kokum plantation but leaf miner and mealy bugs cause damage to the tender leaves and these can be controlled by spraying the plants with 0.03% phosphomidon or 0.03% dimethoate. Another pest by name leaf-eating beetles (*Podontia congregate*) (Coleoptera: Chrysomelidae) are also noticed in kokum plant in Chettalli conditions. This pest is present on the host throughout the year and causes large-scale defoliation of the plant by larvae and adult beetles. Sometimes diseases such as pink disease and drying back of twigs are noticed on branches at Central Horticultural Experiment Station, Chettalli where it can be kept in control by removing the diseased portion of a branch and smearing of Bordeaux paste on the wound.

3.8.6 HARVESTING AND POSTHARVEST TECHNOLOGY

All the fruits on kokum tree are not ready to harvest at the same time and hence periodical plucking is done where the numbers of pluckings vary from tree to tree and the high yielding plants generally takes six to eight pluckings. Spraying of ethrel at the 300 ppm at the stage of full maturity of kokum fruits facilitate harvesting by reducing the number of plucking and improving the yield as well as chemical composition of fruits. The fully ripe fruits are plucked by hand and the skilled persons even climb on the tree to shake the branches and the ripe fruits which fall down are collected but it leads to considerable loss of fruits of approximately 35–40% which include immature and broken fruits. About 30–50 kg yield per plant is obtained in a seedling population, whereas in well-managed plantation, it is 100 kg/plant and as mixed crop in coconut plantation, it yielded 15 kg per plant. Considerable variability in physicochemical composition of kokum is noticed. The annual fruit yield fluctuation is reported in kokum where the higher yield was reported every alternate year. The harvested fruits are exclusively used for processing.

Kokum fruits are juicy and highly perishable and hence need careful handling after harvest and the fruits are to be collected in bamboo baskets lined with rice straw and stored under shade. Sorting of harvested fruits by removing undersized, damaged fruits and grading as per the size and colour into two grades are done accordingly. Because of poor transportability and highly perishable nature, the fruits are mostly processed into *Amsul* or *Amri Kokam* squash. The fruit consists of three to eight seeds which are embedded in a red acid pulp in a regular pattern such as orange segments, in the white pulpy material. The rind and seed both are economical components of kokum fruit. The expected shelf life of fresh fruit is about 1 week and it is mostly used in the form of dried rind to give acid flavour to curries and the fresh fruit juice for preparing cooling syrup and known to have tremendous potential in South Indian curries and is used instead of tamarind and also has many medicinal properties. Kokum seed is a good source of fat called as kokum butter and has food and non-food applications. Kokum oil is traditionally extracted by boiling the kernel powder in water by skimming off the oil that is collected at the top. The yield of oil (fat) is about 25–30% and has been reported to be used in chocolate and confectionary preparations, in the manufacture of soaps, candle and ointments. White dammar resin (resin exuded by *Vateria indica* tree) is an ointment made out of kokum fat and wax and is said to be effective in treating carbuncles. Italy and some other foreign countries import kokum fat from India and use in confectionary preparations. In view of its interesting chemical constituents Kshirsagar et al. (2015) reported that the kokum fruit appears to be a promising industrial raw material for commercial exploitation.

Kokum value-added products	Process of making
Kokum syrup (Amrit kokum)	The fruits are washed with clean water and cut into two halves. Seeds and pulp are removed from them and the rind halves are cleaned internally and placed in layers filling sugar in the halves and they are put in the food grade plastic drums for 7 days. The rind is extracted by osmosis and the syrup is strained through 1-mm sieve or cloth to separate out the rind portion. It is available in two variants (i) syrup with only sugar and no preservatives, colour, salt or water (ii) syrup with permitted preservatives and salts.
Kokum *amsul* (dehydrated salted rind)	Fresh fruits are washed properly and cut into two halves separating the seeds, pulp and rind. The seed and the pulp are mixed with around 10% salt and the salt solution leached out from this mixture is used for dipping the separated rind. The rind is then sun-dried in trays and is dipped again in the salt solution, the next day that is leached on the second day from the salted seed mixture and again placed for drying. Repeating this process of dipping and drying for four to five times, we get the *amsul* that has souring qualities like tamarind and adds taste to coconut based curries, vegetable dishes and fish curries. Generally, three or four rinds are enough to season an average dish.
Kokum *agal* (salted juice)	It is a locally prepared product where salt is added to pulp at four concentration levels (14, 16, 18 and 20%) and the mixture was stirred daily for 7 days after which the whole mixture is strained through stainless steel sieve of 1 mm and filled in pre-sterilized bottles.
Ready-to-serve (RTS) drink from kokum and other juices	The TSS and acidity of different juices are observed and then required quantity of citric acid and sugar was added to raise its Brix value by 20°Bx and acidity to 0.3%. Juice used is 20% and sodium benzoate (NaB) is added as preservative to the remaining water depending upon the colour of the product and it is boiled sufficiently to dissolve the ingredients. NaB is added to 140 mg/kg of the final product and immediately filled into the pre-sterilized glass bottles, sealed with crown corks and pasteurized for 30 min in boiling water, cooled, labelled and stored in cool and dry place at ambient temperature.
Squash	About 25% juice is used to prepare squash. The TSS is noted and required quantity of sugar is added to juice to raise its Brix value to 45°Bx and the acidity is maintained at 1.2%. The remaining water is used for beverage preparation as RTS except that the NaB added is 610 mg/kg of final product.
Kokum butter	The oil is traditionally extracted by boiling the kernels in water and the oil which collects at the top is skimmed off. At present, oil is obtained by solvent extraction also. The yield of oil (fat) which is greasy and whitish yellow in colour is about 25%. It remains in solid state at normal room temperature and is off-white. It is used in food preparations and is helpful in skin ailments such as rashes, allergies, burns, scalds and chaffed skin and is used for the manufacture of cosmetics, creams, soaps, confectionery, candles, etc.

Table *(Continued)*

Kokum value-added products	Process of making
Kokum rind powder	It is used as raw material for various curry preparations, ingredients in various mixes such as *sarbat*, *solkhadi*, kokum RTS, etc. The rind is dried at a certain temperature in a tray dryer and is ground and sieved to get uniform particle size.
Kokum *sarbat* mixture	It is an instant product (ready to prepare) prepared by adding *kokum* powder, sugar and spices in various concentrations and drying in a tray dryer.
Kokum solkhadi mixture	It is an instant product (ready to prepare) such as *sarbat* mixture and is prepared by adding powder, coconut milk powder, milk powder, salt, sugar and spices in various concentrations and drying in a tray dryer.

3.9 MINT

3.9.1 SYSTEMATIC POSITION

Kingdom: Plantae

Division: Magnoliophyta

Class: Magnoliopsida

Order: Lamiales

Family: Lamiaceae/Labiatae

Genus: Mentha

Species: *piperita*

Mints are the perennial plants which belong to the family Lamiaceae. The essential oil is produced in glands present in the leaves and stem of mint (Hegnauer, 1953). From the harvested hay through steam distillation process essential oil is recovered. It is used as herbal medicines (Kumar et al., 2008) and oil is used in chewing gum, toothpaste, candy, oral hygiene products, pharmaceuticals and cosmetic industries (Singh et al., 2007). The United States is the major producer (more than 30%) of mint oil. This oil is mainly exported to Europe and other parts of the world. Commercially cultivated mint species in the world are Japanese mint (*Mentha arvensis*), peppermint (*Mentha piperita* L.), spearmint (*Mentha spicata*) and Bergamot mint (*Mentha citrata*).

3.9.2 BOTANY

M. arvensis is a perennial herb, spreading by rootstocks, which creep along the ground or just under the surface and root at the nodes. The plants have rigid branches that are pubescent and grown up to 60–90-cm tall. The leaves are oblong to lanceolate, 3.7–10-cm long, sharply toothed and hairy. The flowers, arranged in cymes, are usually sessile or rarely pedunculate. The flowers are purplish and minute. The calyx is 3.0-mm long and acuminate, whereas corolla is white to purple coloured and 4–5-mm long. *Mentha piperata* is grown up to 45–90-cm height. The branching stems are dark green to deep purple colour and bear opposite, broadly lance-shaped, slightly toothed leaves of a deep colour. Flowers are born in terminal spikes of main stem and branches from June to September. *M. citrata* grows up to 30–60-cm height, with erect end and decumbent branches. The leaves are thin and bronzy green colour with 1.25–5.0-cm long and smooth broadly ovate to elliptical with the obtuse apex. The flowers are purple, appearing in the upper axil on short, dense terminal spikes. The calyx is glabrous with subulate teeth while corolla is glabrous. *M. citrata* is obtained by crossing between *Mentha aquatica* and *Menthe viridis*. *M. spicata* is grown up to 30–60-cm long, erect, ascending branches arise. The leaves are sessile, smooth, lanceolate or ovate-lanceolate, sharply serrate, smooth above and glandular below, with the acute apex and grow up to 6.5-cm long. The flowers are sharply pointed, long and narrow. The calyx teeth are hirsute or glabrous and corolla is about 3-mm long.

3.9.3 CLIMATE AND SOIL

Japanese mint can thrive best in tropical and subtropical areas under irrigation. However, it does not tolerate damp winters which cause root rot. For proper vegetative growth, temperature ranges from 20 to 25°C is ideal. Under Indian conditions, the higher temperature of 30°C increases the essential oil and menthol content of Japanese mint. Peppermint and spearmint cannot thrive in tropical and subtropical areas, especially in those areas with very high summer temperatures (41°C). The ideal yield is obtained only in humid and temperate conditions such as Kashmir and the hills of Uttar Pradesh and Himachal Pradesh. Bergamot mint can be grown both in temperate as well as subtropical areas. However, the yield is higher in temperate climate. Planting of mint is usually done during the winter months, whereas in temperate climates, planting is done in autumn or spring from the last week of December to the first week of March or from the first week of January to the third week of February. The late planting is usually avoided as it always gives poor yields. The cultivation of mint soil should be medium to fertile deep, rich in humus. The soil should have a good water-holding capacity but waterlogging should be avoided. A pH range of 6–7.5 is the best.

3.9.4 CULTIVATION AND INTERCULTURE

The three most important mint types popularly known and grown across the world are namely Japanese mint, peppermint and spearmint. Japanese mint has several varieties as follows:

Himalaya (MAS-1): It is a selection released by the Central Institute of Medicinal and Aromatic Plants (CIMAP), Lucknow which contains 0.8–1.0% oil (FWB) with 81% menthol content and a low congealing point.

Kalka (Hyb-77): It is a tall, vigorous variety evolved by the CIMAP, Lucknow.

Shivalik: It was introduced from China and released by the CIMAP, Lucknow.

EC-41,911: It is an interspecific cross developed by a cross between *M. arvensis* × *M. piperita.*

Kukrail is one of the most important varieties of peppermint. It is a high yielding variety developed and released by the CIMAP, Lucknow. *Mentha spicata* selection (MSS)-1 and MSS-5 are two popular varieties of spearmint.

MSS-1: This is a selection from the spearmint cultivars introduced from the United States. This variety was released by the CIMAP, Lucknow.

MSS-5: It is a selection from MSS-1 made at the CIMAP, Lucknow.

Punjab spearmint-1: This variety is a clonal selection made at the CIMAP, Lucknow.

Arka and **Neera** are the recently released varieties from CIMAP, Lucknow.

Mints are usually propagated by the creeping stolons or suckers. In the case of peppermint and bergamot mint, even runners are planted. Stolons are obtained from the previous year's planting. A hectare of well-established mint, on an average, provides enough planting material for 10 ha. About 400 kg of stolons are required for planting 1 ha of land. December and January is the best time for obtaining stolons.

The soil should be thoroughly ploughed, harrowed and brought to a fine tilth. All the stubble of weeds should be removed before the crop is planted. Manuring may be done at the time of land preparation by adding 25–30 t/ha of FYM. Green manuring (*Crotolaria juncea* L.) may also be done before the mint is planted. Mints are planted on flat beds of convenient sizes or ridges are made according to the spacing recommended. The stolons (7–10 cm) are planted in shallow furrows about 7–10-cm deep with a row-to-row distance of 45–60 cm, which is done either manually or mechanically. While planting on ridges, the stolons are planted half-way down on the inner sides of the ridges. Immediately after planting, it is irrigated.

Mint gives a positive response with respect to heavy application of nitrogenous fertilizers. Generally, nitrogenous fertilizers at 80–120 kg P and K at 50 kg P_2O_5 and 40 kg K_2O per hectare are required for a good crop of mint. However, in *M. arvensis* an increase of up to 160 kg N per hectare and in *M. piperata*, 125 kg N per hectare has been found to give increased fresh herbage and essential oil yield (Mitchell, 1998). An amount of 100–120 kg N per hectare is recommended for producing the optimum herb and oil yield in *M. citrata* under Pantnagar conditions. A split application of 75 kg N per hectare in combination with P at 60 kg P_2O_5 per hectare is recommended under Kodaikanal conditions. In *M. spicata*, the maximum herb yield is obtained with the application of 100–120 kg N per hectare. The application of nitrogen is usually done in three split doses at 1, 1.5–2 and 3 months after planting and the third dose after the first harvest of the crop. Boron deficiency reduces both the yield of green herb and the essential oil in peppermint. Increased yields of herb, menthol content and essential oil content in peppermint have been obtained by using a combination of boron and zinc fertilizers. Visual symptoms have been documented for some cultivars of Japanese mint towards Fe and Zn deficiencies. With respect to Zn, the crop response is 20 kg/ha if Zn applied at planting. Similarly, experiments conducted at the CIMAP, Lucknow, have shown that the application of 20 kg/ha of sulphur will increase the herb and oil yield in *M. spicata*. Among the different sources of sulphur, calcium sulphate was best followed by ammonium sulphate and elemental sulphur.

The water requirement of mint is very high. Depending upon the soil and climatic conditions, the crop is irrigated six to nine times before the first monsoon. The crop requires three irrigations after the monsoons during September, October and November. Sometimes another irrigation is required during the winter, if the plant is dormant and there are no winter rain to encourage the proper growth of the underground stems. Experiments conducted at Pantnagar have revealed that 15 irrigations are required to get the maximum herb and oil yield in Japanese mint. When mints are grown in temperate climates, only three to four irrigations during the period from July to October are required.

In mint cultivation, 60% reduction in herb and oil yields results from the weed completion. Hence, mints require weeding and hoeing at regular intervals in the early stages of crop growth. One hand weeding is required after the first harvest. Sinbar is usually supplied to control a large number of weeds effectively, when applied as a post-emergence spray at 1 kg/ha. However, combining organic mulch with a combination of 0.5 kg/ha of oxyfluorfen herbicide and weeding or application of pendimethion herbicide at 1 kg/ha and weeding is found to give excellent weed control throughout the crop growth. Dalapon (4 kg/ha) or Gramaxone (2.5 kg/ha) as postemergent spray; Diuron (2 kg a.i./ha) or terbacil treatment (2 kg a.i./ha) as pre-emergent treatment are also recommended for chemical weed control in mints. In order to give a perennial crop (of 3 years only) in peppermint, recultivation is done either in autumn (November to December) or in spring (March to April). When peppermint is grown as a perennial crop, the first-year crop is called *Row mint*, while the second and third-year crop is called *Meadow mint*.

Crop rotations forms an effective practice to maintain a reasonable control on weed growth and also reserves the fertility of the soil and gives higher returns from the land. The following crop rotations are in practice in Uttar Pradesh (a) mint–maize–potato (b) mint–early paddy and potato and (c) mint–late paddy and sweet pea, whereas in Punjab, the farmers practice mint–maize and rapeseed/mustard and mint–maize and potato or mint and paddy rotation. The recommendation for the Terai region of Uttar Pradesh is a 2-year rotation of mint-summer fallowing or millet (fodder) followed by mint on poor fertility lands and mint–wheat–paddy and mint on medium fertile lands.

3.9.5 PLANT PROTECTION

Verticillium wilt (*Verticillium dahliae*) is caused by soilborne fungus, is most serious disease of mint. The fungus spreads readily through diseased planting stock and soil-contaminated equipment. Once established in the soil, the fungus persists for many years. Infected plants are stunted and have smaller, twisted top leaves; the affected plants may be found in spots or portions of a field. The plants first yellow, then die progressively from the bottom up. Infected plants usually die before harvest, but infected stolons and roots may survive to infect the next year's crop. Certified and disease-free planting stock is used. Peppermint variety such as Black Mitcham is most susceptible, while native spearmint is highly resistant and Scotch spearmint is less susceptible. Once the wilt disease is present, peppermint can usually still be profitably grown by using one of the more wilt-resistant varieties (Murray Mitcham, Robertís Mitcham or Toddís Mitcham) in a shorter rotation of at least 3 years, followed by 3 years in corn or soybeans. There is no chemical control for Verticillium wilt and *Puccinia menthae* the mint rust is noticed first on older plants as orange to reddish brown spots on the underside of the lower leaves. In late summer and fall, the rust disease appears as dark brown spots on the leaves of plants in regrowth. Both peppermint and the spearmints are susceptible to mint rust. Careful ploughing of the fields in the early winter to bury crop debris prevents or delays infections. When the crop is 3–4-in tall, 1 kg/ha of chlorothalonil (Bravo) is applied, followed by a second application 10–14 days later. The timing of these treatments is essential for control. If delayed, control will be less effective. There are a number of minor diseases of both peppermint and the spearmints, such as anthracnose, septoria leaf spot and *Fusarium* crown root that rarely occur at levels that require direct control. Most are held in check by disease-free planting stock, crop rotation and crop sanitation such as clean fall ploughing.

The flea beetle (*Longitarsus waterhousei*) damage appears first in the spring, shortly after plant emergence. Affected plants are stunted, usually red in colour and occur in spots in the field. Stunted plants may die. The fine roots of affected plants are clipped back to the underground stem, and the stem is furrowed and brown due to tunnelling of the larvae. Adults feed on the lower leaves in the inner canopy and cause little financial crop damage. Insecticide applications may be necessary if scouting just prior to harvest shows the presence of adult beetles. Within 3 days after the mint is harvested, the stubble is sprayed to eliminate the remaining adults and interrupt egg-laying. Malathion and methomyl (Lannate) are registered for use on mint for flea beetle control. Reducing volunteer mint on roadsides and in ditches helps to control populations. Mint bud mite (*Floridotarsonemus* sp.) causes stunting and distortion of the upper plant late in the season. The mite is spread by infested planting stock, and damage increases in older stands. There is still much that is not known of the life cycle of this mite. Mints effective control depends on careful scouting. If more than 10 mites are found in 30% of the terminals, direct control with the miticide propargite (Omite 6E, Comite) is recommended. Two applications at intervals of 10–14 days are required for suppression of mite populations. Other important control practices include the use of planting stock from mite-free areas, crop rotations of 4 years or less, and early winter ploughing. There are a number of other insects and mites that may occur sporadically in mint, such as mint looper, cutworms (especially the variegated cutworm), two-spotted spider mite and European corn borer.

3.9.6 HARVESTING AND POSTHARVEST TECHNOLOGY

The Japanese mint becomes ready for harvesting 100–120 days after planting when the lower leaves start turning yellow, whereas the delay in harvesting results in loss of oil. Further, harvesting should be done in bright sunny weather. Harvesting consists of cutting the green herb by means of a sickle 2–3 cm above the ground. A second harvest is obtained about 80 days after the first harvest and the third one after about 80 days from the second harvest, whereas in peppermint, spearmint and bergamot mints which are grown in temperate climates, the first crop is ready by the end of June and the second in September or October. A good crop of Japanese mint can give as high a yield as 48 Mt/ha of fresh herb. However, the average yield of mints from three cuttings is 20–25 Mt/ha. The fresh herb contains 0.4% oil.

Mint oil is obtained by distilling either the fresh or the dry herb. The distillation is done both in primitive and modern stills; in the former, the principle of water and steam distillation is followed. While in the later steam generated in a separate boiler is employed. The stems are removed from the dried material prior to distillation because they constitute 30–50% of the material and contain only traces of the oil. The average yield of oil is 50–70 kg/ha. Although bergamot mint as well as Japanese mint give an average yield of 70–100 kg/ha, the yield of peppermint oil is lower with an average of 50 kg/ha. Mint oil is light and golden-coloured, motile liquid and before storage mint leaves should be completely free from moisture It is stored in large steel, galvanized steel or aluminium containers, filled up to the brim to protect against any air remaining inside and placed in a cool storage godown, away from light and humidity.

3.10 PARSELY

3.10.1 SYSTEMATIC POSITION

Kingdom: Plantae

Division: Magnoliophyta

Class: Magnoliopsida

Order: Apiales

Family: Apiaceae

Genus: *Petroselinum*

Species: *crispum*

Parsley (*Petroselinum crispum*), also known as *Ajmood* in Hindi, *Achu Mooda* in Kannada and *Seema malli* in Malayalam, belongs to the family Apiaceae and is a hardy biennial herb. It is widely cultivated as a spice and vegetable and mostly the leaves and seeds are used as spice. There are mainly two types of parsleys where one is found in India and cultivated for leaves and the other grows for its turnip-like roots. It has a characteristic aroma due to the presence of volatile oil. Its name is derived from Greek where 'petroselinum' from *petros* means stone referring to the plant growth habit and *selinon* which was used for parsley in ancient history and is commercially cultivated as an annual for its attractive and aromatic leaves in many parts of the world.

3.10.2 BOTANY

Parsley is a bright green plant that grows as biennial in temperate and as an annual herb in subtropical and tropical areas with bright green fern-like leaves that are finely divided and curled. It reaches a height of 60–100 cm with the numerous stems that grow up from a central crown and vertical roots that are thin or thick and fusiform to tuberous. The leafy stems have a stronger flavour than the leaf and bear greenish yellow flowers in compound umbels and the seeds are 2–3-mm long, smooth, ribbed and ovate with prominent style remnants at the apex. Leaves provide a fresh taste and are known to be a rich source of vitamins A, B and C and the minerals iron, calcium and magnesium. It possesses antiseptic qualities due to the presence of high amounts of chlorophyll.

3.10.3 CLIMATE AND SOIL

It grows best in cool summer climates and sometimes also prefers the hot and humid summers of the Deep South. Good crop growth can be achieved in well-drained soils of 6–7 pH that are consistently moist and rich with organic matter and good levels of nutrients. If it is growing in containers it should be ensured with adequate drainage holes that are not blocked, whereas in indoors normal potting compost is sufficient. Mulching is done to conserve soil moisture and reduce weed competition. Fertilizer application around the soil is to be carried out for every 4 weeks throughout the growing season to ensure sustainable growth.

3.10.4 CULTIVATION AND INTERCULTURE

There are different varieties of parsley as discussed under:

Varieties	**Special features**
Curled leaf (*P. crispum*)	It is very finely divided leaf type whose leaves are bright green with a toothed leaf margin that curled into small frilly leaflets. It is less flavoured than Italian type but used for garnish.
Italian (or plain leaf), (*P. crispum* var. *neapolitanum*)	It resembles original non-curly plants of Europe and is less decorative but flavourful and has flat, dark green leaves with a strong succulent stems. The leaves have stronger flavour hardier than curled leaf.
Hamburg (*P. crispum* var. *tuberosum*)	Consists of thick white roots that resemble parsnips in appearance and use and has a nutty taste when boiled as a vegetable.
Turnip-rooted parsley	It has flat leaves, but the economic part used as a vegetable is the large edible root.
Neapolitan (or celery leaf)	It is dwarf and suitable for both ornamental and culinary purposes whose leaf stalks are eaten like celery.

It is a cool season biennial but grown as an annual from September to May in Florida and is propagated by seeds that are soaked for 24 h before planting to get better germination. If the planting is in September under glass, the seeds should be sown in August in peat blocks. The spacing between rows is 20–30 cm and between plants it is 10 cm apart. Once after seed sowing at ¼-in depth, it is to be covered with a thin mulch layer until the seedlings appear. The germination may take from 7–12 days after which the seedlings may be transplanted. The soil should be well watered as parsley requires very moist soil and is kept weed free. Complete fertilizer application at the time of planting time followed by monthly nitrogen feeding is best on most Florida soils.

Generally, sowing is done in March to April or July to August; but it can be done all around the year for pots under glass. Seed rate of 60–80 g/100 m^2 is sufficient for direct sowing and for press pots 6 g/1000 press pots and for pots 20 g/1000 pots are required. NPK of 1.2 kg, 0.4 kg and 1.7 kg are required per 100 m^2. Seed sowing can be done in spring either in cold frames or window boxes or directly in the garden. The plants grown indoors or under glass shows best results and later these are transplanted to the garden. Before sowing, the seeds are soaked in warm water for 24 h or boiling water is poured into seed drills immediately and if they are sown directly in the garden, they should be sown early. In southern sections of the country, parsley can be grown as a winter crop where it can be kept green under glass in a cold frame during the winter where freezing is not too severe. It bloom and produce seed in the second season. As soon as sufficient growth has been made, the green leaves can be harvested at any time during the season. Parsley leaves are generally used in the fresh state, but both leaves and roots retain their flavour when dried. When the plants attain harvesting stage, they are cut and the stems are removed and the leaves are dried mechanically causing a very dark green colour and seeds are collected when capsules are ripe.

The volatile oil of parsley is very low (0.05%) and can be extracted and sold as an essential oil. Seeds are sown at 45 cm apart in rows which are thinned to 15 cm and are slow to germinate. The process can be speeded up by watering in with warm water. The leaf is harvested in early summer and roots in autumn, whereas the seed is collected in second year of growth when plump and green. The well-formed seedbed prepared by ripping and disking, tilling with rotary hoe fitted with bed formers is used for sowing parsley like the other vegetables and is spaced 50–70 cm apart and two rows of parsley (15–25 cm between rows) sown on the bed at 5–8-mm deep. The soil is kept moist until germination and to avoid crusting on soils peat or a fine mulch should be spread over it.

Depending on the soil type and fertility status, the fertilizers are to be scheduled. In the United States, NPK fertilizer at a rate of 135:135:135 kg/ha is broadcasted before sowing and after each harvest; it is side dressed with high nitrogen fertilizer that improves the production and crop colour. It is very cautious to the environment and is not recommended by NSW Agriculture Department where it will waste money and pollute waterways. The suggested limit is in between1200 and 1500 kg/ha. About 75 kg/ha of ammonium nitrate is sufficient as side dressing at each harvest and if the pH is less than 5.5, lime should be applied to the soil before sowing. However, high N application increases yield and quality (greenness) of parsley without affecting the aroma. According to Antonopoulos et al. (2014) partial dehydration of parsley leaves during storage does not adversely affect its colour or aroma and therefore may be preferable to total drying.

Parsley should be treated as a leafy vegetable when considering watering. During seed germination, the seedbed should be moist but not wet and during growth, the plants should be in moist, well-drained soil. Production on formed beds is considered to be the most suitable. Allowing parsley to experience water stress reduces yields and regrowth while once parsley has 'bolted' to seed its functional life as a foliage plant is over. Removal of the seed heads will not encourage vegetative growth. Plants exposure to water stress during the vegetative growth (before flowering stage) exhibits a decrease in plant height and leaf area (Abbaszadeh et al., 2008; Taheri et al., 2008), dry matter production (Aliabadi et al., 2009) and number of leaves (Misra and Srivastava, 2000). Plant growth (foliage and root weight, leaf number) was significantly reduced by water stress, even at 30–45% deficit. Water stress increased the yield of essential oil (on a fresh weight basis) from the leaves of plain-leafed and curly-leafed, but not turnip-rooted, parsley (Petropoulosa et al., 2008).

Parsley responds well to both organic and inorganic forms of mulch which are good for reducing weeds. The material may be woven weed matting that is cost-effective or plastic film that is cheaper. The increased temperatures produced after mulching can improve the length of the production season.

3.10.5 PLANT PROTECTION

Parsley is known to be infested by many pests such as the carrot weevil whose grubs bore into the crown of plants and cause symptoms and death. The adult weevils overwinter in weed hosts (e.g. Queen Anne's lace) and can be successfully controlled after they become active in spring (mid-May in the north) but before significant egg-laying. Jar traps baited with carrot baby food have been effective as an intregrated pest management tool to determine when adults become active and migrate to fields and corn earworm that can be looked for its larvae in July and August in north and overwinter in the south, whereas flea beetles, leafhoppers or tarnished plant bugs also cause infestation. Root-knot nematode is also noticed in parsley where it is controlled by fumigating the soil with Vapam (metham) or Vorlex (methyl isothiocynate) and coming to the diseases, the most important foliage disease is septoria leaf spot caused by *Septoria apiicola* that is either seed-borne and/or splash disseminated and is most destructive. The disease can be best prevented with the purchase of quality seed and use of drip or trickle irrigation rather than overhead sprinklers. Aster yellow is a viral disease of parsley that is best controlled by eliminating leafhoppers that act as vector.

3.10.6 HARVESTING AND POSTHARVEST TECHNOLOGY

Parsley is best when harvested and used fresh. Leaves are cut and used either whole or chopped, depending on the dish and if preferred for stronger flavour, stems are also used along with the leaves. It can be dried and stored in the dark in an airtight container to retain colour and for later use or it can be placed in the freezer cutting into pieces and placing in ziplock bags which are pulled out and used as needed. Freezing maintains its freshness for later usage. The leaves were first harvested 2 months after seed sowing and harvested again for three additional harvests at monthly intervals.

When grown for oil purpose, it is sown directly as drilled seed with the aim of establishing a good stand and the sowing time is dependent on the climate of the site. Areas with cold winters can plant in autumn but it results only in one leaf harvest and a seed harvest in spring. Murtagh et al. (1990) reported that warmer areas will get a number of leaf harvest during a warm winter and a seed harvest in spring, whereas spring sowing in cold winter areas allows a number of harvests during the summer, a possible winter harvest and a spring seed harvest. The difficulty with summer planting is weed control and avoiding water stress that stimulates flowering and seed production. Parsley leaf oil distilled solely from the plant leaf demands the highest price. Simon and Quinn (1988) recorded leaf yield between 0.04 and 0.15% on a fresh weight basis. Parsley herb oil is derived from the leaves and immature seeds and in some areas, it is classed as leaf oil and is not of same quality as pure leaf oil. Parsley seeds have the highest essential oil content (about 7%) (Prakash, 1990), but produce the lowest quality seed oil that has a characteristic flavour and is a deep green semi-viscous liquid containing 12–15 ml of volatile oil per 100 g. Prakash (1990) reported that 150 g of oleoresin is equivalent to 45.45 kg of fresh parsley.

According to Murtagh et al. (1990), the oil yield and its components will change during the year but when they are at their maximum, harvesting should commence. Monitoring of the crop will enable the optimum harvest time to be determined. The actual process of harvesting is by utilizing standard forage harvester which will cut the stems and deposit the cut material into a trailing distillation bin and it is not necessary to remove the stems or macerate the plant material as this may cause seed and oil loss. Standard combine harvester modified to receive small light seed is suitable when it is desirable to get pure seed oil.

When harvesting for dried product, more care is required than when harvesting for oil. Hand harvesting is the best method, where only good quality green leaves are collected. Yellow leaves will reduce the quality of the final dried product and the collected plant material may be washed to reduce soil and dust impurities as again these will reduce the quality of the final product. If plants are grown on mulched beds, the need to wash the product is reduced and when harvested for fresh market, these are cut with a sharp knife or clippers and bunched with a rubber band and placed in cool boxes to remove field heat.

3.11 SAFFRON

3.11.1 SYSTEMATIC POSITION

Division: Spermatophyta

Subdivision: Angiospermae

Class: Monocotyledonae

Order: Liliales

Family: Iridaceae

Genus: *Crocus*

Species: *sativus*

Saffron is a commercially grown bulbous crop which is perennial in nature. Saffron is commonly used to refer both to the spice and the plant itself. Some archaeological and historical studies indicate that domestication of saffron dates back to 2000–1500 years B.C. (Grilli Caiola, 2004). The dried stigmas are the prominently commercial part used for making saffron spice. Dried stigmas of saffron flowers are the most expensive spice which has been valuable since ancient times for its odoriferous, colouring and medicinal properties (Plessner et al., 1990). The genus Crocus includes native species from Europe, North Africa and temperate Asia, and is especially well represented in arid countries of southeastern Europe and Western and Central Asia. Among the 85 species belonging to the genus *Crocus*, *C. sativus* L. (saffron) is the most fascinating and intriguing species (Fernández, 2004). The origin of saffron crop is southern Europe and it is mainly cultivated in some of the Mediterranean countries such as Spain, Austria, France, Greece, England, Turkey, India and Iran. In other studies, the origin of saffron is obscure, but the plant is believed to have originated in the eastern Mediterranean, (Winterhalter and Straubinger, 2000). In India, it is a legendary crop of Jammu and Kashmir, produced on well-drained karewa soils where ideal climatic conditions are available for good shoot growth and flower production. Saffron cultivation is a tedious job and it requires huge number of labour. Harvesting is done by hand. Hence, it is a labour intensive crop; it is regarded as most expensive spice crop in the world and is popularly known as the *Golden Condiment*.

3.11.2 BOTANY

Saffron plant grows up to a height of 20–30 cm and bears up to four flowers, each with three vivid crimson coloured stigmas. The dried stigmas are used mainly in various cuisines as a seasoning and colouring agent. The saffron crop is multiplied by means of corms. The corms are globular ranging from

0.5 to 5.0 cm in diameter, globose, flattened at the base, the fibres finely reticulate and extended at the apex of the corm. Cataphylls are three to five in number and are white membranous. The corms produce 6–15 narrow, needle-like channelled leaves which are 10-cm long, 1.5–2.5-mm wide, glabrous and surrounded in the lower region by four to five scales. The vegetative phase lasts from November to May. The funnel-shaped flowers (7–8-cm long) are borne either singly or in two or three. The perianth is made up of six segments in two series, violet or reddish purple in colour. Perianth tube is 4–5-cm long, segment subequal, 3.5–5-cm long, 1–2-cm wide, bracts are membranous with long tapering tips. Flowers arise along with aerial shoots by mid-October to early November. The flowers possess a sweet, honey-like fragrance. Upon flowering, plants average less than 30 cm in height. A three-pronged style emerges from each flower. Each prong terminates with a vivid crimson stigma 25–30 mm in length along with 35–50 mm portion of style.

3.11.3 CLIMATE AND SOIL

Saffron thrives well in the cold region and is a temperate season crop. It can tolerate snowfall and frost easily. However, the cultivated area should be free from snowfall during October to November. Saffron thrives well in a sub-temperate climate at an elevation ranging from 1500 to 2400 m in Kashmir valley. During flowering, sunny days and an annual rainfall of 300–400 mm are favourable for a good yield. Spring rains are favourable for the promotion of new croms while the second spell of rain at the end of summer or at the beginning of autumn encourage profuse flowering. The time of flowers blooming is dependent on the temperature prevalent in spring and autumn and also upon the amount of rainfall. Early flowering is conducive by warm spring and long autumn. An average day temperature of saffron ranges from 15 to 20°C during the flowering period with an average night temperature of 6–8°C.

A well-drained, sandy loam to clay loam soil, which is free from clay, is an ideal soil for saffron cultivation. Saffron grows well in medium grade, well-drained light soil rather than rich fertile soil, because fertile soil increases vegetative growth and affects reproductive growth, and it is necessary to have adequate surface drainage for corm development. The soil pH ranges from neutral to slightly alkaline. Saffron is susceptible to calcium carbonate deficiency (Mollafilabi, 2004). Saffron thrives well in salty soils.

3.11.4 CULTIVATION AND INTERCULTURE

SKU-S-86-WAT, SKU-S-86-KIS, SKU-S-86-Sona, SKU-S-86-Lath, SKU-S-86-QUIL, SKU-S-86-Chand and SKU-S-86-Chad are the indigenous varieties and EC-37,328, EC-221,039, EC-217,010, SKUAST, Nag C8708 and Badi c86606 are the exotic varieties of saffron. It is mainly propagated vegetatively by using the underground croms. Corms of 2–5-cm diameter are selected for planting. About two to five cormlets are produced by each mother corm. New corms keep developing each year to replace older ones. The bigger sized corms formed in a year produce flowers during the following season. The corms remain dormant from May to August (Aga et al., 2006).

Land should be ploughed two to three times at a depth of 25–30 cm and should be free from weeds. The land should be rested from 2–3 weeks to the entire winter. Corms should be treated with benomyl, captan or copper-based solution for around 5 min to avoid the fungal attack. The perennial crop is sown in depth of 10–20 cm and spaced wider of 10–15 cm between corms and 20–25 between rows. While annual crops are sown at a depth 8–10 cm, 3–8 cm between corms and about 15 cm between rows. In intensive cropping systems, the shorter spacing is used and planting depth has shown no effect with respect to rooting and flowering in saffron (Negbi et al., 1989). Planting time for saffron is middle of July and end of August in India (Gresta et al., 2003); it has been shown that better quality saffron stigmas are produced when corms sown in the end of July when compared with corms sown in end of August.

In Chaubattia (Ranikhet) region, generally 15–22 Mt of FYM is used (Srivastava, 1973) and 10–50 kg of N, P and K as ammonium sulphate, single super phosphate and potash per hectare in next year is recommended. After harvesting, soil should be immediately applied with manure along the rows which are dug at a depth of 15 cm. In Kashmir, saffron can be grown as a rainfed crop. It requires less amount of water, it can be cultivated in semiarid regions even with a scare rainfall. Hence, July to September rain is favourable for the crop growth. In Spain, it is grown under irrigated condition. Irrigating the crop in early autumn months helps in accelerates blooming and increase crop production.

Information regarding crop rotation in saffron is very scanty. However, for 3–8 years, the land should be left before growing saffron on the same field. Tammaro (1999) observed that when saffron was cultivated on the same soil, it causes decrease in stigma production and increased weed growth. However, in central Italy, saffron is rotated with legumes and wheat with good returns.

The corms planted in July and August should not be disturbed up to 3–4 months. However, one light hoeing in November is helpful for removal of weeds and restoration of aeration in the soil. In subsequent years, second hoeing and weeding are done in May to June for necessary aeration in the soil and development of corms. A deep hoeing (8–10 cm) should be given in August, that is about a month earlier to flowering followed by a light hoeing (4–6 cm) in early September before the flowers appear (Aga et al., 2006). Altogether three to four hoeing operations are required for raising a better production of corms and flowers.

Saffron is the world's most expensive spice, but a poor competitor of weeds particularly during the period between vegetative stage and reproductive stage. Being a very slow and short-statured crop, dense weed growth at any stage of crop growth will have an adverse effect not only on its yield but on quality of the produce. The major weeds of saffron are hoary cress (*Cardaria draba*), downy brome (*Bromus tectorum*), bulbous blue grass (*Poa bulbosa*), hare barley wild barley (*Hordeum spontaneum*), redroot pigweed (*Amaranthus retroflexus*) and Canada thistle (*Cirsium arvense*). In Italy, weed management is done by hands in annual crops, but for perennial crops, weedicides are used, namely Simazine or Atrazine (Goliaris, 1999). While in Iran, broad-leaved weeds are controlled by pre-emergence and post-emergence treatments of Sencor (metribuzin) and narrow-leaved weeds are controlled with Gallant (Haloxy fopetoxy-ethyl) treatments after flower harvest. In summer months, usually Roundup or Buster is applied. Inadequate research is done on indirect weed control methods. Surprisingly, good results are obtained in reducing weeds with the help of wood chips and sawdust mulch (Galigani and Garbati Pegna, 1999). Research trails are under processing to control weeds with plastic which is used as dead mulch in southern Sicily. Manual weeding and herbicide application are the two dominant approaches in controlling established weeds in saffron. But in terms of expenditure manual weeding is costly and time-consuming. Herbicide application is also a costly and non-ecologically sound method of weed control (Soufizadeh et al., 2007).

3.11.5 PLANT PROTECTION

The worst enemies of saffron are rodents and fungi (Tammaro, 1999). Moles, rats and rabbits also can easily damage corms or eat leaves. Fungal attacks are mostly promoted by humid conditions. High moisture percentage together with high temperature creates ideal condition for the rapid development and spread of nematode and fungi (Fusarium, Penicillium, Rhizoctonia, etc.) and consequently corm rot. These conditions generally occur in the hot and rainy spring. Tammaro (1999) indicates that temperature above 10–12°C with rainy weather is a favourable climatic combination for the establishment of fungal disease on saffron. On the contrary, the hot and dry Mediterranean summer inhibits the spread of parasites. To avoid fungal infection, the best practices are crop rotation, the removal and burning of infected plants and corm treatments with antifungal products before planting, such as benomyl or copper-based solution.

3.11.6 HARVESTING AND POSTHARVEST TECHNOLOGY

Harvesting of saffron is very sensible practice. Utmost care should be taken while harvesting stigmas. The flowers are usually a few centimetres away from the ground; the flowers are surrounded by several leaves which should not be damaged, if so daughter corms will not be produced. Usually, manual harvesting is practiced by cutting the base of the flower stem with finger nail. Nearly about 350–450 man hours are required to produce 1 kg of spice which includes 200,000 and 400,000 stigmas altogether, depending on the unitary weight. Saffron flower is with short life hence it should harvest on the same day when it flowers, before the opening of the corolla otherwise the colour and quality will be lost (Zanzucchi, 1987; Tammaro, 1990). The stigmas should be harvested by separating stamens and sepals by gently opening the corolla. Stigmas are cut below the branching with fingers where the style changes colour (from red to yellow).

Lack of mechanization in saffron is due to its fragile nature of flowers and corms and it requires careful handling. The other possible reasons may be due to low wages for labour in some countries which curtails the need for mechanization and also may be due to less amount of land high labour cost countries. A modified onion planter is used to plant saffron, but this does not acquire prominence because of its inability of placing the corms with the apex in upward direction and this delays the emergence of plant and because of this production decrease. A normal hoeing machine can be used to control weeds by following specific distance in the rows. This can be carried out successfully in the first year until the drought corms arise. In the second year, this method failed to give better results due to rising of daughter corms. If hysteranthy is manifested in saffron, flower harvests can be easily done with the help of calibrated mowing or grass cutting machines, without cutting the leaves. Adapted bulb and tuber-picking (such as a potato digger) can be used instead of manual labour for corm lifting with better results.

Drying and storage methods are very important steps in maintaining the quality of the saffron stigmas (Carmona et al., 2005). According to the International Organization for Standardization (ISO) norms, ideal moisture content ranges between 10 and 12% (ISO-3632, 2003). For dehydration of saffron, many methods are followed. In Italian method of saffron production, the stigmas are dried at room temperature in the sunlight or with forced air by spreading them over a large area. In Navelli, stigmas are traditionally dehydrated by placing stigmas on a 20-cm sieve above a charcoal fire (Tammaro, 1999). While in Sardinia drying of stigmas are carried out under sunlight or at room temperature for several days or in the oven the temperature of 35–40°C for less time until the moisture content is reduced to 5–15%. In India and Iran, saffron is sun-dried and in Spain stigmas are toasted over hot ashes. In Greece, they are dried slowly at 30–35°C in dark rooms. But among the procedures, no method has given 100% results (Carmona et al., 2005; Gregory et al., 2005). Saffron pigments are light-, oxygen- and temperature-sensitive hence it should be stored in dark and modified atmosphere. The best method to store saffron is to keep it hermetically closed in darkened glass containers, at low temperature (5–10°C) (Alonzo et al., 1990).

Yield in saffron is essentially a cumulative function of many agronomic, biological and environmental factors, which influence the production. The production is mainly influenced by dimension and storage conditions of corms (Molina et al., 2004), climatic conditions (Tammaro, 1999), sowing time, cultural techniques (annual or perennial), crop management (irrigation, fertilization and weed control) and disease. The production of saffron takes a peak from the first to the third to fourth years of cultivation. Generally, 1 ha of saffron may produce 10–15 kg of dried stigmas, but it can range widely, depending on the above-mentioned factors, from 2 to 30 kg. The yield of 2.5 kg/ha are reported in Kashmir, India and Morocco (Bali and Sagwal, 1987) in rainfed conditions, while it can reach 15 kg/ha in Spain under irrigation and fertilization. In irrigated Moroccan areas, yields of about 2.5–6 kg are obtained (Ait-Oubahou and El-Otmani, 1999). In New Zealand areas, yields of 24 kg/ha of dried stigmas is obtained (McGimpsey et al., 1997). A production of 29 kg/ha was recorded in Navelli (Tammaro, 1999), but this is not comparable with the other yields because an annual cropping system is used and only the biggest corms are replanted every year (Gresta et al., 2003).

3.12 VANILLA

3.12.1 SYSTEMATIC POSITION

Kingdom: Plantae

Subkingdom: Viridiplantae

Division: Tracheophyta

Class: Magnoliopsida

Subclass: Magnoliidae

Order: Asparagales

Family: Orchidaceae

Genus: *Vanilla*

Species: *planifolia*

Vanilla (*Vanilla planifolia*), a member of the family Orchidaceae, is a tropical, herbaceous, climbing monocot perennial crop possessing a stout, succulent stem and is grown for its fruit which yields the vanilla flavour extensively used in foods and beverages (Bailey and Bailey, 1976). It is the most expensive spices after saffron and is the only spicy orchid of the tropics valued for its cured fragrant beans. Though over 50 species have been described, only three are important as source of vanillin. They are *V. planifolia* Andrews, formerly known as *Vanilla fragrans* Saslisb Ames (Bouriquet, 1954); *Vanilla pompona* Schiede and *Vanilla tahitiensis*, Moore. Of these, *V. planifolia* is the most preferred commercially and therefore, widely cultivated as it contains high content of vanillin (Ranadive, 1994).

3.12.2 BOTANY

Vanilla is a tropical, fleshy, herbaceous perennial vine, climbing by means of adventitious roots on trees or other supports. The roots are long, whitish, aerial, about 2 mm in diameter and are produced singly opposite the leaves. The roots at the base ramify in the humus or mulch layer. The long, cylindrical, monopodial stem (1–2-cm diameter) is simple or branched, succulent and brittle. It is dark green and photosynthetic with stomata. The internodes are 5–15 cm in length. Large, flat, fleshy, subsessile leaves are alternate, oblong-elliptic to lanceolate and are 8–25-cm long and 2–8-cm broad. The veins are numerous, parallel and indistinct. The petiole is short and thick. They are borne toward the top of the vine and are 5–8-cm long with up to 20–30 flowers, opening from the base upwards. The flowers are large, waxy, fragrant, pale greenish-yellow and are about 10 cm in diameter. Two upper petals resemble the sepals in shape. The lower petal is modified as a trumpet-shaped labellum or lip whose lower part envelops a central structure called column (gynostemium) which is less shorter than perianth and also have a tuft of scales (Bhat and Sudarshan, 2004). The tip of the lip is obscurely three-lobed and is irregularly toothed on its revolute margin. Dark coloured papillae form a crest in the median line. The gynostemium is long, hairy on the inner surface, bearing at its tip the single stamen. The concave sticky stigma is separated from the stamen by a thin, flap-like rostellum because of which self-pollination is impossible. The fruit is known as bean and is pendulous, cylindrical and a three-angled capsule. It contains ripe myriads of very minute globose seeds of about 0.3 mm in diameter.

3.12.3 CLIMATE AND SOIL

V. planifolia originates from the Mexican tropical rain forest. Its cultivation requires the same climatic conditions. Vanilla thrives well from the sea level up to 1000 m MSL under hot, moist, tropical climate with adequate well-distributed rainfall. Natural growth is obtained at latitudes, 15° north and 20° south of the equator. The optimum temperature ranges from 21 to 32°C and rainfall 2000 to 2500 mm annually. Rainfall must be evenly distributed, with a minimum of 2000 mm/year. The dry period of about 2 months is needed to restrict vegetative growth and induce flowering. It grows best in light, porous and friable soils with pH 6–7. Partial shade is essential for successful cultivation. An ideal soil for vanilla is light, rich in humus and porous, allowing the roots to spread without encountering high moisture. However, vanilla can grow in many types of soils, if they are not heavy and have good drainage. Volcanic soils, sand and laterites are also suitable.

3.12.4 CULTIVATION AND INTERCULTURE

Commercial vanilla is always propagated by stem cuttings. The cuttings are taken from healthy, vigorous plants and may be cut from any parts of the vein. The length of the cutting is usually determined by the amount of planting material available. Short cuttings, 30 cm in length will take 3–4 years to flower and fruit. Cuttings, 90–100 cm in length are preferable and have almost become the standard length of cuttings and will flower for about 2–3 years. Length of the shoot cutting used for propagation has a direct bearing on the growth of vine and yield (Suryanarayana, 2004). Since the longer cuttings bear flowers much sooner than shorter ones, it is more economical and profitable to plant longer cuttings whenever material is available. It is customary to remove two or three leaves from the base of the cutting which inserted into the humid layer and mulch. With short cuttings, at least two nodes should be left above ground. The portion above ground should be tied to support until aerial roots have obtained a firm grasp. Cuttings are usually planted directly in the plantation grounds but they can be started in nursery beds also. Because of its nature, cuttings can be stored or transported for maximum 2 weeks, if required.

Vanilla being a climbing orchid needs some support to grow. It also requires about 50% shade. The support trees can also be used for providing shade. Low branching trees with rough bark and small leaves are preferred as support trees. Some commonly used support trees are *Gliricidia*, *Plumeria*, *Casuarina*, Mulberry and *Erythrina lithosperma*. The cuttings for support trees should be planted at least 6 months prior to planting of vanilla. Cuttings of 1.5–2-m length with 4–5-cm diameter are to be used. They should be planted in the corner of the pits. The size of the pits should be 40 cm × 40 cm × 40 cm and the spacing of 2.5 m between rows and 2.5 m within a row should be maintained. The pits are to be filled with fertile soil before planting the supports.

The ideal time for planting in Indian conditions is August to September when the intensity of the south-west monsoon is low and where by this time the support trees should have grown well. Good quality vines from disease-free plants, sufficiently grown rooted cuttings or secondary hardened tissue-cultured plantlets can be used for planting. Stem cuttings selected for planting should be kept in shade for about a week prior to planting. Generally, three to four basal leaves of the cutting are clipped away before they are put in the shade. It is recommended to dip the basal tip in 1% Bordeaux mixture or Bordeaux paste or pseudomonas paste before planting the cutting. While planting the cutting, the defoliated basal portion is to be placed in the loose soil near the base of the support, just below the surface, in a half loop in such a way the basal tip is above the soil surface. The top end of the cutting is to be tied to the support. Mulching the base of the support tree with partially decomposed organic matter is recommended. It takes about 4–8 weeks for the cutting to take root and to show signs of initial growth.

The vines must be trained to grow within easy reach of workers who must be able to pollinate the flowers, pruned the vines and harvest the fruit without difficulty. The vines should never be exposed to direct full sunlight but rather, should always have checkered shade provided by the supporting trees. It is important to keep a layer of vegetable mulch on the surface of the ground, especially over the roots of the plant. Cultivation of the soil is risky since the roots grow at the surface level and can be easily damaged. Weeds and other vegetable growth should be clipped and spread over the ground as additional mulch. About 9 or 10 months before the flowering season, the lip of the vine is cut off to induce vine to flower. The blossoms are produced in the axils of the leaves on long hanging branches. When the plants are in flower, they need daily attention. After flowers are selectively pollinated and the desired numbers of pods are set on each vine, the remaining flowers and buds must be removed. This prevents any loss of plant vitality through the production of useless blossoms, avoids pollination of too many blossoms and saves workers from having to examine superfluous flowers. All undesirable or malformed pods should also be removed. After flowering and fruiting, old stems should be trimmed away. This will be replaced the following year with new and productive stems. Vines reach their maximum production about 7 or 8 years and if given proper care, continue to produce for several years more.

For optimum growth of these plants, a controlled environment is created by establishing suitable greenhouse/shade net house which provides the appropriate amount of light, temperature and humidity which are essential for the commercial production of vanilla. High-density polyethylene net providing 50–60% shade can be supported with stone pillars of 12 ft height to provide the required shade.

The main source of nutrients for the crop is from organic sources, namely decomposed leaf mould or dry/decomposed FYM/vermicompost. A thick layer of easily decomposable organic matter is applied around the plant base at least three to four times in a year. Besides that, Spices Board recommends spraying 1% 17:17:17 NPK that gives full coverage of the foliage and stem to enhance the growth of the vines and presently the farmers are getting very good response in growth and yield by spraying vermiwash to the foliage.

The flowering commences from the third year after planting during January to February and can be naturally pollinated only by a specific *Melipone* bee found in Mexico. Self-pollination never happens in *vanilla* because a structure called rostellum prevents stigma coming into direct contact with the pollen grains. Hence, artificial hand pollination is resorted to by pushing back the rostellum with the help of a toothpick or a pointed bamboo splinter. The pollen sac is further pressed to spread pollen over the stigma. In its natural home of vanilla, pollination is done by some *Melipone* bees and some hummingbirds. The ideal time for pollination is 7 a.m. to 12 p.m. (Artificial pollination starts from early morning and completed before noon as the flower closes in the afternoon.) On an average, a skilled worker can pollinate 1200–2000 flowers a day. First pollination formed 8–10 flowers on the lower side of the inflorescence are ideal and it is also recommended to maintain only 10–12 inflorescences per vine in order to get beans with maximum length, girth and of high-quality standards. Generally, one flower in an inflorescence opens in a day. The flowering is spread over a period of 3 weeks. Pods take about 8–9 months to attain maturity.

3.12.5 PLANT PROTECTION

No serious pests are noticed in Vanilla in India. But according to the survey of Varadarasan et al. in 2003, pests such as leaf-feeding beetles, caterpillars and sucking bug (*Halyomorpha* sp.) whose adults and nymphs infest tender shoot tips and emerging inflorescences resulting in their drying and rotting that can be controlled by spraying quinalphos 0.05% are noticed. Snails and slugs are also found which feed and damage tender shoot tips and leave especially in moist and shaded areas in the plantation where hand-picking and poison baiting help in preventing the pest.

However, diseases caused by viruses, fungi, bacteria and nematodes are found where the viruses and fungi cause the most serious problems for vanilla. Premature yellowing and bean shedding disease is of relatively recent origin and is noticed in all vanilla plantations of Karnataka and Kerala especially during summer months and it initiates as dried corolla drop off from the tip of immature beans which otherwise remains attached to the beans till half way through maturity. As the dried corolla drop off, exudates from the beans accumulate at the tip and the beans turn yellow followed by brown discoloration from the tip. High temperature (more than 32°C) and very low relative humidity (less than 70%) prevailing during the months of February to May predisposes the plants to infection. Overcrowding of the beans may also play a key role in immature bean shedding. The intensity of the disease is low under conditions of high altitudes where temperature and humidity are maintained under forest cover. Constant association of *Colletotrichum vanillae* and insect larvae inside the flowers are noticed. This can be kept in control by spraying dimethoate or quinalphos 0.05% during flowering period thrice at 15–20-day interval and fungicides such as thiophanate methyl 0.2% or carbendazim–mancozeb (0.25%) at 15–20-day interval thrice from February to May.

Other diseases such as stem and root rot can also be noticed in vanilla where the earlier one caused by *F. oxysporum* sp. *vanillae* usually appears during the post-monsoon period of November to February as yellowing and shrivelling of the internodal area extending to both sides of the stem and when the basal or middle portions of the vines decay and shrivel, the remaining distal portions of the vines show wilting symptoms. Stem rot and drying are observed at the basal portions above the ground level and the later one caused by different fungi: *Phytophthora jatrophae, F. oxysporum* or *Fusarium vanillae* var. *bulbigenum* shows symptoms where during early stages shallow roots turn brown and die. The damage is generally not perceptible at first and vanilla plants can survive without showing symptoms. New roots may form, but they become infected as soon as they touch the soil. As a result, the plant cannot obtain water, minerals and turns yellow. The leaves become pendulous and abscise and at the end, the vines become flaccid and die. In order to control these diseases, some management practices should be done such as using of uninfected cuttings, limited pollination to strong vines particularly at first blooming and regular looping. Removal and burning of all diseased parts and damaged roots and moreover, use of resistant cultivars would be the best solution and fungicidal sprays can also be used.

Viral diseases such as vanilla mosaic virus (VanMV) and cucumber mosaic virus (CMV) are identified in vanilla where VanMV causes leaf mosaic and severe malformations on *Vanilla tahitensis* and *V. pompona* in French Polynesia with symptoms of severe mosaic, blistering and leaf distortion with over 30% of the fields (Wisler et al. 1987). This virus is serologically related to dasheen mosaic virus, extremely common in taro (*Colocasia esculenta*), but has a different host range, whereas CMV was recently found in vanilla causing leaf distortion and stunting. In French Polynesia, 23% of plants surveyed showed virus-like symptoms and contained CMV. The *V. tahitensis* plants were severely stunted with distortions of the leaves and stems. Flowers were sterile and the production was greatly impacted. This virus has an extremely large host range and infects over 800 plant species, with the vanilla isolates able to infect Chenopodiaceae (*Chenopodium*), Cucurbitaceae (cucumber), Fabaceae (mung bean) and Solanaceae (pepper, tomato and tobacco). CMV is divided into two major subgroups (I and II) and subgroup I is further divided into two groups (IA and IB) by phylogenetic analysis. The *V. planifolia* isolates belong to subgroup IB in India (Madhubala) and subgroup I from French Polynesia and Reunion Island (Farreyrol) for *V. planifolia* and *V. tahitensis*, respectively. An unknown rhabdovirus-like particles or 'bullet shaped' virus has also been found in vanilla from Fiji and Vanuatu.

A viral infection means that the entire plant has the virus and it is incurable. The management includes detection of virus and any plants with CMV, VanMV or WMV-II should be destroyed immediately and new plantings should be started with virus-free stock plants. All ornamental orchids and crops such as cucumber, mung bean, pepper, tomato and tobacco should be kept away from vanilla fields. Older vines should be tested for viral diseases before using for propagation.

3.12.6 HARVESTING AND POSTHARVEST TECHNOLOGY

Fruits mature at about 8–9 months after pollination. If harvested before maturation capsules have reduced aroma after postharvest treatment and may deteriorate. If harvested too late, the capsules split. The best stage for harvesting the beans is the so-called canary tail. At this stage, the fruit is green, but its base is canary yellow. Harvesting at this stage is a requisite for producing vanilla beans of high quality and with good aroma. Capsules are harvested one at a time as they mature. At the time of harvesting, the capsules do not have any aroma. The aroma forms only after a long process that spans for 9 months, during which the green capsules become black 'vanilla beans'.

Vanilla is harvested when the pods are mature and split longitudinally. A hectare of land can hold about 4000 productive vines. The average yield is 1.5–2 kg green beans per vine. A mature good quality bean weighs between 15 and 30 g. Fresh beans get the characteristic aroma due to enzymatic action during curing. The enzyme β-glucosidase act on the precursor glucovanillin which result in the harvested beans subjected to a process of nightly sweating and daily exposure to the sun for about 10 days until they become deep chocolate brown in colour. Then they are spread on trays in an airy shelter until dry enough for grading. The best grade may be covered with tiny crystals of vanillin. This coating is known as givre.

After pollination and fertilization, the beans develop very quickly and obtain full size in about 56 weeks but it takes 9–11 months for the same to mature. Around 75–90 mature beans make 1 kg. The beans are harvested when the distal end turns pale yellow in colour. The aroma and flavour develop only after the curing process. The different stages of curing include killing (by dipping the beans in hot water at 63–65°C for 3 min), sweating (through exposure to sunlight for 1–11/2 h by spreading them on a raised platform every day for 5–7 days), drying (by keeping the beans spread on racks in an airy room for up to 30 days) and conditioning (keeping the dried beans bundled and covered in butter paper, in wooden boxes for about 2–3 months).

3.13 LONG PEPPER

3.13.1 SYSTEMATIC POSITION

Kingdom: Plantae

Division: Magnoliophyta

Class: Magnoliopsida

Order: Piperales

Family: Piperaceae

Genus: *Piper*

Species: *longum*

Long pepper (*Piper longum*) belongs to the family Piperaceae and its genus contains more than 700 species grown in tropical and subtropical rainforest. It reached Greece in the sixth or fifth century B.C. and was an important and well-known spice before European discovery of the New World. It is known as Pippali (Thippali) and is a close relative of *Piper nigrum* that gives black, green and white pepper and has a similar but generally hotter flavour and Hippocrates is the first person who described it as a medicament rather than a spice. It consists of dried fruits and is a slender, aromatic, creeping and perennial under shrub, native of the hotter parts of the country and found wild as well as cultivated extensively in Assam, lower hiss of Bengal, evergreen forest of Western Ghats, along west coast of southern states and also recorded from Car Nicobar Islands (Chauhan et al., 2011).

3.13.2 BOTANY

P. longum is a slender creeping under shrub spreading on the ground and rooting at nodes. Young shoots grow downwards with 5–9-cm long leaves of 5-cm wide, ovate and cordate with broad rounded lobes at the base, subacute, entire glabrous bearing unisexual flowers in solitary erect spikes during the rainy season or just after that. The male spikes are larger and slender and are about 2.5–7.5-cm long, whereas the female spikes are 1.25 to about 2-cm long when in flower and growing to about 3 cm when in fruit. The berries are ovoid, yellowish orange, sunk in the thick rachis about 0.25 cm in diameter (Sumy et al., 2000; Banerjee et al., 1999; Viswanathan, 1995).

3.13.3 CLIMATE AND SOIL

It is cultivated as a commercial crop widely in areas having high rainfall, high humidity and moderate temperature of about 15–35°C such as West Bengal, Assam, Meghalaya, Maharashtra (Akola region), Orissa Andhra

Pradesh (Vishakhapatnam area), Uttar Pradesh, Tamil Nadu (Anaimalai Hills) and Kerala. The crop thrives well on a variety of soils and light porous well-drained soil rich in organic content is most suited for its cultivation. Since it is a shallow rooted crop it requires high humidity and frequent irrigation. The plant should be grown under partial shade for good growth. Thus, it can be successfully cultivated as an intercrop in irrigated coconut and areca nut gardens. It is highly sensitive to drought and also waterlogging conditions.

3.13.4 CULTIVATION AND INTERCULTURE

The field should be prepared with two to three ploughings, followed by one or two harrowings and levellings. Considering the slope of the fields, drainage should be provided for excess water. Long pepper is propagated vegetatively by rooted vine cuttings. It is recommended to take three-nodded cuttings from any part of the stem to serve as planting material. Rooting takes about 15–20 days after planting. Cuttings can be directly planted in the field or after induced rooting in the nursery before finally transplanting in the field.

It is propagated by suckers or rooted vine cuttings. Vine cuttings and suckers are transplanted soon after the setting in of monsoon rains. The best time for raising nursery is during March to April and to avoid mealy bug attack on roots, 10% detergent powder is to be mixed with the potting mixture. Normal irrigation may be given on alternate days. Excess moisture in the nursery can cause Phytophthora wilt and by the end of May, the cuttings will be ready for planting. About 60 cm × 60 cm spacing can be maintained between row to row and plant to plant. If plants are to be raised first in the nursery, the best time for nursery raising would be 1 month earlier to actual planting.

Long pepper needs heavy manuring. In soils with low fertility, the growth of the plant is very poor. In the first year, application of about 20 t/ha FYM at the time of land preparation is recommended. Since it comes to economic yield after 3 years, the manuring has to be done each year. In subsequent years, manuring is done by spreading it in bed sand covering with soil. Application of organic manure increases the water holding capacity of the beds. In the subsequent years, apply FYM before the onset of monsoon. No chemical fertilizers are recommended for use. During the first year, weeding may be undertaken as and when necessary. Generally, two to three weeding are sufficient. Once the crop grows and covers the field, no serious problems of weeding are faced.

Ensure irrigation during summer months. Irrigate once or twice in a week depending upon the water holding capacity of the soil. Even during the monsoon period, if there is a failure of rains for quite some time, apply irrigation. As irrigated crop, spike production continues even in summer months. When the crop is not irrigated, it is necessary to give mulch with dry leaves or straw during summer months. If the crop is irrigated during summer, it continues to produce spikes and off-season produce will be available.

3.13.5 PLANT PROTECTION

Long pepper is known to be attacked by mealy bugs that infest the roots and suck its sap resulting in stunted growth. The severity is more in summer and can be controlled by spraying Rogor.

Wilt and pollu are the two important diseases of this crop. The wilt is characterized by the death and decay of the roots, yellowing and shedding of the leaves and the ultimate drying of the plant. Pollu not only causes hollowness of the fruit but also leads to their complete destruction. Spraying Bordeaux mixture (0.1%) in the month of May and two to three times in rainy season found effective in reducing the extent of damage. *Helopeltis thivora* is a pest which causes destruction of leaves. To control this pest, decoction prepared out of neem (0.25%) has to be sprayed.

3.13.6 HARVESTING AND POSTHARVEST TECHNOLOGY

Vines start fruiting 6 months after planting. The female spikes take about 2 months to mature from its inception. A full-grown mature spike should be harvested before ripening. In Kerala, three to four pickings can be taken depending upon the maturity of spikes. The spikes are harvested when these are blackish green in colour. The yield of dry spikes in the first year is about 400 kg/ha and up to 1000 kg/ha in the third year. After the third year, yield declines and after the fifth year gradually becomes uneconomical. Besides spikes, thicker roots and basal stem portions should also be cut and dried before the crop is abandoned, as these are used as important drug constituents in the Ayurvedic and Unani systems of medicine. On an average, 500 kg roots are obtainable per hectare. Dry the harvested spikes in the sun for 4–5 days. Green spike to dry spike ratio is about 5:1. Dried spikes should be stored in moisture-proof container. The produce should not be stored for more than a year. Stem and roots are cleaned, cut into pieces of 2.5–5 cm-length, dried in shade and marketed as *Pippali mool*. There are three grades of *P. mool*, based on the thickness (Joy et al., 1998). The grade I with thick roots and underground stem fetches higher price than grade II and/or III which may comprise their roots, stem or broken fragments (Wealth of India, 2003).

3.14 SWEET FLAG

3.14.1 SYSTEMATIC POSITION

Kingdom: Plantae

Division: Tracheophyta

Class: Magnoliopsida

Order: Acorales

Family: Acoraceae

Genus: *Acorus*

Species: *calamus*

Sweet flag is a semiaquatic plant mostly seen in temperate and subtemperate regions. It is an uncommon, widespread but commonly used for drugs in the traditional system of medicine. It is grown in wetlands of North America, India and Indonesia. In India, it is grown plentiful in marshy tracts of Kashmir, Himachal Pradesh, Manipur, Naga Hills and in some parts of Karnataka. *Acorus calamus* is commonly known as sweet flag in India. The plant parts that are used in the medicine traditionally are scented leaves and rhizomes. The leaves and roots have lemony scent and sweet fragrance, respectively. Powdered rhizome which is having a spicy flavour is used as substitute for ginger, cinnamon and nutmeg for its odour (Balakumbahan et al., 2010).

3.14.2 BOTANY

Sweet flag is a semiaquatic marshy perennial aromatic herb grows to a height of about 1.5 m. The plant has long leaves which are thick, erect, 1-cm long, sword-shaped with single prominent mid-vein and slightly raised secondary and tertiary veins on both sides. The margin is curly edged or undulate. When the leaves are crushed it emits strong scent. Flowers and fruits are rarely seen on the plants. It consists of a leaf-like spathe and a spike-like spadix that is densely covered with yellow and green flowers which are 3–8-cm long, cylindrical in shape. At the time of expansion, spadix grows to a length of 4.9–8.9 cm. The season of flowering starts from early to late summer. The fruits are small green angular berry like with one to three seeded. Rhizomes are long, cylindrical extensively branched, about 19–25 mm in diameter and 10-cm long. Outside of the rhizome is light brown in colour, whereas it is white and spongy inside. Its leaf scars are brown white and spongy and it possesses slight slender roots.

3.14.3 CLIMATE AND SOIL

It is a hardy plant and requires well-distributed rainfall throughout the year. It requires plenty of sunshine for the growth of the plant and drying of rhizomes after harvesting. Temperature ranging should be 10–38°C and annual rainfall between 70 and 250°cm. It grows well at an altitude of 2000 m in the in the Himalayas, Manipur, Naga Hills and in some parts of South India. Cultivation should be avoided wherever there is no irrigation facility. It thrives well in waterlogged marshy soils, clayey loams and light alluvial soils. River banks of clayey loams and light alluvial soils are best suitable for sweet flag cultivation and the pH range should be 5.5–7.5.

3.14.5 CULTIVATION AND INTERCULTURE

CIM-Balya is developed through clonal selection with medium long and broad leaves. It has cylindrical rhizomes in light cream colour having the length and diameter of 115.120 and 15–20 mm, respectively. The yield of the rhizome is 9–11 t/ha and oil content is about 0.60–0.65% (Chauhan, 2006).

Sweet flag can be propagated through seed and vegetatively by plant tops and rhizomes. The plant rarely produces seeds, because it is propagated mainly by vegetative means. In greenhouse, seeds are planted during winter in a tray of 2-in depth which is filled with organic soil mixture. The seeds are scattered sparsely on the tray surface and press firmly into the soil to the depth of 4–5 cm. It does not require any stratification but the soil should be kept from moist to saturation state. Germinates are observed within 2 weeks. They are transplanted in to pots with 10-cm size when the plant reaches to the height of 3–4 in in the tray. To maintain moist to saturated conditions, the pot should be placed in shallow water or irrigated frequently. During spring season, the potted plants are transplanted to the main field with a spacing of 30 cm apart and should be planted where it will be in full sun to partial shade. Avoid flooding in seeded area or newly established plants in the main field. The tops of the previous year are used as planting material. These are kept for 2–3 weeks in a moist place or puddled place and later they are planted in the main field at 30 cm × 30 cm spacing. Within a year, the crop is ready to harvest. Sweet flag is mainly propagated by rhizomes. Rhizomes of the previous planting should be taken and kept preserved in soil rich is moist in condition. After emergence of sprouting, the rhizomes are cut in to small pieces without disturbing the bud and planted in the main field at a depth of 4 cm and with a spacing of 30 cm × 30 cm while planting the leafy portion should be should be little above the soil so that the bud can be seen from outside. The best time of planting is second fortnight of July. The plant population is about 111,000 plants/ha. For the alternate use of rhizomes sprigs from clumped plants can also be used as a planting material. Prepare the land prior to the onset of rains by ploughing twice or thrice like paddy fields. Best season of planting is June to August. Recommended spacing for rhizome pieces is 30 cm × 30 cm and presses into mud to a depth of 5 cm (Krishnamurthy and Avani, 2015). According to the studies of Tiwari et al. (2012), closer planting (20 cm × 20 cm, 30 cm × 20 cm) gave significantly higher rhizome yield (8620 kg/ha, 8120 kg/ha) than wider spacing. Manuring should be done in the field with compost 15 t/ha or with green manure 10–12 t/ha. According to Krishnamurthy and Avani (2015), chemical fertilizers per hectare are recommended at the rate of 100:40:40 kg NPK. For the first 4–5 months, weeding should be done every month. Duration of the sweet flag is 10–11 months. As long as crop is in the field, water is needed. It is a water-intensive crop as paddy, so water should not be scarce. Where the land is saturated with water such as canal or river banks are suitable for its growth. About 5 cm of water level is required during the initial stages of the crop and later it is increased to 10 cm. The level of water standing in the field should be 10 cm till 20 days before harvest. In the rainy season, irrigation can be avoided but during dry spell, at an interval of 2–3 days, irrigation must be given to the field.

3.14.5 PLANT PROTECTION

Mealybugs are the major pests observed in this crop. It can be controlled by spraying the shoots and drenching the roots of plants with 10-ml methyl parathion or 20-ml quinalphos in 10 l of water. Leaf spot is the major disease that is observed and it is controlled by spray of captan 10 g with chlorpyrifos 20 ml/10 l of water.

3.14.6 HARVESTING AND POSTHARVEST TECHNOLOGY

The symptoms of maturity in the sweet flag are that the leaves begin to turn yellow in colour and drying of the leaves is observed and the spadix will turn brown in colour. From the sides of the stalk flowering heads are produced which consists of a fleshy spike of about 9–10-cm long and 1-cm thickness. The flowers are closely covered, small greenish yellow in colour appear from May to July. In October to November, the rhizomes are harvested. The crop is ready to harvest after 1 year of planting. The rhizomes are usually collected during autumn (September to October) till early spring (March to April) seasons, and sufficient moisture should be necessary for deep digging. Carefully harvest the rhizomes that are located at 60-cm depth. After the rhizomes are uprooted they are cleaned and washed with water, fibrous roots are removed from the rhizomes and they are cut in to the lengths of 5–7.5-cm pieces. The cut rhizomes are dried in sun and rubbed with gunny bags to remove the scales. In case of the large-scale cultivation, the harvesting is done by ploughing. In order to retain the oil, the dried rhizomes are spread under the shade. The yield of 4.22 t of dry rhizomes or 10 t fresh rhizomes is expected per hectare. Indian roots contain 3.1% oil in the plains, whereas fresh aerial parts contain 0.125% oil. To a lesser extent, the leaves also contain aromatic properties of the rhizome but they are not used in medicines.

3.15 HORSERADISH

3.15.1 SYSTEMATIC POSITION

Kingdom: Plantae

Division: Magnoliophyta

Class: Magnoliopsida

Order: Capparales

Family: Cruciferae

Genus: *Armoracia*

Species: *rusticana*

Horseradish is one of the oldest herbaceous condiments and is well-known as large-leaved hardy perennial. It is a perennial crop which is cultivated mainly in Europe and Asia because its roots are used in the human diet as a pungent spice. The roots are also rich sources of biological compounds beneficial for humans (Veitch, 2004 and Jiang et al., 2006). It is grown up to 2–3-ft tall and wide. It has a distinct rosette growth habit with numerous erect and long-petioled leaves originating from a central crown. Plants have a deep root system of multiple branches and many finer rootlets. The roots are fleshy tan and thick to medium brown and smooth to corky on the outside and pure white on the inside. The edible roots are very pungent are grated and used fresh or added as an ingredient to commercially processed condiment products, such various mayonnaise-based formulations as and seafood sauce. The pungency of horseradish is due to isothiocyanate (sulphuric) compounds and the naturally occurring enzyme myrosinase which is found in the plant. Pungency is developed when chemicals become volatile in the chopped roots exposed to air, resulting in opening of sinuses and watering in the eyes.

3.15.2 BOTANY

Horseradish is one of the oldest herbaceous condiments and stout glabrous perennial herb. It is grown up to 2–3-ft tall and wide. Its growth habit is distinct rosette with numerous erect and long-petioled leaves originating from a central crown. Plants have a deep root system of multiple branches and many finer rootlets. The roots are fleshy tan and thick to medium brown and smooth to corky on the outside and pure white on the inside. The edible roots are very pungent are grated and used fresh or added as an ingredient to commercially processed condiment products, such various mayonnaise-based formulations as and seafood sauce. The pungency of horseradish is due to isothiocyanate (sulphuric) compounds and the naturally occurring enzyme myrosinase which is found in the plant. When the chopped roots are exposed to air, then these chemicals become volatile, resulting in pungency that opens sinuses and makes the eyes water (Gilbert and Nursten, 1972).

3.15.3 CLIMATE AND SOIL

Horseradish is a hardy plant. Horseradish is grown as temperate crop and it is grown well in cooler weather with adequate sunshine; however, it cannot tolerate extreme cold. It also grows well in hotter climates but it may require afternoon shade.

Horseradish is grown in well-drained, loose, sandy loam and alluvial soils for the healthy development of horseradish roots. Prior to planting horseradish be sure to work the soil deeply for the best results; a depth of 10–12 in or deeper is recommended. Clay, rock and thin layer of soil are filled in the heavy soil which can restrict root development and result in a poor harvest. The soil has a pH of 6.0–6.5. The soil fertility can be enhanced by the application of manure preliminary to planting.

3.15.4 CULTIVATION AND INTERCULTURE

Horseradish is divided into two types: ‘common’ and ‘Bohemian’. ‘Bohemian’ type is a Maliner Kren and local selections have been made from this type. Improved Bohemian and Bohemian form on the basis of the current industry. Superior quality and broad crinkled leaves are found in common types, while low quality and narrow smooth leaves are found in ‘Bohemian’ types, but have better disease resistance.

Horseradish is propagated through crown and root cutting. It is revealed that 6–10-in long section of root that includes at least one bud and has a circumference of ¼–1 in should be selected. Horseradish is also propagated through crown division. Horseradish climb is lifted from the ground and shaken gently to remove excess dirt. Crowns are cut into sections that include the upper leaves of the plant and at least one crown bud. Replant the sections in the desired location. Seeds can be planted, but they are not always reliable. Planting stocks are used from root cuttings that have been trimmed from the crop's main roots at harvest. Root pieces with a diameter of 3/8–3/4 in are used. Root pieces are cut into 8–12-in long leaving a square cut at the top and tapered cut at the bottom, so those planting will orient the root properly in the ground. Space rows 30–36 in apart within-row spacing of 15–24 in. About 8700–9700 root cuttings will be required per acre.

Horseradish plants may be produced through tissue culture. Although more expensive, the rapid increase of the desired planting material through tissue culture may be possible by contracting with plant propagators having tissue culture capabilities. It is advisable to have a soil test done on each field to be planted. Fields should be limed to a pH of 6.0–6.5. Manure may be ploughed at 12–20 tonnes per acre in the fall. A fertilizer containing nitrogen about 100–200 (N) pound per acre, phosphorus about 100–150 (P_2O_5) pound per acre and potassium about 100–150 (K_2O) pound per acre is ideal depending on soil type. Although early irrigation is not required in the growing season, if irrigation should be given in dry period in August and September then more yields will be obtained. The irrigation will be greater on lighter soils where crops are more subjected to moisture stress. The amount of total water needed does not affect the soil type but does dictate the frequency of water application. Lighter soils required more frequent water applications, but less water applied per application.

3.15.5 PLANT PROTECTION

Horseradish can tolerate some pest which damage its leaves without affecting yield and root quality. Flea beetles, caterpillars, false chinch bugs and diamondback larvae have all been known to defoliate horseradish. Growers are often more concerned with insects that cause root damage. One of these pests is the imported crucifer weevil (*Baris lepidii*). In horseradish fields, adult weevils lay eggs overwinter and larvae bore into the roots. Crop rotation is to be done for the control of wild horseradish, and the use of clean root sets can help to control this pest. The beet leafhopper (*Circulifer tenellus*) can indirectly harm the plant because it is the vector for the brittle root virus, a pathogen called *Spiroplasma citri*. Leaves are yellow curled.

Sometimes it is known as curly top, show up within weeks after the plant is infected and daytime wilting can occur. As the disease moves underground, resulting in brittle, discoloured roots that produce lower yields. The disease can be controlled by avoiding the used of infected rootstock.

Armoracia rusticana is susceptible to many pathogens. The turnip mosaic virus is transported by aphids and infected rootstock. The virus is characterized as mottled, streaked or spotted leaves. The fungus *Albugo candida* causes discolouration of the leaves as well as streaking on the leaf stalks. The directly or indirectly damage root may suffer as a result of damage to the top growth. *A. candida* can be controlled by using clean rootstock and crop rotation. The fungus *V. dahliae* may cause the root discolouration. This pathogen is difficult to eradicate because microsclerotia, tiny dark masses of cells in which the fungus overwinters, can be harboured in the soil for several years. Horseradish is also susceptible to other foliar diseases such as bacterial leaf spot and *Cercospora.*

3.15.6 HARVESTING AND POSTHARVEST TECHNOLOGY

Harvesting of horseradish can be done at the end of the first year or 150 days after sowing and also in the autumn. The root is stored up to next spring when plants are replanted. It is important during harvesting that all the roots are dug, as any small roots from the horseradish can become a weed in the garden. The plant is dig up, the tops are trimmed off and side roots and scrub roots are cleaned. Intact roots will retain their flavour for up to 3 months if stored properly. Horseradish roots are stored in moist sand or sawdust in a dark root cellar, or they can be put in a moist sand plastic bag and stored in the refrigerator. The side roots are saved for planting the next spring.

3.16 ASAFOETIDA

3.16.1 SYSTEMATIC POSITION

Kingdom: Plantae

Division: Tracheophyta

Class: Spermatopsida

Order: Apiales

Family: Apiaceae

Genus: *Ferula*

Species: *asafoetida*

Asafoetida is gum resin used as condiment in India and Iran. *Ferula asafoetida* is a Latin name, where *ferula* means 'carrier' or 'vehicle', asa means resin and *foetidus* means 'smelling, fetid'. Origin of *asafoetida* is from Afghanistan and Iran. It has been distributed throughout the Mediterranean region to Central Asia. In India, it is abundant in Kashmir and in some parts of Punjab. During early age, the hollow stems of the plants are used to transport fire between their camps. In ancient Rome, it was used as a flavouring agent in the kitchens. The useful part of the plant is underground rhizome or tap root from which dried latex (gum oleoresin) is extracted. According to (Kareparamban, 2012), asafoetida is referred as 'Food of the God'.

3.16.2 BOTANY

It is a herbaceous, monoecious, perennial plant, 2–3-m tall and with a circular mass of 30–40 cm. Leaves are compound, two to four pinnate very large and bipinnate, pubescent; segments oblong entire, obtuse with wide sheath 'Heeng' petioles. Flowers are produced on hollow 10-cm thick stem, with a number of schizogenous ducts in the cortex region containing the resinous gum. Flowering type is compound umbel with pale greenish yellow in colour. The inflorescence is densely pubescent. Petals are persistent. Fruits are 1-cm long, 8-mm broad, oblong to suborbicular, flat, thin, reddish brown in colour and also and have a milky juice. Roots are useful parts to extract oleogum resin. They are large thick pulpy and carrot shaped with 12–15-cm diameter. All the parts of the plant have foetid odour. Hilteet or hing (asafoetida) is an oleogum resin which is obtained from the rhizome and root of the plant *F. asafoetida* (Mahendra and Bisht, 2012). It is acrid and bitter in taste, strong onion-like smell because of its organic sulphur compounds (Alqasoumi, 2012).

3.16.3 CLIMATE AND SOIL

It can grow with a slope of 15–70% at an altitude of 1900–2400 m with an annual average precipitation of 250–350 mm. The crop is equally adapted to full and partial sunlight. The most suited soil types are clay, loam as well as sand. Soil having acidic, neutral or even alkaline pH with an average drainage capacity is ideal.

3.16.4 CULTIVATION AND INTERCULTURE

Conventionally, asafoetida is propagated through seed, but the germination capacity is low due to seed dormancy in Apiaceae family (Nadjafi et al., 2006; Joshi and Dhar, 2003). By the combination treatment of cold stratification (60 days) with 2000 ppm of GA3 solution, maximum germination was reported in the nursery by Zare et al. (2011). From the nursery, it is transplanted to the main field directly. Transfer the seedling to its permanent location and avoid transferring it again because of long taproots.

There are two varieties of asafoetida under different classifications namely 'Hing' and 'Hingra'. Hing is most fancied and rich in odour than Hingra. According to their country of origin, Hing is further classified as 'Irani Hing' and 'Pathani Hing' from Iran and Afghanistan, respectively. Again there are several varieties among them. The most priced and the strongest variety is Hadda. From the Irani asafoetida, the two varieties are sweet and bitter asafoetida. By the horizontal cutting of stem, sweet asafoetida is obtained which is transparent and brown in colour. By the cutting of the roots, bitter asafoetida is obtained which is white and red in colour and no pieces of wood is seen. Because of strong flavour and odour, the pure asafoetida is not preferred and, therefore, it is sold in compound form. In India, mass form of asafoetida is manufactured and marketed. Adulteration is also common.

3.16.5 PLANT PROTECTION

Asafoetida is generally resistant to disease. However, a few diseases such as fungi grow rapidly and consume nutrition from plants. It also causes decaying of roots, infection in leaves, yellowish colour, mushy spots and holes or infection in the stem. All these also reduce the plant's lifespan.

Identification of the disease and curing the plant with the right kind of pesticide is basic thing every grower should know.

3.16.6 HARVESTING AND POSTHARVEST TECHNOLOGY

The plant is ready to yield from fourth year in the main field. The stems are cut down to the ground level during March to April before the flowering stage. The top soil is removed; an incision is made to the stem close to the crown part of the root. The gum resins exudates from the incised area which is translucent, pearly white appearance but it soon becomes pink and finally turns reddish brown in colour on contact to air. The collection process of oleogum resin is repeated until roots stop exudation. Several incisions were made on the roots to extract oleogum resin. From the first incision, the period may extend up to 3 months. The yield of the gum resin is about 1 kg in 3 months. According to (Coppen, 1995), to produce more exudates the root is sliced every few days. Harvesting is different in different countries. In Afghanistan, the roots upper portion is sliced off with sharp knives and exudates are collected after 2 days. The cut portion is protected from the sun heat by guarding against a heap of stones. In market, asafoetida is available in three grades, namely tears, mass and paste. The tears are rounded and flatted in shape with greyish in colour and 5–30 mm in size. These are the purest form of oleogum resins. Most commonly available form in the market is mass asafoetida consists of stuck tears and form a mass mixed with fragments of soil (Tyler et al., 1976; Evans, 2002). Another form of the asafoetida is paste form, which also contains soil and woody matter (Anonymous, 2002).

Since pure asafoetida is not preferred for use because of its strong flavour/odour, it is sold in compound form. Main raw materials used are gum arabic and starch. Compounded asafoetida types manufactured and marketed in India are in mass form. Adulteration is also common.

3.17 BAY LEAF

3.17.1 SYSTEMATIC POSITION

Kingdom: Plantae

Division: Tracheophyta

Class: Magnoliopsida

Order: Laurales

Family: Lauraceae

Genus: *Laurus*

Species: *nobilis*

It is an evergreen small tree belongs to the family Lauraceae. Some other names of bay leaf are sweet bay leaf, and true, Roman, or Turkish laurel (Patrakar et al., 2012) bay laurel, bay tree, Grecian laurel (Brown, 1956) laurel tree or simply laurel. The useful parts of the tree are leaves and fruits. Leaves are used itself or in powder form in culinary items. Leaf essential and fruit crude oil are extracted and used as spice, flavouring and in cosmetic industries.

3.17.2 BOTANY

The tree is multi-branched evergreen with dense canopy. The bark is smooth in texture. Tree grows to a height of about 10–18 m. leaves are dark green in colour. It has alternate leaves with narrowly oblong-lanceolate shape. They are 8-cm long, 3–4-cm wide large in size glossy in appearance, elliptical and pointed. The upper and lower surface of the leaves are glabrous, shiny, olive green and dull olive to brown with a prominent rib and veins, respectively. The aroma of the leaves, when crushed, is delicate, fragrant and tastes bitter. Flowers are four-lobed, small, creamy yellow in colour and flowers during the summer season. Male and female flowers are present with 8–12 stamens and 2–4 staminodes. Fruits are small, ovoid in shape, 10–15 mm in size, one-seeded dark purple berries. When the fruit matures, the colour changes to black (Konstantinidou et al., 2008).

3.17.3 CLIMATE AND SOIL

It requires ample amount of sunlight and it should be protected during the winter season from cold winds and frost conditions when they are in the young stage. Warm and moist climatic conditions as of Mediterranean region are suitable for its growth. The soil pH should be 6–8 and it requires rich fertile soils with good drainage and organic matter.

3.17.4 CULTIVATION AND INTERCULTURE

Bay leaf can be propagated through seeds, cuttings and layering. Seeds are collected during the autumn season for seed propagation. The fleshy outer core of seed is removed and soaked in warm water for 24 h before sowing. The compost seedbeds are prepared and seeds are placed on the surface of the soil covering with dry compost. Seedbed should be kept in the dark at a temperature of 18°C for germination. It takes about 3 months for germination. Care should be taken to avoid rotting during germination. Cuttings are planned for propagation during late summer or early fall. It takes about 1 year or little longer for the establishment. The best cuttings are taken from well-ripened wood with a length of about 7.5–10 cm. The three to four leaves should be remained after trimming. The cuttings are kept in the sharp sand either under bell glasses or in glass cases. Then they are planted in a small pot filled with potting compost and placed in a site without direct sunlight. For getting proper growth of the plant, it is provided with heated propagating frame which increases the success with high humidity in the ambient air. Under Mediterranean regions, rooting for cuttings is better during July to August. By bottom heating, rooting period can be extended from May to September. After rooting, they are planted in the nursery beds with rich, sandy soil and good drainage. From the studies of Fochesato et al. (2006), best response was obtained using semi-woody cuttings with four leaves and treated with 2000 mg/l of indole-3-butyric acid. Layering should be done during the spring season. It is often successful but the rate of growth slower than cuttings. The lower healthy branches are bent towards the ground and a small cut is made on the stem. The cut portion of the stem should touch the ground and cover it with soil. Within 6–12 months, offshoots are produced from the cut region. A couple of ploughings and hoeings are advocated to bring fine tilth in the main field before transplanting. The field should be weed free and supplied with well rotten FYM for better yield of leaves. Nitrogen and potash fertilizers are applied to overcome the nutrient deficiency in the soil. In the main field, bay leaf is planted during the summer season because it requires good amount of sunlight. Spacing should be about 4–6 m. Training pruning and thinning are the necessary steps taken for the proper growth of bay leaf tree. Mulching with dry grass or dry leaves is a very important operation for providing constant moisture in the soil. For efficient use of water, drip irrigation may also be adopted. Frequent watering is needed during dry spell as the roots are shallow. Overirrigation to the plants is avoided but for better growth, constant soil moisture is recommended. During initial 1–2 years, vegetables or pulses can be grown as an intercrop which gives an extra income to the farmers. Afterwards, shade-loving plants can be grown under the bay leaf tree.

3.17.5 PLANT PROTECTION

Black powder coating—fine layer of black soot, is observed on the leaves. Aphids suck the sap of the leaves and excrete a sugary liquid; this attracts the moulds so that the leaves appear black in colour. The leaves are washed with water to remove the black powder so that the leaves appearance is good and photosynthesis is better and it will also avoid the attraction of insects such as wasps, bees, etc. who feed of the excreta of aphids.

Leaf spot is caused due to waterlogged roots or wet weather conditions. Leaf spots are usually seen in container plants, usually indicating that the compost has become old and tired. Repotting should be done in the spring season with well-drained potting compost. In yellow leaf disease, naturally, the older yellow leaves will shed off. In some conditions, due to waterlogged compost or cold weather, nutrient deficiency will occur. Peeling bark is another major problem. Stem cracking and peeling is observed, especially in the lower main stems during winter. The cause of peeling is not known but the winter cold and stress factors such as soil moisture fluctuations may be the problem for peeling. Though peeling is not the major problem, if the remaining part of the tree grows normally from above the damaged area no action is needed. But from the damaged area, if the above part of the tree growth is not normal or fatal in condition then remove the dead parts by cutting back to the healthy wood (i.e. green under the bark) or near to soil level. Recovery occurs from the soil level.

3.17.6 HARVESTING AND POSTHARVEST TECHNOLOGY

Harvesting is generally done after 2 years of planting. Healthy undamaged leaves are collected manually and they are dried under shade and pressed flat slightly without damaging the leaves. The maximum amount of oil is extracted from the shade dried leaves. According to Uzeyir Erden et al. (2003), 2.89% of oil is extracted from shade dried leaves and 2.88% of oil from solar tunnel dryer in October. Soil moisture holding capacity and weather conditions are the primary factors for yield of the bay leaf. An average yield about 6 t/ha by different harvesting time and drying temperatures a variation of 2.02–3.02% of essential oil may be extracted. The maximum percentage of oil is obtained at 35 and 50°C drying temperature for the October harvest. The higher temperature for drying may also cause the loss of essential oil content (Sekeroglu, 2007).

3.18 NUTMEG

3.18.1 SYSTEMATIC POSITION

Kingdom: Plantae

Division: Magnoliophyta

Class: Magnoliopsida

Order: Magnoliales

Family: Myristicaceae

Genus: *Myristica*

Species: *fragrans*

Nutmeg is the dried kernel of the seed, while mace is the portion of the surrounding dried aril. Theories about its origin revealed that Moluccas Islands of Indonesia might be the centre of origin of the said crop (Barceloux, 2009). Indonesia ranks top in the sector of worlds' export contributing about 50% of nutmeg and mace production followed by Grenada in the same sector of production and export. In India, Thrissur, Ernakulam and Kottayam districts of Kerala and parts of Kanyakumari and Tirunelveli districts in Tamil Nadu are well noted for their contribution in nutmeg cultivation and production.

3.18.2 BOTANY

The dried kernel of the seed is known as nutmeg and the surrounding dried aril is called as mace. Nutmeg is an evergreen spreading tree with dense foliage. It grows to a height of 10–20 m, sometimes reaching a 20-m height or more. The branches are spreading with a dark grey bark. The leaves are shiny, oblong to oval in shape. The inflorescence of nutmeg is cymes. It has several branches bearing number of flowers which hang down. The flowers are small, pale yellow and bell shaped and slightly aromatic. The fruit is fleshy, globose in shape and yellow to light brown colour of lemon. Nutmeg is dioecious in nature (Krishnamoorthy, 2013). The monoecious condition is also reported to occur in aged tree at Burliar in which case, double and triple nut are produced and the yield from such tree is low. The male and female trees can be easily identified by a trained eye. The male tree has erect branches, and leaves are smaller in size. The male tree is conspicuously less leafy than the female tree, and the trees' shape is not regular. The leaf content—calcium oxalate—is also taken as a criterion for the identification of sex in nutmeg.

3.18.3 CLIMATE AND SOIL

While considering the climatic factors, rainfall and temperature are regarded as the most influential factors in determining the cultivation of nutmeg. Warm and humid climatic condition along with average annual precipitation of 150 cm and more are effective in the cultivation of nutmeg. The topographical location of about 1300 m above sea level is considered to be optimum for the proper growth and development of nutmeg. Best soil textures are preferably clay loam, sandy loam and red laterite soils. The dry climate and waterlogging (both temporary as well as permanent) aggravate the problem in the cultivation of nutmeg.

3.18.4 CULTIVATION AND INTERCULTURE

Observable variations can be markedly seen in the crop as they are cross-pollinated. Each plant shows its variations not only in all aspects of growth and vigour but also for sex expression, size and shape of fruit quantity and quality of mace. *Konkan Sugandha* is an improved variety of nutmeg released from the Dr. Balasaheb Sawant Konkan Krishi Vidyapeeth, Dapoli, Maharashtra. IISR has also identified a few elite lines such as A9–20, 22, 25, 69, 150, A4–12, 22, 52, A11–23, 70 as high yielders and grafts of these lines are produced for distribution.

The segregation of seedlings into male and female plants is one of the crucial problems leading to 50% unproductive male trees and the reason behind this problem is its nature of being a dioecious plant. In spite of different schools of thought proposing that sex could be easily determined at the early stage of seedling development through observation of leaf form and venation and colour of young sprouts, vigour of seedlings and shape of calcium oxalate crystals on leaf epidermis, till date no promising reports about the solution have been proposed and rectified. One of the available options for this problem can be adoption of vegetative propagation technique either by top-working male plants or using budded or grafted plants. Epicotyl grafting is worth mentioning that it is one of the most used methods of commercial grafting (Haldankar et al., 1999). In the months of June to July, naturally split healthy fruits are harvested with the main view to raise healthy rootstocks. Extraction of the seeds from the pericarp is followed by immediate sowing in the sand beds of proper dimensions (convenient length, 1–1.5 m width and 15 cm height). In order to achieve good and healthy condition, regular watering is being done at an optimum interval. Normally, the germination period commences from 30th day and last up to 90 days after sowing. About 20 days old sprouts are being transplanted in a polythene bag containing an adequate mixture of soil, sand and cow dung in 3:3:1 ratio.

The selected rootstock at the first leaf stage should have a thick stem (diameter of 0.5 cm or more) with sufficient length so as to enable to give a cut of 3 cm length. Scions with two to three leaves are collected from high-yielding trees for grafting. The diameter of stock and scion should be appropriate to have maximum compatibility. A 'V'-shaped cut is made in the stock and a tapered scion is fitted carefully into the cut and then bandaging at the grafted region is done with the help of polythene strips. They are then planted in polythene bags of 25 cm × 15 cm size containing the potting mixture. The scion is covered with a polythene bag and kept in a cool, shaded place protected from direct sunlight. After 1 month, the bags can be opened and those grafts showing sprouting of scions may be transplanted into polythene bags, containing a mixture of soil, sand and cow dung (3:3:1) and kept in shade for development. The polythene bandage covering the grafted portion can be removed after 3 months. The wilting of scions should be prevented in order to complete the grafting as soon as possible. The grafts can be planted in the field after 12 months.

The planting of nutmeg is usually done in June to July at the beginning of the rainy season. Pits of 0.75 m × 0.75 m × 0.75 m size are dug at a spacing of 9 m × 9 m and filled with organic manure and soil about 15 days earlier to planting. For planting plagiotropic grafts, a spacing of 5 m × 5 m has to be adopted. A ratio of 10:1 female and male plants has to be maintained in the field. The plants should be shaded to protect them from sun scorch during early stages. Permanent shade trees are to be planted when the site is on hilly slopes and when nutmeg is grown as a monocrop. Nutmeg can best be grown as an intercrop in coconut gardens that are more than 15 years old where shade conditions are ideal. Coconut gardens along river beds and adjoining areas are best suited for nutmeg cultivation. Irrigation is essential during summer months. The Kerala Agriculture Department recommends 20 g N (40 g urea), 18 g P_2O_5 (110 g superphosphate) and 50 g K_2O (80 g muriate of potash) during the initial year and progressively increasing the dose to 500 g N (1090 g urea), 250 g P_2O_5 (1560 g superphosphate) and 1000 g K_2O (1670 g muriate of potash) per year in subsequent years for a fully grown tree of 15 years or more. About 25 kg FYM is to be applied for 7–8 years old trees and 50 kg for grown-up tree of 15 years.

3.18.5 PLANT PROTECTION

Dieback starts drying of mature and immature branches from the tip to downwards, which is a most visual symptom. *Diplodia* sp. and a few other fungi have been isolated from such trees. The infected branches should be cut and removed and the cut-end pasted with Bordeaux mixture of 1% (Suganthy and Kalyanasundaram, 2010). However, two types of blights diseases are noticed in nutmeg such as white thread blight disease is caused by *Marasmius pulcherima*, wherein fine white hyphae aggregate to form fungal threads that traverse along the stem underneath the leaves in a fan-shaped or irregular manner causing blight in the affected portions. The dried-up leaves with mycelium form a major source of inoculum for the spread of the disease and the second type of blight is called horsehair blight. Fine black silky threads of the fungus form an irregular, loose network on the stems and leaves. These strands cause blight of leaves and stems. However, these threads hold up the detached, dried leaves on the tree, giving the appearance of a bird's nest, when viewed from a distance. This disease is caused by *Marasmius equicrinus*. Both the diseases are severe under heavy shade. These diseases can be managed by adopting phytosanitation and shade regulation. In severely affected gardens, Bordeaux mixture (1%) spraying may be undertaken in addition to cultural practices.

Immature fruit split, fruit rot and fruit drop are serious in a majority of nutmeg gardens in Kerala. Immature fruit splitting and shedding are noticed in some trees without any apparent infection. In the case of fruit rot, the infection starts from the pedicel as dark lesions and gradually spreads to the fruit, causing brown discolouration of the rind resulting in rotting. In advanced stages, the mace also rots emitting a foul smell. *Phytophthora* sp. and *Diplodia natalensis* have been isolated from affected fruits. However, the reasons for fruit rot could be both pathological and physiological. Bordeaux mixture (1%) may be sprayed when the fruits are half mature to reduce the incidence of the disease. The shot hole disease is caused by *C. gloeosporioides*. Necrotic spots develop on the lamina which are encircled by a chlorotic halo. In advanced stages, the necrotic spots become brittle and fall off resulting in shot holes. A prophylactic spray with Bordeaux mixture (1%) is effective against the disease.

The black scale (*Saissetia nigra*) infests tender stems and leaves especially in the nursery and sometimes young plants in the field. The scales are seen clustered together and are black, oval and dome-shaped. They feed on plant sap and severe infestations cause the shoots to wilt and dry. The white scale (*Pseudaulacaspis cockerelli*) is greyish white, flat and shaped like a fish scale and occurs clustered together on the lower surface of leaves, especially in nursery seedlings. The pest infestation results in yellow streaks and spots on affected leaves, and in severe infestations, the leaves wilt and dry. The shield scale (*Protopulvinaria mangiferae*) is creamy brown and oval and occurs on tender leaves and stems, especially in nursery seedlings. The pest infestation results in wilting of leaves and shoots. The scale insects mentioned above and other species that may also occur sporadically on nutmeg can be controlled by spraying 0.5% monocrotophos (Pruthi, 2001; Rema et al., 2003; Anandaraj et al., 2005).

3.18.6 HARVESTING AND POSTHARVEST TECHNOLOGY

Delay in harvesting results in splitting of the fruit while still on the tree and the seed surrounded by the red aril falls to the ground after 2 days. Harvesting involves collecting the seed or seed with aril from the ground. Sometimes fruits with partially-opened pods may be picked from the tree using a long pole 'rodding'.The latter method affords a better quality aril and pods that could be used in agro processing. This procedure may also lead to excessive dropping of flowers and young fruits. The frequency of harvesting in nutmeg is highly influenced by the location of the field, the availability of labour, the level of production and the price offered to farmers. Most farmers collect the fallen seeds daily during the two peak production periods, namely January to March and June to August and every 2–3 days during the rest of the year. Once the field is readily accessible nutmegs are harvested with a higher frequency. In the cases where farmers are part-time, fields located in distant areas or when the farm is comprised of several plots of land at different locations, then the collection rate may be as low as once per week. Observations show that a larger proportion of women are usually involved in harvesting.

In the case where rodding was used, open fruits may fall to the ground intact. The seed with the surrounding red aril is removed from the pod which oftentimes is discarded. The collected seeds and seeds with mace are transported from the field by workers, on their heads or assisted by animal (donkey) or vehicle to the Boucan or farmer's residence where the mace is carefully separated from the seed, graded and allowed to dry directly in the sun. Care is taken so that drying mace does not get wet. Wetting will encourage mould growth and such mace will have to be discarded.

The seeds are usually delivered green (fresh), within 24 h after harvesting to the receiving station. However, depending on the distance from the receiving station and the quantities of nutmegs involved, deliveries maybe made once weekly or at a much later period if the nutmegs are being delivered in the dry state. This is usually the situation with large estates with adequate drying facilities. Mace is always delivered to receiving station dried. A tree from seedling usually 'declares' in 5–8 years. Trees propagated vegetatively by marcots may fruit as early as in 3 years. Yields increase gradually and at 25–30 years the plant may have peaked to its maximum production level. It continues to bear up to 100 years and over. However, after age of 70–15, yields tend to decline. On an optimum level, a healthy tree can produce about 2000 fruits on an average per year; however, variations in the yield may range from a few hundreds to about 10,000 fruits. After 8 years of planting, an improved variety IISR Viswashree can produce about 1000 fruits. From a population of 360 plants/ha, an average yield of approximately 3122 kg dry nut (with shell) and 480 kg dry mace per hectare could be obtained. The nut and mace dry recovery of IISR Viswashree are 70 and 35%, respectively. The nut has 7.1% essential oil, 9.8% oleoresin and 30.9% butter, while the mace has 7.1% essential oil and 13.8% oleoresin. The nutmeg value-added products are nutmeg oil, mace oil, nutmeg oleoresin, mace oleoresin, myristicin, nutmeg butter and the volatile oil of the bark.

3.19 BASIL

3.19.1 SYSTEMATIC POSITION

Kingdom: Plantae

Division: Tracheophyta

Class: Magnoliopsida

Order: Lamiales

Family: Lamiaceae

Genus: *Ocimum*

Species: *basillicum*

Basil (*Ocimum basillicum*) also known as French basil or sweet basil or tulsi, belonging to the family Lamiaceae, is one among the species of the genus *Ocimum* that includes around 30 plant species from tropical and subtropical areas, which are much differentiated in respect of morphological and chemical features (Vina and Murillo, 2003; Telci et al., 2006). It is a pleasant-smelling perennial shrub which is native to India and other Southeast Asian countries and also grows in several regions all over the world (Akgul, 1993; Bariaux et al., 1992). It is well known for its medicinal properties and also for economically important essential oils. It is used as a spice and medicinal herb.

Basil is an annual aromatic plant, widely grown because of its pleasant spicy odour and taste. It is grown primarily in Egypt and the United States and is one of the ancient spices whose remnants were found in Egyptian burial chambers. Both the leaves and the essential oils that are distilled from the flowering plants are used as flavouring agents.

3.19.2 BOTANY

Basil is a bushy aromatic annual of 60-cm height with brittle branched stems. There are large and dwarf types with oval, shiny, fleshy and fragile, dark green, or purple, variegated leaves, some of which are ornamentals. The fresh leaves are petiolate and about 5-cm long and 2-cm wide and have numerous oil glands with aromatic volatile oil. The freshly picked bright green leaves turn brownish green on drying and become brittle and curled. Creamy white flowers are in whorls with fruiting calyx, very shortly pedicelled, two lower teeth ovate-lanceolate awned longer than the rounded upper, lateral smaller than the lower and with a corolla of about 0.85–1.25 cm long. Seeds are dark brown to black, ellipsoid and mucilaginous (Hooker, 1885; Bhasin, 2012).

In view of the great diversity, the various species and varieties have been classified in accordance with their chemical composition and geographical sources into four major types as hereunder.

1. European or sweet basil whose oil consists of mainly methyl chavicol and linalool, but no camphor. Due to high quality and finest odour, the French and American sweet basil oils are in demand. It is distilled in France, Italy, Bulgaria, Egypt, Hungary, South Africa and occasionally in the United States.
2. Reunion basil has main constituents of methyl chavicol and camphor, but no linalool. The cultivation of oil from Ocimum Directorate of Medicinal Aromatic Plants Research, Boriavi 19 is of lower quality. It is distilled in the Reunion Island, Comoros, Malagasy Republic, Thailand and occasionally in Seychelles.
3. Methyl cinnamate basil whose oil consists of methyl chavicol, linalool and methyl cinnamate that is present in substantial amounts. It is distilled in tropical countries like India, Haiti, Guatemala and a few African countries.
4. Eugenol basil with eugenol as the principle constituent is distilled in the Union of Soviet Socialist Republics and North African countries like Egypt and Morocco.

3.19.3 CLIMATE AND SOIL

Basil can be cultivated on a wide range of soils from moderately fertile well-drained loamy to sandy loam with a pH range of 4.3–9.1. Clayey, waterlogged soils are unsuitable for cultivation. This crop tolerates higher concentration of copper and zinc but is susceptible to cobalt, nickel and frost.

It comes up well in warm and humid climate up to an altitude of 1800 m. The weather conditions such as long day with high temperature and high humidity have been found favourable for plant growth and high oil production. It shows poor growth in areas which receive heavy and continuous rainfall where it can be raised prior to the onset of monsoon and care should be taken that the rain water does not stagnate in the field. Waterlogging condition causes root rot and it results in stunted growth.

3.19.4 CULTIVATION AND INTERCULTURE

Promising varieties of basil include:

It is highly cross-pollinated and quality oil yield gets deteriorated over generations. Hence, the grower has to collect fresh seeds from the pedigree stock that are in good condition and free from pests for planting. About 125 g seeds are required for raising seedlings in 1 ha. As they are highly cross-pollinated, certain amount of heterozygosity is essential for vigour and yield attributes that are mostly controlled by polygenes whose effect is considered additive.

Seed sowing is done in the third week of February on well-prepared seed beds. It extends to the middle of March depending on the season. According to Sedigheh Sadeghi et al. (2009), the first week of March is the best time of sowing where it gives the highest yield. In the plains of North India, the seeds may be sown in the nursery in the months of April to May or August to September and in the hilly regions, seeds are sown in April. The seeds start germinating 3 days after sowing and the germination will be completed in about 8–12 days. The seedbed should be kept weed free. The seedlings will be ready for transplanting in about 6 weeks after attaining a height of 10–15 cm. A spray of 2% urea solution on nursery beds 2 weeks before transplanting provides vigour to nursery growth.

The land is well prepared with 2–3 ploughings until a fine tilth of soil is obtained. FYM/compost at 10–15 t/ha is to be applied before the second and third ploughing. Transplantation is done in March end or the first week of April with a spacing of 45–60 cm that is found suitable for most of the *Ocimum* sp. Cloudy weather and fine drizzle are considered ideal for transplanting

The recommended dose of fertilizers at 60 kg N with 40 kg P_2O_5 and 40 kg K_2O per hectare is uniformly applied in splits throughout the growth

Name of the species	Source	Fresh herb yield (t/ha)	Oil yield (kg/ha)	Oil recovery (fresh basis) (%)	Major oil constituents
Ocimum Basilicum (RRL-07)	RRL, Jammu	40	200	0.50	Citral (75–80%)
RRL-011	RRL, Jammu	50	220	0.50	Linalool (40v) and methyl chavicol (35%)
CIM-Soumya	Central Institute of Medicinal and Aromatic Plants, Lucknow	Fresh 290 q/ha Dry 197 kg/ha	–	–	Methyl chavicol 62.54% Linalool 24.61%
Vikarsudha	Hybrid: exotic basil from Australia (EC-331,886) × land race (Badaun)	–	–	–	–
MC-05	–	–	–	–	Methyl chavicol

period. Among them, the nitrogen fertilizer helps in increased vegetative growth and oil yield in aromatic plants due to the enhancement of the leaf area and photosynthetic rate. Laura Frabboni etal., 2011 found that 160 kg N/ha showed the best performance. Basil is not tolerant to water stress. A regular and even supply of moisture via trickle or overhead irrigation is necessary and care must be taken during harvest so that no damage is done to the irrigation line (De Baggio and Belsinger, 1996). To maintain quality and yields, adequate soil moisture should be maintained throughout the growing season. Hence, mulching the field with organic material between the rows is to be practised to control weeds and conserve moisture which reduces soil splashing on to the leaves that simplify the washing of foliage at harvest time and helps reduce the incidence of certain diseases.

3.19.5 PLANT PROTECTION

It is found to be infested with few insect pests and diseases. Pests such as lacewing bug (*Cochlochila bullita*) affects the crop where the adult and nymphs feed gregariously on leaves and younger stems leaving their excreta making it unsuitable for use. Due to this feeding, the leaves initially get curled and later the whole plant gets dried up. They can be managed by the spraying of 10,000 ppm at 5 ml/l of azadirachtin. Other pests such as leaf rollers stick to the undersurface and then fold of the leaves backwards lengthwise and webbed together.

The plant is susceptible to powdery mildew (*Oidium* sp.) that can be controlled by spraying wettable sulphur at 4 g/l of water; seedling blight (*R. solani*) and root rot (*Rhizoctonia bataticola*) that can be managed by improved phytosanitary measures and by drenching the nursery beds with 1% Bavistin.

3.19.6 HARVESTING AND POSTHARVEST TECHNOLOGY

Time of harvest plays an important role in qualitative and quantitative oil production and is done at the flowering stage or initiation of flowering, that is 90–95 days after planting in a stage when the plant is in full bloom and the lower leaves start turning yellowish. It is done with the help of sickles. There are two grades of oil like herb oil and flower oil of which the flower oil has a superior note. The flowering tops are only harvested to get high-quality oil. Usually, three to four floral harvests are obtained in this crop. The first harvest is when it is in full bloom and the subsequent harvests are at 65–75 days interval. The whole plant is harvested leaving about 15 cm from the ground level for regeneration and the harvested produce will be allowed to wilt in the field for 4–5 h to reduce the moisture and bulkiness.

An average yield of 3–4 t of flowers and 13–14 t of herbage can be obtained per hectare. The inflorescence contains 0.4% while the whole herb contains 0.10–0.25% which gives an oil yield of about 30–35 kg/ha from the flower (flower oil) and 18–22 kg/ha from the whole herb.

The most critical stage in determining the end quality of the aromatic plant material is the postharvest processing. Oil from sweet basil can be obtained either by hydrodistillation, which is cheaper and used for small plantations or steam distillation that is preferred for larger plantations from young inflorescence or the whole herb, where the latter is better as it takes less time and improves the oil recovery. Distillation unit should be clean and free from rust and other odours. The oil obtained is then decanted and filtered. To remove moisture from the distilled oil, it should be treated with anhydrous sodium sulphate or common salt at the rate of 20 g/l. The oil is then stored in sealed amber-coloured glass bottles, stainless steel containers, galvanized tanks and aluminium containers and kept in a cool and dry place.

3.20 POPPY SEED

3.20.1 SYSTEMATIC POSITION

Kingdom: Plantae

Division: Tracheophyta

Subdivision: Spermatophytina

Class: Magnoliopsida

Order: Ranunculales

Family: Papaveraceae

Genus: *Papaver*

Species: *somniferum*

Opium poppy is commonly known to the world as opium. Botanically, opium is an erect annual glaucous herb which yields latex from fruits, gives a commercial product when dried, which is the main cause for the popularity of this crop. The origin of opium is southeastern Europe and Western Asia or the Mediterranean region. Stepping back into the history of opium, Portuguese first exploited the opium poppy but in later stages, Dutch and English were made to regulate its production and trade. In the global scenario, opium is grown in many countries such as Netherlands, Germany, France, Romania, Hungary, Turkey, Iran, India and the United States. Opium is a herb which grows up to 1.5 m in height. The colour of the opium flower is white to pink or purple and has good medicinal properties.

The latex of opium poppy is viscous and possesses many strong alkaloids such as morphine, codeine, papaverine, noscapine and thebaine. Latex should be collected 2–3 weeks after flowering by giving a sharp cut into the surface of the skin of pod. Then the latex should be collected when the pods or capsules stop secreting the latex and then the coagulated latex should be scraped and collected gently. Alkaloids are absent in the mature poppy seeds. Hence, these seeds are used for flavouring food as spice since ages. These are also used in rolls and other pastry goods in baking.

3.20.2 BOTANY

Papaver somniferum is an erect, scarcely branched, glaucous annual, growing to a height of 60–120 cm. The leaves are ovate, oblong or linear oblong; flowers are large, usually bluish with a purplish base or white, purple or variegated. It produces capsular type of fruits from which the latex known as opium is obtained on lancing. The fruits are about 2.5 cm in diameter, globose in shape. Seeds are small, kidney-shaped and white or black in colour. Though nearly all parts of the poppy plant contain milky-white latex, the unripe capsules contain larger amounts (Stace, 2010; Blamey et al., 2003).

3.20.3 CLIMATE AND SOIL

Opium grows well and gives higher yield when grown in a long spell of cool weather (20°C) with adequate sunshine during the early stage of vegetative growth. Basically, it is a temperate crop, but it can grow successfully during winter in subtropical regions. Cool climate favours higher yield, while the yield is affected by the higher day and night temperature. Heavy rains after sowing cause losses in seed germination. Warm dry weather with the temperature of 32–38°C is needed during the reproductive period. The quality and quantity of opium are drastically reduced in the desiccating and freezing temperature. It grows well in deep clay loam, highly fertile light black or loam soil with an optimum pH of around 7.0. Such soil containing adequate organic moisture does not need irrigation during lancing. Heavy clay or fine sandy soils generally pose a problem since they remain wet during the rain and are too difficult to cultivate during the dry periods.

3.20.4 CULTIVATION AND INTERCULTURE

The duration of the crop is 140–160 days. In India, only a few varieties are being cultivated. The following varieties are recommended for heavy black soils.

i. Talia: Early sown crop, duration 140 days, flowers are pink in colour with large petals. The capsule is oblong, ovate, light green and shiny.
ii. Ranghatak: A medium-tall variety, takes 125–130 days to mature for lancing after sowing, bears flowers of white and light pink colour, medium-sized capsules (7.6 cm × 5.0 cm) which are slightly flattened on the top.
iii. Dhola Chota Gotia: It is a dwarf cultivar, bearing pure white flowers and light green capsules which are oblong-ovate in shape, ready for lancing after 105–115 days of sowing, matures for seeds in 140 days.
iv. MOP-3: Developed by Jawaharlal Nehru Krishi Vishwa Vidyalaya, Mandsaur, bears pinkish-white flowers comprising of large non-serrated petals. Capsules ready for lancing after 20 days, recommended for areas with adequate irrigation facilities in the later part of the season.
v. MOP-16: Developed by Jawaharlal Nehru Krishi Viswa Vidyalay, Mandsaur. Flowers are with serrated petals and capsules are round and flat-topped. It is drought tolerant, ready for lancing after 105–110 days of sowing, early maturing.

vi. Shama: Released by the CIMAP, Lucknow in 1983. The main alkaloids are morphine (14.51–16.75%), codeine (2.05–3/24%), thebaine (1.84–2.16%), papaverine (0.82%) and narcotine (5.89–6.32%), etc. It yields 39.5 kg of latex and 8.8 kg of seeds per hectare.
vii. Shweta: Released by the CIMAP, Lucknow, along with Shama. The main alkaloids are morphine (15.75–22.38%), codeine (2.15–2.76%), thebaine (2.04–2.5%), papaverine (0.94–1.1%) and narcotine (5.94–6.5%). It gives an average yield of 42.5 kg of latex and 7.8 kg of seeds per hectare.
viii. BROP 1 (Botanical Research Opium Poppy-l) (NBRI-3): A synthetic variety, developed at the National Botanical Research Institute (NBRI), Lucknow, by crossing selections from Kali Dandi, Suyapankhi and Safaid Dandi. This variety is moderately resistant to diseases, yields about 54 kg/ha of opium and 10–13 q/ha of seeds.
ix. Kirtiman (NOP-4): Developed at the Narendra Dev University of Agriculture and Technology, Faizabad. Selection from local races, moderately resistant to downy mildew, yields 35–45 kg/ha of latex and 9–10 q of seeds per hectare. The morphine content is up to 12%.
x. Chetak (UO-285): Developed at the Rajasthan Agriculture University, Udaipur, moderately resistant to diseases. The opium yield is up to 54 kg/ha, the seed yield is 10–12 q/ha and contains up to 12% morphine.
xi. Jawahar Aphim 16 (JA-16): Developed at the Jawaharlal Nehru Krishi Vishwa Vidyalaya College of Agriculture, Madhya Pradesh, moderately resistant to downy mildew, yields 45–54 kg/ha of latex, 8–10 q/ha of seeds and contains up to 12% morphine.

NBRI, Lucknow, RRL, Jammu and CIMAP, Lucknow recently released few varieties like 'NBRI-3' of opium, 'Sujatha' an opium-free poppy for the production of oil and seed and 'Shubhra' for high morphine and seed yield.

The land should be thoroughly ploughed at least three or four times and soil is pulverized well. Care should be taken to free the field from weeds. Convenient sized beds are prepared with ample spacing. Seeds are treated with Dithane M-45 at 4 g/kg to get rid of pathogens and they are either line sown or broadcasted evenly. Seeds should be mixed with fine sand for even broadcasting. Line sowing is mostly preferred compared to broadcasting because line sowing ensures low seed rate, good crop stand and ensures easy intercultural operations. Seed sowing is carried out during October-early November. Seed rate in the broadcasting method is 7–8 kg/ha and 4–5 kg/ha for line sowing. The spacing of 30 cm × 30 cm is usually followed. Depending on the moisture content of soil, germination of seeds takes place in about 5–10 days after sowing. Thinning is done in order to ensure a better plant growth and development. Thinning should be carried out until the plants attain 14–15 cm with a time period of 3–4 weeks after sowing. Opium poppy shows positive response to manures and fertilizer application. After the field preparation, field should be broadcasted with FYM at 20–30 t/ha. Along with FYM, 60–80 kg of N and 40–50 kg of P_2O_5 per hectare is recommended. Potassium is not applied. Initially, half N and full dose of P are applied at the time of sowing and at the rosette stage, the remaining half of N is applied.

Irrigation plays a vital role in ensuring good growth of poppy. Light irrigation is given twice in an interval of 7 days after sowing when the seed starts germination. And then irrigation is given at an interval of 12–15 days till it attains the flowering stage. After that, irrigation frequency is reduced to 8–10 days interval during flowering and capsule formation stage. Altogether 10–15 irrigations are given in the entire crop growth period. Yields are affected because of moisture stress during the stage of fruiting and latex extracting, hence care should be taken.

3.20.5 PLANT PROTECTION

The most prominent pests of opium are cutworm and weevil. Among them, cutworm (*Agrotis sulfusa*) causes great harm to the plant by remaining burrowed in the soil during the day and cutting as well as eating the leaves during the night. Flooding the field with water and dusting the crop with 2% carbaryl can control this insect. Weevil (*Stenocarpus fuliginous* root weevil and *Ceutorhynchus macula* alba-capsule weevil) is known to infest the young seedling and leaves. As the plant matures, the insect eats it up to the stem and ultimately bores up to the poppy heads and proceeds to destroy the capsule. The capsule borer (*Helicoverpa armigera*) and aphids (*Myzus* sp.) also infest the shoot and capsules, respectively.

Downy mildew (*Peronospora arborescens*) diseases may appear at the seedling to maturity stages at any time. The infested plant develops chlorosis, curling, stunted growth. If the plant is attacked in the later stage, the capsules are affected. The application of Dithane Z-78 (0.4%) can control this disease. Powdery mildew (*Erysiphe polygoni*) appears 90 days after sowing with white growth of mycelia on the stem and lower side of the leaves. Later on, the large patches of white powdery masses turn to black. This disease can be controlled by uprooting and burning in the early stage. Spraying Dithane Z-78 or Dithane M-45 at the beginning of the flowering stage and repeating it two to three times at 15 days intervals also prevent the diseases (Alam et al., 2014).

3.20.6 HARVESTING AND POSTHARVEST TECHNOLOGY

Opium attains flowering stage after 95–115 days of sowing. The petals start shedding after 3–4 days of flowering. Capsules attain maturity after 15–20 days of flowering. This is the stage where more latex is obtained. This stage can be identified by the compactness and change in the colour from greenish to light green coloured ring in the capsule. The stage is called as industrial maturity. Physically, the opium latex is found within laticiferous vessels, which lie just beneath the epicarp of the seed capsule. It is harvested by making a series of shallow incisions through the epicarp, which allows the latex to 'bleed'onto the surface of the seed capsule. The latex is allowed to partially dry on the capsule surface and is then removed by scraping the capsule with specially designed hand tools. Lancing may be done with a knife having three or four equispaced pointed ends which do not penetrate the capsule more than 1–2 mm. It is done early in the morning (before 8.00 a.m.) at 2 days interval in each capsule. The length of the incision should be one-third or less than the full length of the capsule. The dried latex is a malleable gum, which is light to dark brown and is known as raw opium. The major constituents of raw opium are plant fragments, resins, sterols, triterpenoid alcohols, fatty acids, alcohols, polysaccharides and more than 30 alkaloids.

When the last lancing on the capsules stops the exudation of latex, opium is left for drying for about 20–25 days. The capsules are then picked up and the plants are removed with the help of sickles. These capsules are then dried in an open yard for days and seeds are collected by beating with a wooden rod. Raw opium varies from 50 to 60 kg/ha. On an average

25–30 kg of crude opium and 400–500 kg of seeds are obtained per hectare in India. Well matured seeds are cleaned and packed in gunny bags or in friction top tins, in smaller scale.

Poppy seeds contain 50% of edible oil which is extracted by using cold or hot expression method. The oil obtained from opium is odourless and has a pleasant almond-like taste. The colour of the raw opium oil is pale to golden yellow in colour. Poppy seed oil has the characteristics: specific gravity 15°/25°:0.924–0.927; refractive index: 1.467–1.470; iodine value: 132–142; saponification value: 188–196; acid value: 3.13%. It contains vitamin F, which is used for skin treatment.

3.21 ROSEMARY

3.21.1 SYSTEMATIC POSITION

Kingdom: Plantae

Division: Tracheophyta

Class: Magnoliopsida

Order: Lamiales

Family: Lamiaceae

Genus: *Rosmarinus*

Species: *officinalis*

Rosmarinus officinalis is an aromatic, medicinal and condiment plant that belongs to the family Labiatae, commonly known as rosemary. It is native to the Mediterranean region. It is widely spread in Algeria and broadly used in traditional medicine. It is a perennial herb, evergreen woody plant with a characteristic fragrance. It bears needle-like leaves and flowers that are in various colours such as white, pink, purple and blue. It belongs to the family Lamiaceae. Useful parts of rosemary are the aromatic leaf and flowering tops. Rosemary is mainly cultivated in Europe and California and is native to Mediterranean region. It is also grown in Algeria, China, Middle East, Morocco, Russia, Romania, Serbia, Tunisia, Turkey and to a limited extent in India. It grows above altitudes of 750 m MSL. It requires warm climates and well-drained soils for successful cultivation. Yield and the composition of rosemary oil is influenced by the soil properties.

3.21.2 BOTANY

Rosemary grows up to the height of 2 m with small (2–4 cm) pointed, sticky, hairy leaves. It is an evergreen, hardy, dense, perennial, aromatic shrub. The economically important part of this plant is the leaf. It contains fibrous root system. The upper and lower surfaces of the leaves are dark green and white in colour, respectively. The leaves are resinous and contain 0.9% essential oil (Johnykallupurackal and Ravindran, 2006). Branches having fissured bark are rigid in nature. The stem is square in shape, woody and brown in colour. Small pale blue flowers appear in cymose inflorescence. The calyx contains maximum amount of oil (Ravikumar, 2002). It can be grown as an indoor plant or field crop.

It is used as a medicinal, stimulant, food preservative and, of course, as a flavouring agent in cooking and as a memory enhancer. This ancient strewing herb and Romany charm was revered in ancient Greece for its association with memory and became a symbol of faithfulness to lovers. During medieval times, it was thought to be a protector from evil.

3.21.3 CLIMATE AND SOIL

Rosemary requires sandy to clay loam soil and the pH range is from 5.5 to 8.0. It can tolerate maximum of 30% clay in the soil. Rosemary is a hardy, temperate plant. Frost-free tropical and subtropical summer monsoons are suitable. However, places where frost occurs frequently should be avoided as the plant is susceptible to it. Temperature less than 30°C is optimal. 20–25°C day temperatures are favourable for better growth of the plant. It is grown in almost all parts of the country in South Africa. Rosemary prefers the Mediterranean type of climate with low humidity, warm winter and mild summer for its successful growth. In India, the climates of Nilgiris and Bangalore are found to be suitable for its cultivation.

Ooty-1: It is a perennial variety with a high yield potential developed by seedling progeny selection. Leaves contain high oil content (0.9%). And it is resistant to leaf blight disease and pests such as whiteflies and aphids. Crop yields about 12.4 Mt of green leaves/hectare. It grows well in the temperate zone of the Nilgiris from 900–2500 m above MSL and similar areas. It is suitable for cultivation in the well-drained loamy soil with a pH of 5.5–7.0.

3.21.4 CULTIVATION AND INTERCULTURE

Rosemary is propagated through seeds, cuttings, layering or division of roots. Germination of seeds is very slow. As there is always a problem of cross-pollination, growing true-to-type plants from seeds is not a good practice unless controlled properly. Cuttings of 10–15 cm length from the actively growing stem tips are the best way to propagate new plants efficiently. The bottom two-third of leaves is stripped off from cuttings and they are inserted in a proper growing medium, half to two-thirds of its length. Rooting hormones enhance root formation within 2–4 weeks. For better root formation, a mist bed with a heated floor will give best results. Layering is done in the summer season by pegging some of the lower branches under sandy soil. After formation of the roots, they are separated from the host plant and used for planting in main field.

The main field is prepared by digging and thoroughly ploughing twice and in the last ploughing well-decomposed FYM of 25 Mt and neem cake of 500 kg should be applied and mixed well. About 5 kg of *Azospirillum* and 5 kg *Phosphobacterium* should be applied to the soil at the time of planting and mixed well. Rosemary plants are planted in east–west direction for exposure to maximum amount of sunlight. When it comes to field spacing, beds of convenient length, 30-cm height and 1.5-m width are to be prepared and 40–50 cm of row spacing within the bed is effective for machine cutting, where plant to plant distance should be 25–50 cm. Planting density is about 50,000–60,000 in 1 ha. Rosemary cuttings prepared in a greenhouse or nursery are transplanted to the main field during spring to midsummer. Tip of the shoot or central shoot should be removed after 6 months for getting lateral shoots.

Usually, rosemary is grown in dry lands. It requires irrigation mainly during the establishment of roots in the soil. The field should not be overirrigated. Irrigation is not necessary if annual rainfall is more than 450 mm, check whether water is stagnated during rainy season and drain out the water immediately from the main field. In India, a fertilizer dose of 40 kg P, 40 kg K and 20 kg N with 20 t of FYM per hectare is required for rosemary cultivation (Farooqi and Ramulu, 2001). Oil yield can be increased by further application of N to 300 kg/ha in different splits. Anuradha et al. (2002) reported that biofertilizers such as *Azospirillum*, *Azotobacter* and vesicular–arbuscular mycorrhiza applied along with inorganic fertilizers has shown a beneficial effect. Yield and quality of essential oil is affected by weeds in the field. Hand weeding and hoeing are very important cultural operations to control the weeds. About two to three weedings are necessary per year. Weeding can be done by tractor-drawn cultivators or hand hoe. Care should be taken that roots should not be damaged during inter cultivation. Rosemary is very sensitive to root damage and causes dieback and care should be taken while inter cultivation. By increasing the canopy and plant density weeds will be reduced effectively.

3.21.5 PLANT PROTECTION

Spider mites, mealybugs, whiteflies and thrips commonly occur in rosemary. Application of insecticides along with crop rotation and careful monitoring will assist in keeping the foliage free of pests. Whiteflies usually suck sap from the leaves of plants with the help of their piercing/sucking mouthparts. They excrete large quantities of honeydew which serves as a growth medium for sooty mould. Spider mites normally feed on the lower stem and then move to feed on the upper section of the plant and on leaves. Affected leaves turn yellow and drop. Under heavy infestation, silk webbing may be seen on the plants. Female mealybugs feed on the plant sap. They attach themselves to the plant and secrete a powdery, white, waxy layer used for protection while they suck the plant juices. With the help of their piercing and sucking mouthparts, thrips feed on leaves and damage the plants, causing browning and leaf drop. They also act as vectors for other diseases. Fungal problems may arise in the plants due to overirrigation, so plants should not be overirrigated. Late blight disease is also reported in rosemary. It can be controlled by drenching the plant with Maneb (1%) at 8–10 days interval.

3.21.6 HARVESTING AND POSTHARVEST TECHNOLOGY

Depending on the geographical area and plant material used, rosemary is harvested once or twice a year. Harvesting should be done 6 months after planting when the crop produces flower tops. However, in some areas, harvesting should be done in the second year in the month of August after the full flowering, which commences in May to June. For the preparation of dried rosemary, the crop is harvested frequently before flowering commences. The dried product contains only leaves. For the fresh market, harvesting should be done early in the morning and the herb is cut frequently at a young stage. After harvesting, fresh rosemary is kept cooled at 5°C before packaging for the market. With a temperature of 5°C, a minimum shelf life of 2–3 weeks can be expected. These fresh shoots are used in culinary preparations. After temperature, prevention of the excess moisture loss is the second most important postharvest factor affecting the quality and shelf life of herbs.

For the extraction of essential oil, the harvested product is subjected to different steps of postharvesting such as sorting and distillation. After harvesting, drying of the product can be done by different methods from sunlight to sophisticated driers. The dried product should be processed to remove the leaves from the stems and then sieved to remove dirt and to produce a uniform product. Sun-drying may result in poor quality whereas artificial drying gives better quality of the product. A forced airflow drier is a suitable system to dry better quality leaves. Rosemary oil extraction is done by steam or water distillation, though supercritical fluid extraction using CO_2 is also in practice. In distillation process, it takes about 120 min for the full recovery of oil whereas most of the oil (90%) comes out within the first 60 min of distillation. Oil extracted from leaves and flower tops is of better quality than the oil extracted from the whole plant distillation. Steam distillation was found to be better than water distillation in terms of yield and quality profile of rosemary oil. Blanching is observed to have a positive effect on the retention of the antioxidant principles, green colour and the texture of rosemary, though blanching leads to total loss of volatile oils. The oil content of fresh rosemary leaves is 1% and in shade-dried leaves it increases to 3%. From a hectare one can harvest approximately 10–12 t herbs per year, yielding 25–100 kg oil. Close spacing coupled with increased nitrogen dose results in higher herbage and oil yield.

3.22 THYME

3.22.1 SYSTEMATIC POSITION

Kingdom: Plantae

Division: Tracheophyta

Class: Magnoliopsida

Order: Lamiales

Family: Lamiaceae

Genus: *Thymus*

Species: *vulgaris*

Thyme (*Thymus vulgaris* L.) is a herbaceous perennial subshrub that belongs to the family Lamiaceae (Farooqi et al., 2005) and is native to the western Mediterranean region (Stahl- Biskup, 2002) and southern Italy. There are 350 species of thyme cultivated all over the world (Cronquist, 1988; Zaide and Crow, 2005) of which three of these species, that is garden thyme, European wild thyme and lemon thyme are most prominent (Ortiz, 1996). It has got many decorative and variegated forms for which it is the best culinary hedging herb in gardens. It is used for its leaves and essential oil and is widely cultivated in France, Germany and Spain. It was used as medicinal herb and a flavouring agent in some food items by the ancient Greeks and Romans and also found associated with courage and sacrifice.

3.22.2 BOTANY

It is an aromatic perennial small shrub that grows up to 30–45 cm in height with the stem about 0.5 cm in length, greyish green, square shaped and woody at the base. Leaves are very narrow with aromatic warm and pungent flavour. The dried brownish green curled leaves of 6.7 mm long are marketed in the whole or ground form and it contains 5 mm long, small lilac or white fragrant flowers with a hairy glandular calyx.

3.22.3 CLIMATE AND SOIL

Thyme grows well in a temperate to warm dry sunny climate without shade. It needs full sun to grow to its best potential and a rainfall of 500–1000 mm per year in the Mediterranean region. This crop is drought-tolerant (Herb File. Global Garden. global-garden.com.au.) and does not like excessive moisture because of its susceptibility to rot diseases.

It prefers light, fertile and well-drained soils with a pH of 5–8 and does best in coarse rough soils that would be unsuitable for many other plants. Although it grows easily, especially in calcareous light dry stony soils, it can be cultivated in heavy wet soils but it becomes less aromatic.

3.22.4 CULTIVATION AND INTERCULTURE

There are about 215 species of the genus Thymus that has numerous hybrids as well. Three principal varieties such as broad leaved, narrow leaved and variegated are usually grown for use. The narrow-leaved types also known as winter or German thyme are small with greyish green leaves and are more aromatic than the broad-leaved thyme. The fragrant lemon thyme that ranks as a variety of *Thymus serpyllum*, the wild thyme has a lemon flavour and rather broader leaves and is not curved at the margins than the ordinary garden thyme, whereas hardiest of all is the silver thyme that has the strongest flavour.

There are many different types of thyme available as it hybridizes easily and the most cultivated ones used for culinary and essential oil extraction are:

Variety	Features
Thymus vulgaris	It is the common thyme that is prostrate yellow with silver and variegated foliage and is used in cooking.
Thymus zygis	It is similar to *T. vulgaris* and is mostly distilled for essential oil.
Thymus citriodorus	It is a lemon thyme that is upright golden with variegated silver foliage and has a strong lemon scent.
Varico	It is a robust cultivar which has an upright growth with greyish blue foliage and excellent herbage yield. It is seed propagated and yields 50% thymol and more than 3% essential oil levels and is resistant to frost. Other promising new cultivars are currently developed in various countries for the colour of the leaves and flowers are used mainly as ornamental shrubs.

Thyme is best propagated from seeds that are sown early indoors or under glass in an outdoor bed. When the seedlings attain 5–8-cm height, they are planted at intervals of 30–45 cm between plants and 45 cm in rows. Few plants were used for permanent flower bed as ornamental and that furnish herbs sufficient for flavouring purposes. Every 3–4 years, new plants are started by sowing seeds or planting cuttings or layering as the old ones become too woody to produce tender leaves for culinary use. The source from where the seed has been taken should be checked properly as there are possibilities of hybridization. Plants raised through cuttings give homogeneous individuals and the cuttings are taken during spring at about 5–10 cm. Root-promoting hormones seem beneficial but care has to be taken if any detrimental soil organisms are present.

Thyme comes up well in very shallow soils where other crops cannot survive. When herbal plants are grown on natural soils they yield high-quality essential oil that is in global demand. They are planted at a spacing of 15–30 cm between plants and 60 cm wide between rows. Beds of 1–2-m width are made where machinery can easily enter and the seedlings are transplanted in three rows on each bed. Care is to be taken not to squash thyme as it will break the woody stem at soil level and the plants may die off. Special planters can be used for small seeds and the seed rate is 3300–4000/g. According to the size of mature plants and moisture availability, plant density is to be maintained. Spring is the best time to sow or plant seeds and transplant cuttings.

Thyme should not be heavily fertilized because overfertilized plants tend to show tall, spindly and weak growth. According to the soil analysis results, basal fertilizer dose containing nitrogen, phosphorus, potassium and sulphur should be applied annually. To promote new shoot growth, an additional application of nitrogen is given after each harvest. But the dosage should not be more as it affects the oil quality. Organically cultivated thyme fetches better price in markets and is produced globally on a larger scale than the dried herb. Drip and overhead irrigation are suitable for its growth. The crop is to be kept weed-free to avoid contamination of the end product and it establishes successfully only by hand weeding without using any herbicides and by planting it in a land covered with weed suppressing plastic and organic mulches (Fraser and Whish, 1997).

3.22.5 PLANT PROTECTION

Pest infestation is not very frequent in thyme as their volatile oils have pest repellent properties. However, whitefly, scale and spider mites may sometimes infest wherein spider mites can be controlled by spraying a solution mixture of three tablespoons of dish soap to 1 gal of water for 1 week. The plants are affected by a few diseases like Rhizoctonia root rot which is prevalent in wetter environments with imperfect soil drainage. Alternaria blight, another fungal disease, also causes harm to the crop which is detected by the appearance of round yellowish or black spots with concentric rings on the lower leaves. It causes lesions or holes followed by drying and dropping of leaves and death of plants in severe cases. Removal and destruction of infected leaves, use of wider plant spacing is recommended to check the disease. Rust and botrytis are also a problem in thyme.

3.22.6 HARVESTING AND POSTHARVEST TECHNOLOGY

Thyme grows best in well-drained sunny locations and the harvesting of leaves and flowers that are used for culinary or medicinal purposes is done 5 months after sowing or planting. It is done either by plucking from the plants or cutting the shoots of about 15-cm length. The harvested produce is dried immediately in shade or in mechanical drier and then stored in an air-tight container to prevent the loss of flavour. Sometimes two or more crops can be harvested the same season.

Under favourable and good management condition, the yield in dry form varied from 1.6 to 2.2 t/ha. In the first year, the yield is very low and the plants need to be replanted after 3–4 years as they become woody. The plant is not damaged by any serious pest but wilt disease is a major problem.

The dried product is to be processed to remove the leaves from the stems and then sieved to remove dirt and to produce a uniform product. There are several methods for drying from the sun to sophisticated driers; however, essential oil obtained through sun drying results in poor quality, whereas artificial drying allows better control of product quality while forced air-flow drier is suitable to dry better quality leaves. In order to reduce flavour loss through volatilization of essential oil and to maintain a good green colour, it should be dried at temperatures lower than 40°C and after drying, the leaves should be separated further from the stems, sieved and then graded. The fresh product should be free from any foreign material and look fresh and crispy with a good colour and flavour.

Quality requirements for dried thyme were prescribed by an International Standard (ISO 6754:1996), that set down certain requirements of the finished product such as the essential oil content that contributes to the flavour intensity should be minimum of 0.5% essential oil, which equals 5 ml/kg dried herb in whole leaves and the ground thyme should contain at least 0.2%. There is a diverse range of chemotypes for essential oil production occurring in thyme of which the most important six different chemotypes are thymol, carvacrol, linalool, geraniol, thuyan-4-ol and α-terpinyl acetate and the most frequent are thymol and carvacrol, extracted from low altitude plants near the sea and linalool extracted from plants occurring at higher altitudes while geraniol, thuyan-4-ol and α-terpinyl acetate are rare and found mixed with the first three. The essential oil contains 20–54% thymol (Montvale, 1988) and owing to this powerful disinfectant property, there is a demand for thymol in the pharmaceutical industry.

The packaging of essential oils is to be properly handled as they are volatile in nature. It can be done either in bulk or in smaller quantities, where smaller packing has higher prices owing to extra handling and for bulk handling of fresh thyme, the packaging is in crates or in clear cellophane sachets that can be marketed directly in shops and supermarkets. Cardboard boxes or glass–plastic containers are usually used for dried thyme. Dark glass is preferred for preservation as factors such as moisture, heat, oxygen and light destroy the oils. Thyme should be stored under 18°C for extended shelf life and its essential oil should be kept in dark, airtight glass bottles in cool dry area without exposing them to heat or heavy metals until it is used. If it is opened once, then it should be refrigerated by tightly closing the cap to prolong its shelf life.

3.23 ANISEED

3.23.1 SYSTEMATIC POSITION

Kingdom: Plantae

Division: Magnoliophyta

Class: Magnoliopsida

Order: Apiales

Family: Apiaceae

Genus: *Pimpinella*

Species: *anisum*

Anise is an annual spice having medicinal values and belongs to the family Apiaceae. It is also known as sweet cumin and has the flavour of liquorice. It is chewed after meals and used as mouth freshener and is native of Iran, Turkey, Egypt, Greece, Crete and Asia Minor. The active substances of anise are used in food products and pharmaceuticals (Nabizadeh et al., 2012). Useful plant part is fruits commonly called as seeds. Oil is extracted through distillation process which is used as flavouring agents, cosmetic, and also having medicinal values.

3.23.2 BOTANY

Plant height is about 30–75 cm. Basal and stem end leaves are long-stalked and shorter stalked, respectively. Flower type is loose umbels. Flowers are small yellowish white in colour. Anise is a cross-pollinating species and is genetically heterogeneous. The fruits are 3.5-mm long ovoid in shape and compressed at it sides. Seed type is schizocarp and has five longitudinal dorsal ridges. The fruit consists united carpels two in number, each containing an anise seed. Seeds are small, curved, greyish brown, 0.5-cm long and contain hairy protrusions from each side.

3.23.3 CLIMATE AND SOIL

It requires a temperature of 8–23°C with 400–1700 mm of precipitation for production. Soil pH should be of 6.3–7.3. The anise plant requires well-drained sandy loamy soil. It requires at least 120 days of frost-free season and uniform rainfall during growing season is essential because the plant is adversely affected by fluctuating soil moisture. The temperature should not fluctuate during growing season with adverse hot periods and rainfall. The alternate wet and dry periods during seed maturity causes it to become brown, reduces the seed quality and makes harvesting difficult.

3.23.4 CULTIVATION AND INTERCULTURE

Anise is propagated by seed. It is a slow-growing, annual plant and requires a sunny, warm, sheltered location with deep, well-drained, light soil. Sowing is done directly in main field on a warm day in April or May. It is not suitable for transplanting. Thinning should be done when seedlings come up to keep the area weed free. If sowing is done too late in hot weather, the plants will become lanky and small, and they will bloom and produce the seeds too quickly. Therefore, seeds should be sown early in the spring with a spacing of 60–90 cm in rows. The depth of sowing is 2 cm. Plant spacing is an important factor in determining the microenvironment in the anise field and can lead to a higher yield in the crop by favourably affecting the absorption of nutrients and exposure of the plant to the light. Habib et al. (2014) reported that significantly higher fruit yield was produced at narrow row spacing of 15 cm, whereas wider row spacing produced higher essential oil content. Sowing on 15 April gives the highest essential content in both years and row spaces of 20 and 25 cm give the highest yield, yield components and essential oil content both in 2013 and 2014. (Nimet, 2015). Only light cultivation is needed for weed control. Seed and essential oil yield were the highest in the case of the application of vermicompost and plant densities of 50 and 25 plants/m^2, respectively (Mahdi et al., 2013). Plants grown under 15-cm row spacing with plant density ranging from 200 to 300 plants/m^2 gave the highest fruit yield due to reduced competition among plants (Habib et al., 2015). Fertilizers are applied at a rate of 80–100 kg of K_2O and 50–75 kg of P_2O_5 per hectare. Excessive nitrogen application results in luxuriant vegetative growth with reduced yields, and increased vulnerability to lodging. As irrigation regimes increased from 6 to 14 days, seed yield was reduced from 734 to 566 kg, respectively. Essential oil is affected by drought stress and percentage of essential oil highest in 14 days irrigation regime and lowest in 6 days irrigation regime (Rasoul et al., 2013). Majid et al. (2014) found that the highest grain yield (636.07 kg/ha) and essence yield (23.76 kg/ha) were obtained in 7 days irrigation interval.

3.23.5 PLANT PROTECTION

Anise is resistant to most pests, and will rarely see disease. Semi-looper is caused by *Thysanoplusia orichalcia*. It is a defoliating insect. Besides this, the other storage pests of anise are cigarette beetle, *Lasioderma serricorne*.

Alternaria leaf blight is caused by *Alternaria* spp. This is a fungal disease characterized by the brightening of leaves. It creates small round yellow, brown or black spots on the leaves, concentric ringed pattern and holes in leaves where lesion has dropped out. This disease is spread by seed and poor air circulation favours the spread of disease.

The management of the plant is done by treating the seed with hot water before planting, providing sufficient irrigation to the field to keep the plants away from this disease, removing and destructing the affected plants and cleaning of all plant debris from the soil as fungi can survive on pieces of plant.

3.23.6 HARVESTING AND POSTHARVEST TECHNOLOGY

Flowering takes place during midsummer and seed maturation occurs 1 month after pollination. When the seeds turn brown, the whole plant is cut to the ground level. In a dry place, the harvested plants are hanged upside down. The heads are covered with a paper bag, so that whenever the umbels shed the seeds they are collected in the bag itself. Allow the paper bag in well-ventilated place. The seeds will lose their flavour quickly, so keep them in an airtight container to retain its flavour even for 2 years. By delaying in harvesting time from hard dough to full maturity stage, number of seed in umbrella, grain yield, grain essence content and essence yield significantly decreased. The results also showed that interaction effect of irrigation intervals and harvesting time on grain yield was significant. The highest grain yield was obtained with hard dough stage (Majid et al., 2014). The seed yield of aniseed is about 5–7 q/ha.

3.24 STAR ANISE

3.24.1 SYSTEMATIC POSITION

Kingdom: Plantae

Division: Mangoliophyta

Class: Mangoliopsida

Order: Austrobaileyales

Family: Iliciaceae

Genus: *Illicium*

Species: *verum*

Illicium verum (star anise) is a Chinese spice. It is a small tree belongs to the family Iliaceae. Native to southern China and also grown in Indonesia and Japan. It is a spice that closely resembles anise in flavour, obtained from the star-shaped pericarp. In Vietnam, it is grown near to the border of China. Star anise was known beyond China long before the Christian era as one of the few familiar spices such as cinnamon. However, it was not until the late 16th century that this spice was first brought to Europe by an English navigator, Sir Thomas Cavendish (Kybal and Kaplicka,

1995). Commercial production of star anise is limited today to China and Vietnam. Growing areas in China are southern and south-eastern provinces, particularly mountainous elevations of Yunnan. China has the largest area of star anise cultivation. Star anise oil is extracted through steam distillation process, which is substituted for European aniseed (*Pimpinella anisum* L.) in commercial drinks (Morton, 2004). The fruit is star-shaped and consists of 8–13 carpels joined centrally and is a well-known spice used in Vietnamese cuisine (Loi and Thu, 1970). The name is derived due to the stellate form of its fruit.

3.24.2 BOTANY

Star anise is a small- to medium-sized evergreen tree grows up to a height of 8 m. Leaves are lanceolate in shape, evergreen and aromatic. Flowers are large, bisexual, yellowish green, sometimes flushed pink to dark red in colour. The fruits are star-shaped, reddish brown. Fruits consisting of 8–13 carpels joined centrally. Carpel is boat-shaped, hard and wrinkled with 10-mm length, light brown, smooth, shiny and ovoid shaped seed within.

3.24.3 CLIMATE AND SOIL

Other countries were not allowed to attempt the cultivation of star anise due to the specific agroclimatic conditions available in traditional growing areas. Woodlands, sunny edges and dappled shade are required for cultivation of star anise. The plant is well adapted to humus rich, light to medium soils with mild acidity and neutrality besides good drainage. It tolerates low temperature of about −10°C, but is not very cold hardy.

3.24.4 CULTIVATION AND INTERCULTURE

Vietnam and China are the confined areas for natural spread and cultivation of star anise. Commercial worthwhile crops are failed to grow star anise crop in spite of repeated attempts. This is due to the requirement of specific agroclimatic conditions which are prevalent only in traditional growing areas (Pruthi, 2001). Propagation is by fully matured brown large seeds which are collected from vigorously grown high yielding tree fruits.

As the seeds quickly loose germinating power, there are at the need to be planted within 3 days of fruit harvest at 3–4 cm apart in well-prepared bed. After harvesting of fruits, they have to be planted within 3 days because the seeds will lose quickly its germination power (Anon, 1991). The produced seedlings with the fourth leaf are transferred to the secondary nursery at 25 cm apart. When they are sufficiently grown and become strong at the age of 3 years they are planted in the main field. Layering is one of the best and successful propagating methods but has yet to become popular. Spacing in the main field is about 5 m. Special care is not required for the young plants except weeding. The flowering is unusual. Flowers are bisexual, scented with colour ranges from white to red. Flowering starts at about 10 years old with three seasons of flowering. The first blossom is from March to April end which is sterile in condition and do not develop into fruits. The second blossom is from July to August and lasts for 2–3 weeks only, as they produce the fertile flowers and developed into fruits. But some of the fruits at their premature stage were lost during November to January. The third immediately after second but sometimes partly dovetailed with it and developed in to fruits by August to October of the following year. The flowers are relatively small but help to produce a bigger harvest. Thus, flowers and fruits are available all the year round with seasonal variations.

3.24.5 PLANT PROTECTION

Star anise are having with antibacterial and pest repellent properties, hence it has no specific pest or disease. Because of the high moisture content required for storage, fungal problems are often encountered in the stored spices.

3.24.6 HARVESTING AND POSTHARVEST TECHNOLOGY

Approximately 80% of product is harvested during August to October. Premature fruit drop due to strong winds and sudden fluctuations in temperature occurs (Anon, 1991). Essential oil is valued. Essential oil in star anise and its maximum quantity can be obtained just before maturity hence fruits should be gathered at this stage. Fruits are harvested by shaking branches or climb the trees and gather the fruits by using long poles with hooks. Initial yield of fresh fruit is only 0.5–1.0 kg/tree and it increases with age reaching 20 kg/tree by 15th year and 30 kg/tree at 20th year. Fruits are sun-dried and are deep reddish colour when dried. This drying process develops characteristic aroma and flavour to star anise. Dried star anise is cleaned first by removing the stalk, leaves and other extraneous matter. The main criterion for grading is the diameter of fruit. About 85% of fruits with 2.5-cm diameter comprise first quality. Partly broken dried pieces with less diameter comprises second quality. Natural colour is also a factor in grading (George and Sandana, 2000). Throughout the storage the freshness of star anise can be retained hence fruits are stored in cool place. The freshness of fruit can be determined by breaking one segment, squeezing it between the thumb and forefinger until the brittle seed pops and then sniffing for the distinct aroma. If aroma is weak, fruits have probably passed their optimum storage life or been kept in undesirable hot and open conditions. Normally, dried fruits can be kept for 3–5 years in airtight containers away from heat, light and humidity (Hemphill, 2000).

KEYWORDS

- **systematic position**
- **botany**
- **climate and soil**
- **cultivation and interculture**
- **plant protection**
- **harvesting and postharvest technology**

PHOTOGRAPHS OF A FEW UNDERUTILIZED SPICES

Ajmund/Celery Plant Seeds

Ajmund/Celery Seeds

Ajmund/Celery Flowers

Asafoetida/Hing Plant

Asafoetida/Hing Flower

Asafoetida/Hing Powder

Horseradish Plant

Horseradish Root

Horseradish Flowers

Thyme plant

Thyme Flowers

Thyme Flowers

Thyme Seeds

Marjoram Plant

Marjoram Flower

Marjoram Seed

CHAPTER 4

RECENT APPROACHES ON IMPROVED PRODUCTION TECHNOLOGIES OF UNDEREXPLOITED SPICES AROUND THE WORLD

CONTENTS

4.1 INTRODUCTION

With increased health consciousness, modern people are more inclined to low-carbohydrate, low-sugar, low-calorie food products packed with bioactive compounds, nutraceuticals and antioxidants. Herbs and spices linked to numerous health benefits include major, minor as well as underutilized spice crops. They not only impart aroma, colour and taste to food within defined cuisines, but sometimes mask undesirable odours as well. Scientific research works on various facets of underutilized spice crops have become an exciting field nowadays to unveil many mysteries of this particular group of spices. The following sections will describe glimpses of those outstanding efforts as reviewed hereunder.

4.2 PROMISING UNDERUTILIZED SPICE CROPS OF THE WORLD

Several spice crops around the world are brought into attention for their intrinsic qualities impacting human health. Ferrao (1993) discussed on spices cultivation technology and trade in his book. An introductory chapter is followed by chapters on old world spices (cinnamon, cardamom, cloves, turmeric, ginger, nutmeg and pepper [*Piper* spp.]) and new world spices (vanilla, allspice and peppers [*Capsicum* spp.]). For each spice, information is included on botanical characteristics, ecology, propagation, cultural methods, harvesting, yields, products, active principles, commercial classification and grading, import specifications, utilization and trade, and processing methods used in various countries. The appendix contains statistics on trade information (total spice production by country, principal spice importing and exporting countries and spice trade and consumption in Portugal).

Large cardamom (*Amomum subulatum*) is an important cash crop in Sikkim, parts of West Bengal, north-eastern states of India, and also in Nepal and Bhutan. Three main cultivars of *A. subulatum* (Ramsey, Golsey and Swaney) are grown in Sikkim. The morphology, cytology, flowering and pollination, genetic variation, crop management in relation to pre-planting and post-planting practices, plant protection, curing methods, chemical composition including volatile oil content, of *A. subulatum* and six other *Amomum* spp. are presented by Rao et al. (1993).

Thankamani et al. (1994) reviewed the agronomy of tree spices, namely, clove, nutmeg, cinnamon and allspice. Climatic and soil requirements, propagation and planting, care after planting, and harvesting and yield are discussed for *Syzygium aromaticum, Myristica fragrans, Cinnamomum verum* (*Cinnamomum zeylanicum*) and *Pimenta dioica*, with respect to their cultivation in India.

Spices research in India is narrated by Peter (1998) with research, crop management and cultivar development for spice crops in a summarized form. Ten high-yielding cultivars and hybrids in black pepper (*Piper nigrum*), 7 in cardamom (*Elettaria cardamomum*), 2 in cinnamon (*C. zeylanicum*), 5 in ginger (*Zingiber officinale*), 16 in turmeric (*Curcuma longa*), 12 in coriander (*Coriandrum sativum*), 5 in cumin (*Cuminum cyminum*), 4 in fennel (*Foeniculum vulgare*) and 4 in fenugreek (*Trigonella foenum-graecum*) are recommended. Plant propagation methods have been standardized for nutmeg (*M. fragrans*), cinnamon, cloves (*S. aromaticum*), allspice (*P. dioica*) and vanilla (*Vanilla planifolia*). Tissue culture research for micropropagation and biotechnology has been initiated. High-production technologies for black pepper and cardamom were formulated. Measures to manage phytophthora foot rot in black pepper, rhizome rot and 'katte' in cardamom and soft rot in ginger were formulated and successfully transferred to farmers' plots.

The vegetative anatomy of *V. planifolia*, an orchid that produces vanillin-yielding pods, was studied by Baruah and Saikia (2002). Studies conducted by Bory et al. (2008) showed that interspecific hybridization and perhaps even polyploidization played an important role in the evolution of the genus. The study has demonstrated the urgent need for preservation of the genetic resources of *V. planifolia* (primary and secondary gene pools and cultivated resources). To measure the knowledge level of vanilla growers a Teacher Made Test was prepared by Pankaja et al. (2009). The findings indicated that, out of fifteen variables, nine variables, namely, age, education, family dependency ratio, social participation, risk orientation, economic orientation, innovation proneness, level of aspiration and management orientation were significantly related with knowledge level of vanilla growers.

The opium poppy (*Papaver somniferum*) is a self-pollinating herbaceous annual plant with various rates of outcrossing and is cultivated since prehistoric times. It is used to obtain a number of medicinally important alkaloids, rarely to obtain edible oil and is used as an ornamental plant.

Saffron (*Crocus sativus*) is the most expensive spice of the world. It is native to Asia Minor, and it is cultivated in Mediterranean countries. Saffron predominantly contains certain chemical constituents that are responsible for imparting colour, flavour and aroma. Some of its components have cytotoxic, anticarcinogenic and antitumour properties. Kumar et al. (2009) compiled the recent agronomic research on saffron for commercial flower and corm production. The major constraints influencing the productivity of saffron include poor agronomic practices (preparation of land, method of sowing, seed rate, harvesting method), drought, cold and in some areas, Fusarium wilt and virus in saffron wilt and virus in saffron.

Mollafilabi et al. (2010) depicted the replacement scenario of saffron (*C. sativus* L.) with poppy (*P. somniferum* L.) and its socio-economic results in Afghanistan. At present, the cultivated area of poppy is 193,000 ha in Afghanistan with 8200 t opium production, providing 90% of the world's narcotic materials. Saffron as a replacement crop is being supported and extended and is under sustainable development. An attempt was made to determine the genetic variability, correlation coefficient path analysis with respect to latex yield, seed yield and their components in opium poppy by Dubey (2010).

The agronomy of saffron cultivation including date of sowing, seed rate, method of sowing, plant population, weed control and method of harvesting has been researched and recommendations are developed for different areas. The efforts are being made to transfer these recommendations to farm level with the help of extension specialists. Transfer of these technologies to farmers is in progress and in some areas farmers are getting almost 50% or more productivity with adoption of winter- or entezari sowing (Javadzadeh, 2011).

Vanilla has been utilized in a strictly cultural context: as a flavouring agent, a ritual plant and as a medicine. Although vanilla cultivation was absent in pre-Columbian tropical America, vanilla curing was not. In addition to various Mesoamerican techniques for curing, and the Maya area, the Amerindians of the Guianas have cured vanilla since at least mid-17th century, while Siona–Secoya of Colombia have also developed their own process. A 'Guiana method' and a 'Peruvian process' both employ the use of oil to preserve the vanilla fruits and tying them up into bundles (Lubinsky et al., 2011).

The nutmeg is unique among tree spices as it gives two distinct spices, nutmeg and mace. It is native of Indonesia (Moluccas Islands). In India, it is cultivated in Kerala, Tamil Nadu, Karnataka, Goa, Maharashtra, Northeast India and Andaman. The area, production and productivity in our country is very low and it is mainly due to the non-adoption of improved crop management and postharvest handling technologies, decline in area under cultivation and incidence of pest and diseases. Hence, the innovations made in various crop improvements, production, protection and postharvest handling techniques are reviewed by Thangaselvabai et al. (2011).

Umamaheswari and Mohanan (2011) studied on vegetative and floral morphology of *V. planifolia* and *Vanilla tahitensis*. It has shown that the two species, even though have got almost similar appearance, they show a significant difference in some characters. The analysis of the results on profitability and competitiveness of the vanilla (*V. planifolia* J.) production systems in the Totonacapan region, Mexico by Barrera et al. (2011) suggests that the macroeconomic policy measures such as the overvaluation of the exchange rate and the high interest rate have a negative impact on the production systems studied.

Traditional practices of large cardamom cultivation in Sikkim and Darjeeling have also been described by Gudade et al. (2013). According to them, most of the tribal farmers living in remote places are following traditional methods for large cardamom cultivation, which are eco-friendly, less expensive due to utilization of local resources, knowledge and labour. Large cardamom is cultivated in terraced and sloppy land. Marginal and tribal farmers mostly rely on indigenous traditional knowledge for pest and disease management. A large number of tribal farmers still practice the traditional methods of curing of large cardamom. Royandezagh et al. (2013) was successful in facilitating the *Agrobacterium*-mediated genetic transformation of *P. somniferum* L. using semi-solid-agar-gelled primed seeds as explant.

4.3 MORPHOLOGY AND PHYSIOLOGY

Earlier research findings of Ramin and Atherton (1991) show that the chilling-imbibed seeds of celery cultivar New Dwarf White for 6 or 8 weeks at 5°C induced flowering subsequently at a minimum glasshouse temperature of 15°in 50% of the treated plants. The longer chilling treatment induced flowering at a lower leaf number. Promotion of flowering by chilling-imbibed seeds was only apparent when plants were subsequently grown in a glasshouse at a minimum temperature of 15°. Low-temperature treatments applied in growth rooms during seed production decreased the percentage both of seeds germinating and of seedlings emerging. Developing seeds chilled on the mother plant 10 days after anthesis failed to germinate. Seed yield per umbel was reduced by chilling during seed development as were individual seed weight and seed size.

Percentage germination was highest when celery seeds were primed in *polyethylene glycol* (PEG) 8000 at −12.5 bar at 15° for 20 days. In the second experiment, celery seeds were primed at 15° for 14 days in solutions containing PEG 8000 at −12.5 bar together with 100 ppm benzyl adenine (BA), 100 ppm GA4/7, 500 ppm ethephon, 1000 ppm Enersol (crude humic acid extract), 1% Response Extra (seaweed extract with added nutrients) or 1% BL 9–97 (seaweed extract containing cytokinin). None of these solutions increased germination percentage at 15° above the untreated control value. However, at 25°, 100 ppm BA or 1% BL 9–97 increased it from 20 (control) to >40%. In a third experiment, celery seeds of cultivars Earlybelle and 683-K were primed in PEG 8000 (−12.5 bar) together with 2% Response Extra, or with 100 ppm GA4/7 and either 100 ppm BA, 500 ppm ethephon or 1% BL 9–97. Germination occurred in the dark at 25°. It was negligible in Earlybelle except in treatments including GA4/7 + BA or Response Extra, but in 683-K, it was approximately 80% when BA or BL 9–97 was included with GA4/7, approximately 30% with PEG alone or with Response Extra, and negligible in the other treatments. It is suggested that such cultivar differences demonstrate the need for high-quality seedlots for priming to be effective (Tanne et al., 1989).

The relationship between phytochrome and endogenous hormones in controlling seed dormancy is discussed by Thomas (1990). Mutation breeding studies of Choudhary and Kaul (1993) reported that the seeds of genotype RRL85–1 were irradiated with 10–30 kR γ-ray doses and resulting M1s and controls grown in the field using the bulk method. Selection for early flowering was made in the M3 and M4 generations and confirmed in the M5. One selected mutant showed a 20–25-day improvement over the parent, with a longer flowering period and a mean of 175 days to maturity versus 200 in the parent. Most of the other morphological properties were inferior to the parent (seed yield/plant was 5.1 versus 7.2 g) but oil content was similar.

Phenology of flowering in Vanilla (*V. planifolia* Andr.) has been studied by Bhat and Sudharsan (2002) in high ranges of Karnataka. Development of inflorescence, time taken for the first flower to bloom, morphometrics of flowers, flower blooming period in a season, frequency of flower blooming in an inflorescence and days taken for maturity and harvest has been observed for four seasons. These observations would help vanilla growers to take up flower induction operations, pollination and other cultural operations in time.

The metabolomic analysis of developing *V. planifolia* green pods (between 3 and 8 months after pollination) was carried out by Palama et al. (2009) by nuclear magnetic resonance (NMR) spectroscopy and multivariate data analysis. Older pods had a higher content of glucovanillin, vanillin, p-hydroxybenzaldehyde glucoside, p-hydroxybenzaldehyde and sucrose, while younger pods had more bis [4-(β-D-glucopyranosyloxy)-benzyl]-2-isopropyltartrate (glucoside A), bis [4-(β-D-glucopyranosyloxy)-benzyl]-2-(2-butyl) tartrate (glucoside B), glucose, malic acid, and homocitric acid. Ratios of aglycones/glucosides were estimated and thus allowed for detection of more minor metabolites in the green vanilla pods. Quantification of compounds based on both liquid chromatography–mass spectrometry (LC–MS) and NMR analyses showed that free vanillin can reach 24% of the total vanillin content after 8 months of development in the vanilla green pods.

Mature green vanilla pods accumulate 4-O-(3-methoxybenzaldehyde)-β-D-glucoside (glucovanillin), which, upon hydrolysis by an endogenous β-glucosidase, liberates vanillin, the major aroma component of vanilla. Sites of storage of glucovanillin in the pod have been controversially reported for decades. Odoux and Brillouet (2009) clarified this controversy by providing an anatomical, histochemical and biochemical evidence-based picture of glucovanillin accumulation sites. Glucovanillin is essentially stored in the placentae (92%) and marginally in trichomes (7%). Trichomes store massive amounts of a fluorescing oleoresin (44%) rich in alkenylmethyldihydro-γ-pyranones and synthesize a mucilage made of a glucomannan and a pectic polysaccharide carrying monomeric arabinose and galactose side chains.

Brillouet et al. (2010) also researched on green, ripening and senescent vanilla pod on their anatomy, enzymes, phenolics and lipids profile. They concluded that the green mature vanilla pod is spatially heterogeneous for its phenolics, β-glucosidase and peroxidase activities and that its placentae play an important role in the liberation of vanillin and its subsequent oxidation.

An experiment was conducted by Mini-Abraham et al. (2010a) to study the absorption pattern of 14C-urea by aerial roots of 1-year-old potted vanilla (*V. planifolia*) plants in Kerala, India. The percentage of absorption increased with time up to 96 h. After 96 h of application, the percentage of absorption was 82.57%. The aerial roots were the major nutrient-absorbing plant part and they can absorb ~80–90% of the applied quantity. Thus, spraying of nutrients on aerial roots of vanilla is recommended.

Reis et al. (2011) described some aspects of the floral biology, reproductive biology and vegetative propagation of natural stands of *Vanilla chamissonis*, aiming to generate information for its sustainable management.

4.4 PROPAGATION

There are numerous researches on formulating propagation protocol including that of micropropagation of many underutilized spices. Studies carried out by Haldankar et al. (1991) on the effect of season and shade provision on softwood grafting of kokum (*Garcinia indica*) indicated that October was the best season for grafting. The grafted plants could be successfully maintained either in the glasshouse or outdoors without shade.

There is considerable potential for improving the slow and irregular germination of celery (*Apium graveolens*) seeds by pre-sowing hydration treatments, but poor germination performance as a result of drying seeds after treatment is a recurrent problem. A pre-sowing imbibition treatment temperature of 15° was identified as the most effective for the cultivars Tall Utah 52–70 and Green Giant Hybrid when the seeds were only surface-dried before sowing. However, conventional air-drying of treated seeds at 20° caused severe delays in germination and decreased uniformity. Higher treatment temperatures reduced the adverse effect of drying but resulted in little or no germination advantage because of thermodormancy, which was particularly evident at 25°. Deleterious effects of drying-back were avoided, and in some instances further germination enhancement was gained, by slowing the rate of drying, which was achieved by exposing the treated seed to a series of controlled humidity atmospheres at 80, 68 and 48% RH for various times (Coolbear et al., 1991).

The vanilla plant (*V. planifolia*) could have a higher commercial production but there are problems in the efficiency of propagation. The effects of culture media type (Murashige and Skoog (MS), Woody Plant or Gresshoff and Doy media) and BAP (BA; 0 or 1 mg/l), GA (0.1 mg/l) or $AgNO_3$ (20 mg/l) on shoot proliferation in vitro of *V. planifolia* were investigated in by Ganesh et al. (1996).

In another study by Haldankar et al. (1997) it was, however, found that May was the best month for softwood grafting (80% success) of kokum followed by June (54%) and July (50%). The retention of one terminal leaf on the scion sticks gave 75% success. The retention of the leaves on rootstocks did not influence the success of softwood grafting.

Bhuyan et al. (1997) studied on micropropagation of curry leaf tree (*Murraya koenigii* (L.) Spreng.) by axillary proliferation using intact seedlings. The shoot-forming capacity of intact seedlings was influenced by explant orientation. Maximum shoot proliferation was obtained when the shoot-forming region was in direct contact with the medium surface or slightly embedded into the medium. Proliferating shoot cultures were established by repeatedly subculturing mother seedlings on fresh medium of the same composition after excising all newly formed shoots. Roots were formed on excised shoots following transfer to half-strength MS containing IBA (1 mg/l). Plantlets were acclimatized and established in the soil where they exhibited normal growth.

The influence of seed-conditioning methods on germination, vigour and field emergence in three types of soil (grey-brown, alluvial and black) was studied by Kolasinska and Dabrowska (1996) in parsley (cultivar Berlinska) seeds. There were no differences between germinability of the variously treated parsley seeds. The highest field emergence was shown by PEG-treated and control seeds on all types of soil, irrespective of the place of evaluation (greenhouse, microplots). Low-temperature treatment (10°C) of celery seeds during the summer increased percentage germination, germination vigour and the germination index. Treatment for 8–10 days gave the best results (Zhang et al., 1998).

The effect of the biochemical constituents in the stems and the rooting ability of curry plants (*M. koenigii*) were studied at Bangalore, India, by Reddy et al. (1998). The highest starch content was found in hardwood cuttings (6.83%) and a highly significant positive correlation was found between starch content and rooting. Sugar content was also similarly correlated with rooting. Softwood cuttings were high in N and protein and here there was a highly significant negative correlation with rooting. There was also a high negative correlation between total phenols and rooting.

Rooting was lowest in softwood cuttings, which were high in total phenol content. Maximum dry matter (DM) was found in hardwood cuttings and there was a significant positive correlation between DM and rooting.

Micropropagation of curry leaf was standardized by Mathew et al. (1999) using nodal segments from mature trees on MS media supplemented with 6-benzylaminopurine (BA), kinetin, NAA and IBA. A multiplication ratio of 1:3 was observed over a 5-week culture period on media with benzylaminopurine (1.0 mg/l), kinetin (0.5 mg/l) and NAA (0.1 mg/l). The best response of one to two roots per shoot was obtained on medium containing NAA and IBA (0.1 mg/l each).

An experiment was conducted by Haldankar et al. (1999a) to study the effect of season and rootstock on approach grafting in *M. fragrans*. Approach grafts can be prepared throughout the year. Good graft take was recorded on both edible nutmeg rootstock (40–90%) and wild nutmeg (*Myristica malabarica*) rootstock (30–100%). Mortality after separation of the grafts was as high as 30% on edible nutmeg rootstock and 50% on wild nutmeg rootstock.

In another study, Haldankar et al. (1999b) investigated the factors influencing epicotyl grafting in nutmeg and found that age of scion influenced grafting success, with treatments using 4- and 6-month-old scion sticks recording better success. Prior defoliation was not a prerequisite for grafting. When grafted during the non-rainy season with low humidity, graft union and scions should be covered with a polybag; no such covers are required when high humidity prevails (July). Grafting success was not affected by the location of the scion on the mother plant. September was the best month for grafting.

Effect of maturity and season on rooting of curry leaf cuttings was interpreted by Mohanalakshmi et al. (2000). The results indicated that semi-hardwood cuttings planted during May and July recorded 30 and 20% rooting, while the semi-hardwood cuttings from the current season shoot recorded only 10% in both seasons. The cuttings from the past season shoot in May was found to the best for rooting of cuttings in curry leaf.

M. koenigii (curry leaf tree) is cultivated for its aromatic leaves, which are used as condiment. Nodal cuttings from mature curry leaf plants cultured in woody plant basal medium supplemented with 4.4 μM BA and 4.65 μM kinetin produced 12–30 multiple shoots per node by the 8 week of inoculation. The shoots easily rooted in vitro in woody plant medium containing 1.35 μM NAA. Ninety per cent of the plants survived transfer to a hardening chamber and were transferred to the field after 3 months. In vitro-developed shoots were also rooted ex vitro by dipping in 2.46 μM IBA for 1 min. They were transplanted to sand in a hardening chamber with 70–80% relative humidity and a temperature of 28±2°C. Eighty to ninety per cent of the ex vitro-rooted plants survived and were transferred to the field after 3 months (Babu et al., 2000).

Seeds of parsley (*Petroselinum crispum*) cultivar Moss Curled were primed osmotically in PEG or matrically in fine, exfoliated vermiculite at −0.5 MPa for 4 or 7 days at 20 or 30°C with 0 or 1 mM GA3. All priming treatments stimulated and hastened germination. All matric priming treatments (other than 4-day priming) were repeated to assess seedling emergence in a greenhouse (25°C day/22°C night). Priming increased the percentage, rate and synchrony of emergence, and increased hypocotyl length at 3 weeks after planting. It was concluded that matric priming is a satisfactory alternative to osmotic priming of parsley seeds (Pill and Kilian, 2000).

The success rate of softwood grafting in 22 female, 12 hermaphrodite and 5 male nutmeg genotypes was studied by Haldankar et al. (2003). The variation among genotypes for sprouting, survival and growth parameters was statistically significant. The graft survival has a strong negative correlation with leaf width. Highest graft sprouting was associated with the faster production of new leaves with less breadth and longer petiole.

Greenhouse experiments were conducted by Nicola et al. (2004) to study the cutting techniques on thyme through season-long testing of products registered for organic farming (Cytokin and Radix) and products of synthetic origin (Stimolante 66f, Germon, Naftal, Kendal and Radifarm) to increase rooting and to evaluate the difference on rooting among the products tested. The orthogonal contrasts indicated that the production of adventitious roots was not favoured by the rooting products; on the contrary, the untreated cuttings produced longer roots at 22 days after cutting. The best results were obtained with products registered for organic farming.

An efficient protocol for plant regeneration from stem segments of *M. koenigii* was developed by culturing on MS medium supplemented with 2.5 mg/l BA, 25 mg/l adenine sulphate, 0.25 mg/l indole-3-acetic acid (IAA) and 3% sucrose. The maximum percentage of rooting was obtained on MS medium supplemented with IAA and NAA, each at 0.25 mg/l. During acclimatization, 95% of rooted plantlets survived and were grown normally under greenhouse conditions (Rout, 2005). A comparative research study on the effect of growth regulators and methods of application on rooting of thyme (*Thymus vulgaris* L.) cuttings was undertaken by Chandre et al. (2006).

An experiment was conducted by Khandekar et al. (2006) to find out effect of time of softwood grafting on sprouting, survival and growth of nutmeg grafts. The maximum sprouting (100%) was recorded in July and August, which was followed by June (99.55%). The mean maximum survival (99.24%) was recorded in July, which was at par with August and June, that is, 98.56 and 97.89%, respectively. These months also showed minimum days for initiation and completion on sprouting of nutmeg softwood grafts.

Cutting experiments were carried out with garden thyme with two cutting preparation methods (cutting and tearing), four kinds of rooting medium (perlite, universal flower soil, peat and perlite/peat mixture) as well as three types of hormones (β-indolebutyric acid (IBA): 0.4 and 0.8%, naphthalene acetic acid (NAA): 0.5%). According to the results, it turned out that early autumn (middle of September) is more advantageous period for rooting thyme cuttings. However, in the case of autumn-rooted cuttings, over-wintering is also necessary. The higher amount of thymol and carvacrol has been detected in the oil of the plants originated from normal cuttings. Root formation, plant height and width were also affected by rooting medium: perlite proved to be the best one (Gyongyosi et al., 2008).

They observed good shoot proliferation in the presence of BAP. While WP produced the longest shoots, the combination of BAP + GA was found to promote shoot multiplication and growth. Culture of shoots on MS resulted in rooting. Plantlets were successfully transferred to the soil. A study was conducted to investigate the effect of different substrates in the adventitious rooting of cuttings taken at five heights in the plant stem. Adventitious rooting of cuttings and formation of new vanilla cuttings occurred when the cuttings were obtained from position P2 (between 20 and 40 cm stem height) and planted in the solid substrate in a greenhouse with intermittent misting system (Silva et al., 2009).

Towards a study on in vitro propagation of *V. planifolia* from axillary bud explants, Khunsri and Poeaim (2009) found that the shoots can be induced by this modified MS medium supplemented with 0.113 and 5 mg/l BA in combination with 0.5 mg/l IBA. The result showed that the highest average shoot multiplication about 4.3 shoots/explant was obtained from 1 mg/l BA in combination with 0.5 mg/l IBA for 12 weeks.

Experiment conducted on the seed germination, seedling growth and vigour of nutmeg (*M. fragrans*) revealed that there were significant differences in germination and seedling growth behaviour of nutmeg seeds sown in 21 different combination of growth media (Abirami et al., 2010). The investigation on rooting and growth of vanilla (*V. planifolia*) stem cuttings was carried out by Murthy et al. (2010) under 50% shade net to find out the effect of growth regulators and bio-inoculants on rooting and growth of vanilla cuttings. They opined that *Trichoderma harzianum* could be conveniently used for the commercial propagation of vanilla stem cuttings planted under 50% shade net, which would be more economical and eco-friendly.

Multiple shoots of *V. planifolia* were induced by Tan et al. (2011) from nodal explants under influence of different concentrations of plant growth regulators (PGRs) and coconut water (CW). Ninety-seven per cent of the explants produced a mean number of 9.6 shoots with a mean shoot length of 4.70 cm when cultured on liquid MS supplemented with 1.0 mg/l BAP in a combination of 15% CW. The plantlets were transferred to sand: compost mixture (1:2) with 85.0% survival rate recorded after 4 weeks of acclimation.

The effects of environmental conditions (polyhouse, greenhouse with mist, 50% shade net and natural shade) and type of cuttings (single, three and five nodes) on rooting and growth of vanilla (*V. planifolia*) were evaluated for a year by Umesha et al. (2011). Their results show that greenhouse is more conducive for producing longer sprouts with greater number of leaves, early and high rooting percentage and early *sprouting*.

Later, Abirami et al. (2012) evaluated several nutmeg genotypes for seed germination, seedling growth and epicotyl graft take. The genotype A-4–17 performed best in terms of seed germination and seedling growth characters when compared to the other genotypes. In epicotyl grafting of the variety Viswashree with rootstocks of all the seven genotypes under study, genotype A-4–17 was again found most promising as it recorded higher survival percentage of grafts and maximum values for different graft growth characters like plant height, scion length, leaves per plant, rootstock and scion diameter.

An experiment was conducted by Girma et al. (2012) to determine optimum node numbers of *Vanilla fragrans* for the production of quality planting materials in Southwestern Ethiopia. Mean leaf, shoot and root parameters values were consistently increasing from node number one to four and declined with node number five in a similar pattern. Four-node cuttings resulted in vigorous and quality planting materials and recommended for mass propagation in commercial and small-scale farming.

The genus *Mentha* is cultivated worldwide for essential oil production, with emphasis on its major constituent, menthol, which is used in the pharmaceutical, cosmetics and food industries and for personal care. Santos et al. (2012) investigated the use of different propagation structures and harvesting times of *Mentha canadensis* L. The results suggested that early harvest of the regrowth results in great menthol concentrations in the essential oil. The use of stolons as a propagation structure can be considered a viable alternative for mint, as it provides similar levels of essential oil and menthol productions after the second harvest when compared to stem cuttings, and it also reduces the production cost.

Ramos-Castella et al. (2014) evaluated the efficiency of shoot multiplication of *V. planifolia* using solid medium, partial immersion and a temporary immersion system (TIS) to improve micropropagation in this species. Root initiation was 90% successful in TIS using half-strength MS medium supplemented with 0.44 μM NAA and an immersion frequency of 2 min every 4 h. With this system, the shoot multiplication rate increased threefold compared to that obtained with solid medium. In addition, this system produced good results for the transplantation and acclimation (90% of survival) of in vitro-derived plants. These results offer new options for large-scale micropropagation of vanilla.

4.5 PLANTING REQUIREMENT

Sridharan et al. (1990) investigated the effect of planting density on the damage to cardamom by *Sciothrips cardamomi* in an 8-year-old plantation in Tamil Nadu, India. Plants placed at the smallest distance (1.0 m) away from each other were the most damaged by the pest (92.88% of pods damaged, 85.8% weight loss from pods), with damage decreasing with increased spacing. The herb yield and essential oil content of three ecotypes of *T. vulgaris* were recorded by Rometsch (1993) while studying on the effect of climatic factors on these parameters. Plants were strongly affected by susceptibility to frosts. Soil temperature and humidity were observed to be important by studying the adaptation of the local flora at different sites.

A research work was carried out by Palumbo and D'Amore (1994) on parsley cultivars, grown during the winter-spring season in different plant densities and cold tunnels covered with plastic film. The effects of different cultivars and plant densities were significant, while the plastic film had no effect. In the field, parsley is sown in early April and the first cut is done about 2 months later. The harvest continues until September or October, and produces 20–25 bunches/m^2. In the polytunnel, seedlings are planted in pots in August and the harvest begins 6 weeks later, continuing until May, yielding 30–45 bunches/m^2. Picking and bunching the parsley is labour intensive, but given sufficient expertise, high production and good yields, growing parsley can be economic (Hartmann, 1996).

The effect of different methods of sowing on germination and seedling growth was studied in *E. cardamomum* in a nursery in Mudigere (Karnataka, India). Fewer seeds were required in the dibbling/hill sowing method than in the line sowing/drilling method. Percentage germination, seedling height, number of leaves, leaf area, number of tillers, fresh weight, dry weight and percentage of standard seedlings (4–5 tillers/seedling) obtained at transplantable age were also significantly higher in the dibbling/hill sowing method (Siddagangaiah, 1997). In a trial at Coimbatore, 1-year-old seedlings of *M. koenigii* were planted at spacings of 0.75, 1.0 or 1.5 m. They were pruned to 30 cm 1 year later and given 0, 90, 120 or 150 kg N/ha. Leaf yields declined with increased spacing. They were highest (42.24 t/ha) at a spacing of 0.75 m and with 90 kg N/ha (Lalitha et al., 1997).

The effect of different pretreatments on cardamom seed germination was investigated by Korikanthimath and Ravindra (1998). Seeds (cultivar Malabar) were soaked for 10 min in 20, 40, 60 or 80% of acetic, hydrochloric, sulphuric or nitric acid, washed and then sown in nursery beds. In a separate experiment, seeds were also treated for 12 h with GA3 or Planofix (NAA) at 25, 50, 75, 100, 500, 750 or 1000 ppm. Germination decreased with increasing acid strength. The best acid treatment was 20% nitric acid (44.6% germination). GA3 treatment increased germination up to a concentration of 100 ppm (49.9%) but germination decreased as GA3 concentration increased further. Similarly, with Planofix, germination increased up to 75 ppm (49.3%) and then decreased.

A model field investigation was conducted in Karnataka, India, for converting the low-lying marshy area for cultivation of cardamom by adopting high-production technology. As cardamom is a moisture- and shade-loving plantation crop, an attempt was made to bring the marshy area under cardamom cultivation by providing adequate drainage and raising fast-growing shade tree species. The study revealed that the cultivation of cardamom in marshy areas would be an economically remunerative and ecologically feasible proposition, as the discounted cash flow measures, that is, net profit value (NPV), benefit:cost ratio (BCR) and internal rate of return (IRR), justified the viability of cardamom cultivation. The payback period of 2.14 years indicated that the investment on cardamom can be met in 3 years by successful conversion of marshy areas (Korikanthimath et al., 1999).

Bhagyalakshmi (1999) identified the factors influencing direct shoot regeneration from ovary explants of saffron. The best response towards caulogenesis (28%) with highest shoot numbers per ovary was observed when full-strength MS medium was supplemented with NAA and BA. Incubation in the dark at 20°C was beneficial for the induction of shoot buds. Ovaries of different growth stages having stigmas of pale yellow, pale orange and bright orange colour regenerated a maximum mean number (3.8–4.2) of shoots per ovary. Regenerated shoots produced normal photosynthetic leaves and corms.

Staugaitis and Starkute (2000) conducted a field trial for testing the impact of planting time and agrofilm cover on yield, quality and profitability of plain- and curly-leaved parsley. Up until the second half of August, cultivars planted in May produced 1.8–2.5 kg/m^2 of leaves and an average leaf weight per plant of 400 g. After cutting, leaves grew again until mid-October and the total yield from the two harvests was 3.8–3.9 kg/m^2. Yields from parsley planted in July were significantly lower than those from planting in May. Covering crops planted in spring with agrofilm increased yields by 9%.

Another study was carried out by Esiyok and Ilbi (2000) to determine the most suitable sowing times under two growing conditions (open field and under plastic tunnels) for high yield and quality of parsley grown in Bornova, Turkey. Seeds of parsley cultivar Italian Giant were sown every month in the field, and under plastic tunnels from October to March. The total yield and number of bunches for each sowing time were higher under plastic tunnels compared to open-field cultivation.

Results of a study conducted by Khan and Ahmad (2000) regarding the cultivation, seed production and nutritive value of parsley cultivar Mass curled were reported. The results showed leaves are ready for cutting after 65 days of transplanting. The yield per plant and yield per hectare were recorded to be 290 g and 11 ql, respectively. Crop is ready for seed harvesting after 185 days of transplanting and seed yield was to the tune of 7.90 g/plant and 89.00 kg/ha. A sufficient amount of moisture (81.85%), minerals (3.81%), crude protein (1.69%), crude fat (0.68%), carbohydrate (10.58%), fibres (1.36%), ascorbic acid (132.63 mg/100 g) and total soluble solid (8.50%) was also recorded in the leaves.

A field study with *Mentha arvensis* L. cultivar MAS-1 was carried out to study the effect of planting material and spacing on growth, biomass yield, essential oil yield, oil quantity and economics of cultivation under agroclimatic conditions of Hyderabad (India). The use of stem cuttings as a planting material with 60 cm row spacing recorded a minor increase in menthol, and slight decreases in methyl acetate and menthone contents compared to other treatments. The economics of cultivation showed that highest net returns (Rs. 28,381/ha) were obtained with runner cuttings planted at a 60-cm row spacing with a cost–benefit ratio of 1:1.99 under Hyderabad conditions (Kattimani and Reddy, 2000).

Korikanthimath (2001a) analysed the feasibility of high-density planting in cardamom (*E. cardamomum* Maton.) for Karnataka region of India. The fresh yield of capsules decreased significantly with increasing plant density. Increasing the planting density up to 5000 plants/ ha, spaced at 2m × 1 m, resulted in a significant increase in yield.

Patra et al. (2003) provided descriptions of menthol mint (*M. arvensis*), the development of cultivars and their suitability, commercial cultivation of mint (including land preparation and selecting planting materials), sucker production, direct planting of suckers, transplanting, fertilizer and manure application, irrigation and drainage, interculture and weed control, crop rotation, intercropping, mint under agroforestry, harvesting and yield. Mint pests and diseases, distillation of herbage for essential oil, purification and storage of oil, use of mint oil and its derivatives, are also discussed.

The performance of parsley cultivar Paramount under different transplanting methods (single- or multi-seeded transplants produced at two or three per plot) and plant covers (non-woven fleece, perforated foil and low plastic tunnel) was investigated by Koota and Winiarska (2004). Transplants produced individually in cells of trays were better developed and produced higher total leaf yield than two or three plants grown in one pot. Fleece was the most suitable material for covering, followed by plastic tunnel and perforated foil. Individual planting was preferable for obtaining good DM and ascorbic acid contents.

Under Danish conditions, harvesting in June when the plants are in full bloom gives the highest oil yield (15–20 kg/ha) with an oil concentration of 1% DM. Over 60% of the essential oil from plants harvested at this time is thymol. Later harvesting, in September, gives a higher concentration of oil (1.6% DM), but a lower content of thymol (38%). Drying temperatures also influence the final yield (Grevsen et al., 2004). Horseradish (*Armoracia rusticana*) has been cultivated in Denmark since the 17th century and is also found growing wild. General descriptions are given of the roots, which are used in the preparation of sharp-tasting condiments, soil type (deep, humus-rich), propagation and planting methods, fertilizer use and irrigation, cultivation methods, plant diseases and pests, harvesting methods and details of the various varieties and clones in common use (Henriksen and Bjørn, 2004).

Gruszecki (2004) found out the effect of a few Polish parsley cultivars, that is, Berlinska, Cukrowa, Lenka, Eagle and Omega on early yield grown from the late summer sowing. Berlinska wintered the best (64.8%), however, it was the most susceptible to vernalization. Cukrowa and Omega were the least susceptible to vernalization. The average highest total (0.76 kg/m^2) and marketable yields (0.64 kg/m^2) were obtained from Lenka, while the lowest yields were obtained from Berlinska. Turnip-rooted parsley (*P. crispum* ssp. *tuberosum*, a temperate crop of Northern Europe, was tested by Petropoulos et al. (2005) for its suitability for cultivation under warm, Mediterranean conditions. It is concluded that turnip-rooted parsley may be satisfactorily grown there, preferably adopting an autumn sowing. This species may be exploited as a new, alternative crop for Mediterranean horticulture.

Tort et al. (2005) investigated the effects of temperature, growth media and NaCl concentrations on seed germination and proline content of 3-week-old seedlings of aniseed (*Pimpinella anisum* cultivar Burdur). Germination rate was 58.6% at 40°C (16 h light), while it was 74.6% at 250°C in quartz sand medium. The germination period ranged from 3 to 22 days at a salt concentration of 0 and 400 mM NaCl, respectively. The salt tolerance of Burdur was restricted to only 50 mM NaCl, with a proline content of 0.33000.174 micromol/g.

The effects of increased N rates (0, 100, 200 and 300 kg N/ha) on yield, plant length, number of stalk, phosphorus and potassium content in parsley leaf and stalk during seven harvesting at two sowing time were investigated by Yoldas (2007). The highest yield, plant length, number of stalk were observed with 100 kg N/ha treatment, but maximum P and K content in leaf, stalk were analysed with 100 kg N/ha treatment. In second harvest, the highest yield and yield criteria values depend on ecological condition. In both sowing time, P, K content in leaf and K content in stalk were determined significantly higher in first harvest than others ($p<0.01$). In this research, P content in leaf was observed to be higher than stalk content. In spite of this, K content in stalk was higher than leaf.

Rozek (2007) conducted field experiments in Poland to study the effect of seedling planting density and irrigation on the yield of two cultivars of leaf celery (Afina and Safir). Transplants were planted in May with a spacing of 25×20 and 15×20 cm^2. Harvest of celery leaves was carried out in three turns, that is, in July, September and October. It was shown that higher planting density significantly increased the yield of leaves. Irrigation increased leaf yield, on average, by 55.9%. Depending on the cultivar and planting density, leaf yield of nonirrigated plants was 5.29–6.34 kg/m^2, while for the irrigated plants, it ranged from 7.70 to 11.97 kg/m^2.

In the field experiments conducted by Seidler et al. (2008), the quality of thyme herb and usefulness of Polish cultivar Soneczko for organic cultivation were tested. The following features were tested: dried herb yield, stem content in dried herb, essential oil content, nitrate content, macro- and microelements content and microbiological purity. Only from Sonsk thyme herb yield was higher compared with the yield from conventional cultivation though it contained high amount of stems.

Kolodziej (2009) studied the effects of the method of plantation establishment and foliar fertilizer Resistim application on *T. vulgaris* L. yields and quality. The obtained results indicate that for thyme plantation establishment, transplants produced earlier in multicells should be used on lessive soils. On the other hand, seed sowing was connected with unequal plants distribution and, as a result, it decreased both yields and quality of raw material gathered. Spraying with Resistim positively affected morphological parameters and yields of thyme and could be recommended on commercial plantations of this species.

Al-Ramamneh (2009) found out plant growth strategies of *T. vulgaris* L. in response to population density. An experiment was undertaken to determine the effects of three intra-row spacing (15, 30 and 45 cm) and four harvesting times (vegetative, beginning of blooming, full blooming and fruit set) on plant growth and herbage biomass of thyme in Jordan. Closer spacing resulted in significantly taller plants that exhibited higher shoot: root ratio and, therefore, plants grown using 15 cm intra-row spacing had a better use of light and accumulated more biomass compared to plants in wider spacing. The decrease in specific leaf area between vegetative and full-bloom stage indicated that leaves became thicker as part of an adaptive mechanism to the surrounding environment. Information obtained from growth parameters, leaf area and specific leaf area in particular, could be related to essential oil production in leaves of thyme plants.

Zeinali et al. (2010) studied the effects of planting density and pattern on cucumber yield and yield components in its intercropping with celery. Cucumber was planted as the main crop with three planting spaces (20, 30 and 40 cm), while celery being transplanted as the partner within the rows, between the rows and in a combination. The value of land equivalent ratio (LER) was calculated. For all the treatments, a greater than 1 LER obtained, indicated the intercropping system as beneficial. The treatment of intercropping between cucumber and celery (planting space: 30 cm between the rows) produced the maximum yield. In single-crop system, the treatments with planting spaces of 20 cm for cucumber and 30 cm for celery produced the maximum yields. According to the results obtained, an intercropping culture of cucumber and celery can be beneficially employed by the vegetable growers.

A characterization of production systems of *V. planifolia* under orange trees and under mesh shade (50% brightness) was conducted by Barrera-Rodriguez et al. (2010) through the identification of technical and climatic variables that may affect the yield of vanilla plantations in the Totonacapan region, in the Mexican states of Puebla and Veracruz. Based on the traditional knowledge of producers, the most important factors that determine the proper development of the vanilla pods in the production system under orange trees were nutrition (21%), moisture (19%) and pollination (16%), while for the mesh shade production system, these were nutrition (20%), temperature (15%) and moisture (14%).

Poggi et al. (2010) conducted field studies in Argentina to verify possibilities of extending and modifying the flowering time of saffron, evaluate the effect on flowers and saffron production of corm size and the duration of incubation under summer-like conditions (23–27°C) and compare results from controlled environments with field crop data as well. Chagas et al. (2011a) evaluated the effects of organic fertilization at planting and dressing on the above-ground part dry biomass and on the content and yield of essential oil of mint (*M. arvensis*). The first and second harvests were carried out at 120 days after planting and 120 days after first harvest, respectively. The fresh biomass was dried in the oven at 37°C and the essential oil was extracted in a Clevenger apparatus. Above-ground part dry biomass (g), essential oil content (w/w %) and essential oil yield were determined. The levels of cattle manure applied at planting and dressing affected in a positive linear way the above-ground part dry biomass production and the essential oil yield. Essential oil content was not affected by different organic manure (OM) levels.

Moaveni et al. (2011) conducted a split-plot field experiment in Iran with sowing dates as main factors and planting densities as subfactors. It was concluded that sowing date and planting density are the main factors influencing the quantity and quality features in thyme.

Leaf celery (*A. graveolens* L. var. secalinum Alef.) is a vegetable with spicy and medicinal properties. A study on the effect of seeding rate and irrigation on yield of two leaf celery cultivars, Gewone Snij and Green Cuttnig, was carried out. Seeds were sown in the field in the last decade of April in rows 25 cm apart. Two seeding rates were used: 15 and 25 kg/ha. Plant irrigation was applied during critical periods of soil water deficit. The raw material was harvested twice: in the second decade of August and in the second decade of October. The investigated factors were shown to have a significant effect on leaf celery yield. A significantly higher content of essential oil was found in the leaves of both celery cultivars harvested on the first date. The highest essential oil yield was obtained from irrigated plants of the cultivar Green Cutting harvested in the second decade of August. The main components of the essential oil of leaf celery were limonene and myrcene (Rozek et al., 2013a). Gruszecki and Salata (2013) studied the effect of sowing date on biometrical features of Hamburg parsley plants.

A 2-year study was carried out by Habibullah et al. (2014) to explore the effect of seed rate and row spacing on the fruit yield, essential oil yield and composition of aniseed. The study factors included seed rate (6, 12, 24 g/10 m^2) and row spacing (15, 25, 37.5 cm). A significantly higher fruit yield was produced at narrow row spacing of 15 cm among treatments. Wider row spacing produced markedly higher essential oil.

Plant grown at 37.5 cm row spacing accumulated the highest estragole and trans-anethole concentration among the row spacing treatments. It can be concluded that higher plant density and wider row spacing increased the disease infestation and lodging.

4.6 NUTRIENT MANAGEMENT

4.6.1 INORGANIC: PRIMARY, SECONDARY AND MICRONUTRIENTS

Omidbaigi and Rezaei-Nejad (2000) studied the influence of nitrogen fertilizer and harvest time on the productivity of *T. vulgaris* L. in Iran. The nitrogen fertilizer had a significant effect on the DM production of the species: the herb yield increased. The essential oil yield proved to have a similar tendency because neither the accumulation level of essential oil nor the ratio of thymol, were affected by the nutrient supply. In an unheated greenhouse experiment, celery (*A. graveolens* var. dulce) cultivars Rosati and Golden Boy were given a pre-planting application of N plus a top dressing of N. Marketable yield was higher with the proposed application rate than with the traditional application rate and higher in Rosati than in Golden Boy (Temperini et al., 2000).

The effect of N fertilizer on the quality of celery (cultivar Darklett) was investigated in the region of Bologna, Italy. Fibre content decreased with increasing N rates, as did the levels of soluble solids and essential oil yield (D'Antuono, 2001).

The effect of drip irrigation and liquid N fertilizer application compared to broadcast N application on yield and quality parameters of celery was investigated by Elkner and Kaniszewski (2001) in a field experiment conducted in Poland. The highest total and marketable yield of cultivar Tango F1 was obtained with 200 kg N/ ha applied 50 kg/ ha as pre-plant and 150 kg/ ha supplied as fertigation through surface drip lines. Drip irrigation and N fertilization methods had similar effect on the chemical composition of leaf stalks of both cultivars. The highest ascorbic acid level was observed in plants fertilized with 100 kg N/ha, irrespective of application method. β-carotene level did not depend on N fertilizer application.

Cardamom-growing soils are acidic, rich in organic matter, low in available phosphorus, low to medium in available nitrogen, high in potassium and adequate in available micronutrients, except zinc and boron. Low fertilizer use, low fertilizer use efficiency and micronutrient deficiency, particularly zinc and boron, are the major reasons for the low productivity. The fertilizer dose of 75:75:150 kg of N:P:K/ha is recommended for producing 100 kg dry cardamom capsule/ha, and an additional fertilizer dose of 0.65 kg N/ha, 0.65 kg P/ha and 1.3 kg K/ha is to be applied for every increase in yield of 2.5 kg of capsules. Supply of zinc and boron along with major nutrients as a soil or foliar application will increase the cardamom yield. A fertilizer combination of 150:75:300 kg N:P:K/ha is recommended for high-density planting with 5000 plants/ha (Korikanthimath et al., 2001b).

An experiment was conducted by Madhaiyan et al. (2001) to study the effect of orchid mycorrhizal fungi on growth and nutrient status of *V. planifolia*. Inoculation of orchid mycorrhizal fungus, *Rhizoctonia solani* MSK-01, was found to be superior in increasing the colonization percentage, followed by *Rhizoctonia repens* MSK-02, than uninoculated control. In addition, significant increases in total carbohydrate, starch content, plant nutrient content (N, P and K) and phosphatase activity (acid and alkaline phosphatase) due to inoculation of *R. solani* MSK-01 and *R. repens* MSK-02 were observed when compared to other isolates (*Rhizoctonia* spp. (MSK-03 and MSK-04), and strains MSK-05, MSK-06, MSK-07 and MSK-08) or uninoculated control.

Elkner and Kaniszewski (2001) compared the effect of drip irrigation and liquid N fertilizer application with that of broadcasting N on yield and quality factors of celery (*A. graveolens* L. var. Dulce Mill/Pers.) in Poland. Drip irrigation significantly reduced nitrates and insoluble dietary fibre content but did not influence ascorbic acid and β-carotene contents. The highest ascorbic acid level was observed in plants fertilized with 100 kg N/ha, irrespective of application method. β-carotene level did not depend on N fertilizer application. N fertilizer application affected leaf stalk colour.

The effects of N fertilizer rate on the yield of Japanese mint (*M. arvensis* cultivar MAS-1) were studied by Vikrant et al. (2004). The oil yield increased by 24.22, 65.20, 118.52 and 152.46 kg/ha with the application of 40, 80, 120 and 160 kg N/ha, respectively. The application of 160 kg N/ha and harvesting at 130 DAP gave the highest oil yield (149.59 kg/ha), oil value (Rs. 59,836/ha), profit (Rs. 43,236/ha), BCR (3.60) and link relative index (3.42).

A greenhouse study was conducted by Chenard et al. (2005) to measure the effects of different concentrations of N on growth, elemental accumulation and carotenoid production in parsley cultivar Dark Green Italian. Increasing N in the nutrient solution increased plant biomass, leaf tissue N, phosphorus, potassium, lutein–zeaxanthin, β-carotene, and chlorophyll. Increasing the elemental and carotenoid concentrations in parsley through N fertilizer modification would be expected to increase the nutritional value of this culinary herbal crop.

Xiong et al. (2005) determined the effects of a self-developed coated urea on uptake and balance of nitrogen and on the yield and quality of the crop. The coated urea contained 41.0% nitrogen (N). Compared with the control (urea), coated urea application increased celery yield by 11.5–15.2%, N uptake by the plants by 5.9–9.5%, N use efficiency by 19.2–27.1% and soil-retained N by 32.0–37.3%, and decreased N loss through ammonia volatilization and leaching and denitrification by 14.2–14.9% and by over 200%, respectively.

Cheng-Hua et al. (2006) conducted trials to find out the effects of nitrogen depot with saturated ammonium (NDSA) on N uptake, yield and quality of celery. Compared with the conventional fertilization practice, NDSA in combination with dicyandiamide (DCD) application significantly increased the biomass of celery while the effect of NDSA or DCD alone was non-significant statistically.

Changes in essential nutrients (total NPK) and metabolites (total soluble sugars and free amino acids) were studied by Puthur and Kumar (2006) in the plant parts of *V. planifolia* associated with flowering, shooting and that devoid of any of these. Along with flowering, an increase in total NPK and metabolites was noticed in different plant parts such as inflorescence peduncle, node-bearing inflorescence, node-bearing shoot, as compared to node devoid of inflorescence or shoot. Phosphorus (P) and potassium (K) were found to be required in higher percentage as compared to nitrogen (N) and the influence of sugars was more prominent than free amino acids for flowering in *V. planifolia*.

Nitrogen rates had a significant effect on nitrogen, ascorbic acid and essential oil accumulation in leaf and stalk of parsley during the harvesting times. The highest values except essential oil were observed with the highest nitrogen rate, that is, 300 kg N/ha treatment (Ceylan et al., 2006).

Another research was aimed at evaluating the production of leaves' phytomass, the content and quality of *Mentha* x *gracilis* Sole essential oil with four potassium (K) levels under hydroponics, on nutrient film technique (NFT) system. The K concentrations affected leaves fresh phytomass, content and quality of the essential oil. The K level maximum showed increase of essential oil yield, but decreased the fresh phytomass accumulation, decreasing the oil yield by plant and linalool quantity, the main constituent of this chemotype. The K concentration suggested for the greatest yield of *Mentha* x *gracilis*, under hydroponics, does not surpass 276 mg/l in nutritive solution (Garlet et al., 2007).

By means of quadratic general rotational combination design, the coupling effects of water–fertilizer (N and K) on celery was studied by Chen et al. (2008), and three water–fertilizer coupling regression models demonstrating the relationship between yield and factor were established. The accuracy and reliability of the simulation model were verified by an experiment conducted under greenhouse conditions on celery.

Golcz and Bosiacki (2008) determined the effect of nitrogen fertilization doses and mycorrhization on the yield and essential oil content in thyme (*T. vulgaris* L.). Nitrogen fertilization exerted a significant effect on yield of thyme raw material and on the content of essential oils. On the other hand, substrate mycorrhization brought a positive effect only at low nitrogen content in the substrate.

In order to determine the optimum N level for oil production by parsley, Petropoulos et al. (2009) conducted a research study. The concentration of essential oils in the roots and leaves of plain leaf parsley and turnip-rooted parsley was not affected by N application, but decreased with increasing N rate in curl leaf parsley. Increasing N application caused a reduction in the percentage of β-phellandrene in the essential oils of parsley leaves. In turnip-rooted parsley, increasing N caused a reduction in the percentage of myristicin and apiole. As these three components of the essential oils contribute to parsley aroma, it may be concluded that although application of N fertilizer leads to higher parsley biomass and oil yield per plant, the essential oil components may change and aroma quality may be affected negatively.

The vegetation experiment was carried out by Golcz and Politycka (2009) in pots, in an unheated greenhouse. It was found that nitrogen, irrespective of its dose, significantly decreased the level of chlorophyll a and chlorophyll b and the iron content. The decreased chlorophyll level in thyme leaves with nitrogen application was positively correlated with iron content. Nitrogen fertilization significantly increased both the yield of herb fresh matter and of the dry raw material.

Macro- and microelement deficiencies in curly parsley (*P. crispum* Mill. spp. crispum) were analysed by Graifenberg et al. (2009) in Central Italy under the plastic greenhouse. The lack of macroelements reduced strongly the fresh biomass production, particularly in P deficiency. In all deficiencies, the leaf yellowing was the symptom with differences for extension and localization only. Glen (2009) studied the influence of foliar fertilization on horseradish (*A. rusticana* Gaertn.) yielding and its infection by *Verticillium dahliae* and *Phoma lingam*. It was found that the application of foliar fertilizers in horseradish cultivation contributes both to the increase in marketable and total yield of roots.

Effect of irrigation intervals, nitrogen sources and nitrogen levels on some characters of parsley (*P. crispum* Mill) was studied by Mirdad (2011). The obtained results revealed significant effects of the irrigation intervals, nitrogen sources and nitrogen levels on vegetative growth and yield characters during the two seasons. Kolota (2011) analysed the yield and quality of leafy parsley as affected by the nitrogen fertilization. The most favourable method of leafy parsley nutrition was the application of 80 kg N/ha before seedlings transplanting, and the same amount in two top dressing doses used directly after subsequent harvests. Gradual decrease of DM, vitamin C, and nitrates and the increment of total N contents were observed at the subsequent harvests of leaves during the growing period.

Khalid et al. (2012) carried out an experiment to study the effect of different levels of NP fertilizers, trace elements and their interactions on the morphological and biochemical contents of these three plants under arid regions conditions. The effects of NP and trace elements on the growth (height, leaf number, branch number, umbel number, fresh weight, dry weight and fruit yield per plant) were measured and quantitative analysis of essential oils, fixed oil, total carbohydrates, soluble sugars and nutrient content of anise, coriander and sweet fennel were performed. A greenhouse experiment was conducted by Sharafzadeh (2011a) to evaluate the effect of nutrients on pigment contents of thyme. The treatments were using N, P, K, NP, NK, PK, NPK and Agrihansa (a complete fertilizer, 20–20–20) and comparing them to control (without using fertilizers). The results indicated that maximum level of chlorophyll a, chlorophyll b and carotenoids was achieved on Agrihansa treatment.

Sharafzadeh (2011b) also compared the essential oil yield and components in two parts of garden thyme shoot. The above sections of shoots were divided in two 10 cm parts (top and bottom). Hydrodistillation was used to isolate the essential oils and chemical analyses were performed by gas chromatography (GC) and GC-MS with four replications. The yield of essential oils extracted from top and bottom parts of shoots were 0.87 and 0.75%, respectively. Thirty-three components were identified. The major components of essential oils in top part were thymol (73.54%), carvacrol (4.03%), p-cymene (3.32%), terpinolene (3.13%), β-caryophyllene (2.79%), γ-terpinene (2.22%), and linalool (1.45%). Similarly, the major components of essential oils in bottom part were thymol (60.77%), p-cymene (12.22%), γ-terpinene (4.25%), carvacrol (3.70%), terpinolene (3.13%), linalool (2.44%) and β-caryophyllene (2.24%). Klc et al. (2012) examined the effects of potassium and phosphorus fertilization on green herb yield and some quality traits of *T. vulgaris* L. (garden thyme). The yield generally increased in accordance with the increases in K and P fertilizer rates. The N and P contents of the herb were measured higher under P-rich conditions. The highest three volatile oil constituents of the herb were found to be thymol, p-cymen and carvacrol.

A greenhouse experiment was designed by Kavvadias et al. (2012) to determine the effects of soil amended with cadmium and lead on growth, yield, and metal accumulation and distribution in parsley. The main soil properties; concentrations of the diethylenetriaminepentaacetic acid (DTPA)-extractable metals lead (Pb), Cd, copper (Cu), iron (Fe), zinc (Zn) and manganese (Mn) in soil; plant growth and total contents of metals in shoots and roots were measured. The DTPA-extractable Cd was increased significantly by the addition of Cd. Available soil Pb was increased significantly with Pb levels. The interactive effects of Pb and Cd on their availability in soil and plants and their relation to other metals are also discussed.

Gao et al. (2013) evaluated the effects of foliar fertilizer on yield and quality of celery seed. Foliar spraying during the secondary inflorescence flowering stage could not only increase celery seed yield but also improve 1000-grain weight and enhance germination ability, thus to effectively improve seed quality.

A study was conducted with the hypothesis that vanilla plant nutrient status and growth depends on the substrate composition and fertilization applied. Results indicated that there was a highly significant ($P \leq 0.01$) interaction between substrate and fertilization on plant growth. The presence of coconut fibre produced foliar contents of P, K, Cu, Mg, and Mn significantly higher ($P \leq 0.05$) than those obtained when woodchips were used in the substrate. On the contrary, the presence of woodchips in the substrate significantly ($P \leq 0.05$) increased foliar N and Ca contents in respect to those levels observed with coconut fibre in the growth substrate (Osorio et al., 2014).

Tesfaendrias et al. (2014) studied on the influence of nitrogen and calcium fertilizers on Septoria late blight and yield of celery. Late blight (*Septoria apiicola*) reduces both yield and quality of fresh market and processing celery. The application of both N and Ca can improve yield and also contribute indirectly to improved late blight management in combination with fungicides and other integrated crop management practices. A greenhouse experiment was conducted by Vakili and Sharafzadeh (2014) to evaluate the effects of nitrogen source and level on growth characteristics, volatile oil percentage and oil yield of garden thyme in Iran. The results indicated that the source and level of nitrogen fertilizer altered growth characteristics, oil percentage and oil yield of garden thyme significantly.

Yadegari (2014) worked on the effect of foliar application of micronutrients on growth, yield and essential oil content of thyme. The results showed that applications of micronutrients could affect the growth and yield of thyme, especially when the plant was grown in alkaline soils, and in this, the 200 ppm concentrations of micronutrients were the best. Perhaps, the physiological basis of this effect was immobilization of micronutrients in this soil.

4.6.2 ORGANICS AND BIOFERTILIZER

Experiments were conducted in Karnataka, India, to study the effect of different types of compost applied to the planting hole, on growth and yield of cardamom grown as an intercrop in an areca nut plantation. Coir pith compost prepared with the additive urea, or coir pith compost with farm yard manure (FYM) or coffee cherry husk compost in combination (1:1) showed promising results with regard to sucker and panicle numbers. Coir pith compost prepared with neem cake alone or with urea or ammonium sulphate showed a low response, compared with the other treatments (Moorthy et al., 1997).

Ram and Kumar (1997) investigated on yield improvement in the regenerated and transplanted mint *M. arvensis* (cultivar Hy 77) by recycling the organic wastes and manures. It is recommended that under subtropical conditions, citronella distillation waste mulch should be applied after the sprouting of suckers during the planted crop period to obtain higher yields of herb and essential oil from the planted as well as the regenerated crop harvests. The use of citronella-waste mulch with 160 kg N/ha minimized the cost of transplanted mint production by a factor equivalent to one-third.

Varanashi Farms, a certified organic farm situated in coastal Karnataka, India, is successfully growing vanilla with high productivity in an eco-friendly mixed farming system: with cashew nut, areca nut and small wild tree species; or with *Gliricidia* (in the case of coconut or teak). The farm has recorded an average yield of 265 g dry beans/support plant. Minimum pest and disease problems are encountered in the farm. The paper described the cultivation practices evolved and adopted in the farm (Moorthy and Moorthy, 2004).

A field study was conducted by Madaiah et al. (2006) with an objective of utilizing coffee pulp compost and testing efficiency of biofertilizers along with compost on bean yield of vanilla. The pooled results of 2 years indicated that application of coffee pulp compost along with *Azotobacter chroococcum* and *Aspergillus awamori* resulted in maximum bean yield (1.98 kg beans/vine) followed by the recommended package (1.70 kg beans/vine). The application of coffee pulp compost either alone or in conjunction with P-solubilizer recorded significantly higher bean yield than that of application of FYM alone or in conjunction with P-solubilizer.

Wei et al. (2009) determined the influence of wood vinegar as leaves fertilizer on yield and quality of celery. The results showed that its application increased the yield and quality of celery. Spraying wood vinegar decreased the chlorophyll content and did not increase the nitrate content in celery. The application of compost to agricultural land can benefit the low-fertile sandy soils in Florida and subsequent crop production, while providing an outlet for recycling municipal solid wastes and biosolids. In this regard, Mylavarapu and Zinati (2009) investigated on the improvement of soil properties using compost for optimum parsley production in sandy soils. It was found that the addition of compost resulted in improvement of both physical and chemical properties as well as increased parsley yields.

Shaddad et al. (2009) found out the effect of solarization and organic fertilizer on yield and quality of rocket and parsley fresh herbs. Solarization for 6 weeks reduced significantly total weeds/m^2 after 21 days from rocket and parsley seed sowing compared with non-solarized treatment in addition to increment in yield and quality. Organic fertilizer showed a significant decrease in total weeds/m^2 and significant increase in yield and quality. It can be generally concluded that solarization for 6 weeks after adding organic fertilizer is a successful method to control weeds without using pesticides and increasing yield and keeping quality for rocket and parsley fresh herbs.

Ateia et al. (2009) studied on the effect of organic fertilization on yield and active constituents of *T. vulgaris* L. under North Sinai conditions, Egypt. Results revealed that (20 m^3 compost combined with 10 m^3 chicken or sheep manure) were superior in most cases of growth characters and yields. Moreover, they are leading in oil percentage and oil yield. The highest value for oxygenated compounds, especially thymol, was obtained from 30 m^3 compost combined with 10 m^3 sheep manure treatment (82.84%) compared with the control (42.69%).

Mini-Abraham et al. (2010) assessed the effect of organic nutrition on disease tolerance in vanilla (*V. planifolia*). The effects of organic and inorganic fertilizers (IFs) on the performance of *V. planifolia* were studied in Kerala, India. Plants treated with organic fertilizers had greater average vine length (11 m) and yield (26.0 beans) than plants treated with IFs (9.60 m and 15.40 beans, respectively). Plants supplied with organic fertilizers also showed less incidence of fungal rot caused by *Fusarium*, *Phytophthora* and *Colletotrichum* spp. (10% of the plants were infected) than plants treated with IFs (37.5% of the plants were infected).

A study conducted by Chagas et al. (2011) aimed to evaluate the effects of organic fertilization at planting and dressing on the above-ground part dry biomass and on the content and yield of essential oil of mint (*M. arvensis*). The levels of cattle manure applied during planting and dressing affected in a positive linear way the above-ground part dry biomass production and the essential oil yield. Essential oil content was not affected by different OM levels.

A 4-year study by Filipovic et al. (2012) confirmed that the lowest root yield was found in the control treatment (23.22 t/ha), while the highest was with 1000 kg/ha of Siforga (37.03 t/ha), a certified organic fertilizer. The best market quality of parsley root, that is, the largest portion of first-class roots had 3500 kg/ha rate of Siforga, whereas the poorest quality was obtained at 500 kg/ha rate of Siforga (75.11%).

Sandheep et al. (2013a) demonstrated the results of combined inoculation of *Pseudomonas fluorescens* and *T. harzianum* for enhancing plant growth of vanilla (*V. planifolia*). Combined inoculation of these strains registered the maximum length of vine, highest number of leaves, highest fresh and dry weight of shoots, fresh weight of roots, whereas the highest dry weight of roots was achieved with treatments of *P. fluorescens* alone. Among the inoculated strains, combined inoculation recorded the maximum nitrogen uptake and the highest phosphorus uptake.

Singh et al. (2013) explored the beneficial role of *Bacillus circulans*, a bacterial fertilizer on yield, quality and economics of aniseed (*P. anisum*). The main objective was to increase the quantitative and qualitative production of aniseed on riverine soils of Uttar Pradesh through the integration of bacterial fertilizer with recommended dose of NPK. The instalment of *B. circulans* containing biofertilizer more than 20 kg/ha confined to the further progress in aniseed production. The marketable bright green colour and the chewing taste of aniseed kernels obtained from bacterial fertilizers-treated plots were better than the control plots product.

Tanwar et al. (2013) determined the potential of arbuscular mycorrhizal fungi (AMF) used alone or in combination with the bacterium *P. fluorescens* and/or the nonmycorrhizal fungus *Trichoderma viride* Pers. on celery in a greenhouse. It was found that AMF, or other growth-promoting microbes, for soil inoculation improved establishment and growth of celery transplants. The effects of AMF (*Glomus intraradices*) and vermicompost application on yield and nutrient uptake in garden thyme (*T. vulgaris* L.) were evaluated by Khorshidi et al. (2013). The highest dry yield of thyme was obtained with a combination of mycorrhiza and vermicompost compared with control treatment. There was a positive and synergistic interaction between *G. intraradices* inoculums and vermicompost treatments on Zn uptake in thyme leaves. In vermicompost treatments, Mn and N concentration in leaves of thyme was higher than other treatments.

Mahesh et al. (2014) evaluated the effectiveness of various microbial inoculants on large cardamom suckers. The study demonstrated the beneficial effect of microbial inoculants, namely, *Azospirillum*, *P. fluorescens*, *T. harzianum*, phosphate-solubilizing bacteria (PSBs) and AMF on plant growth. The application of FYM + *Azospirillum* + *P. fluorescens* + *T. harzianum* + AMF recorded significantly higher number of tillers and plant height as compared to other treatments.

4.6.3 COMBINATION OF ORGANIC MANURE AND INORGANIC FERTILIZER

Thimmarayappa, et al. (2000) quantified the effects of OM and IFs on growth, yield attributes and yield of cardamom (*E. cardamomum*) cultivar Mudigere-1. Plants treated with 75% OM + 25% IF were the tallest (225.75 cm) and had the highest number of capsules per clump (342), whereas those treated with 50% OM + 50% IF had the largest leaf area per clump (40,732.75 cm^2) and leaf area index (1.25). Plants treated with 100% IF had the greatest number of bearing suckers per clump (9.96) and of panicles per clump (23.15), and green (707 kg/ha) and dry (184 kg/ha) yields. Treatment with 100% IF gave the highest total (Rs. 5592.00) and net (Rs. 41,637.79) returns, and BCR (4.19).

Kincses et al. (2008) investigated on the effect of nitrogen fertilization and biofertilization on element content of parsley. The N fertilization increased the N content of parsley, however, no relationship was found between N content and biofertilization rate in chernozem soil. The P content of parsley was enhanced by both N fertilization and biofertilization on the chernozem soil.

Gavazzi et al. (2009) compared conventional to organic field production of horticultural crops (celery) in Italy. The crops were given the equivalent of 120 kg N/ha; an organic fertilizer was ploughed in prior to transplanting in the organic plots, while two doses of calcium nitrate were applied in the conventional plots. Yields obtained in the two systems were not significantly different. Contents of nitrates depended on many interacting factors, such as fertilizer dose and type, the period and duration of production, crop species and cultivar. However, the amount of nitrates in both crops was always lower in the organic system, but not always significantly so.

According to Siddappa and Hegde (2011), foliar spray of vermiwash produced vigorous growth with significantly higher fresh leaf yield (13.07 t/ha) compared to control (11.13 t/ha) in curry leaf cultivar Suvasini. Among nutritional treatments, recommended dose of fertilizer (RDF) + FYM (10.00 kg/plant) with vermiwash foliar spray at 50% dilution recorded higher fresh leaf yield (17.74 t/ha) followed by FYM + RDF without vermiwash (15.79 t/ha) and FYM along with vermiwash (15.65 t/ha) compared with control (6.32 t/ha).

Hegde et al. (2012) reported on the response of curry leaf (*M. koenigii* Spreng) Suvasini for foliar spray of vermiwash and nutritional treatments.

Vermiwash obtained from earthworm bed is a foliar spray solution with many growth-regulating substances and beneficial microbes. Significantly higher mean annual fresh leaf yield of 5.79 kg/plant was obtained by foliar spray of vermiwash at 50% dilution compared with 5.05 kg in control. Among nutritional treatments, the RDF + FYM with vermiwash foliar spray at 50% dilution, recorded higher annual fresh leaf yield (8.00 kg/plant) followed by FYM + RDF without foliar spray of vermiwash (6.95 kg/plant) and FYM along with vermiwash (6.60 kg/plant), compared with control (2.92 kg/plant). Foliar spray of vermiwash at monthly intervals also recorded slightly higher essential oil content in fresh leaves (0.44%), compared with control (0.42%).

The effects of bio- and chemical fertilizers on essence, dry and fresh weight production in thyme (*T. vulgaris*) was evaluated in Iran. Significant differences were observed in dry and fresh weight, essential oil production and number of lateral shoots while soils were fortified with one of the chemical or biofertilizers or combination of both. The application of biofertilizers (PSB, nitroxin and frowzy manure) caused the highest biomass and essence production. It was found that soil fortification with biofertilizers make more plant production comparing to usage of chemical fertilizer (Yadegari, et al., 2012).

Yu et al. (2012) investigated on the effects of reduction and optimization of nitrogen fertilizer application on the yield and quality of celery and soil nitrate leaching under greenhouse condition. There was no significant difference in celery yields between the treatments under fertilization reduction with organic and IFs combined and the conventional treatment, and the highest yield was obtained with 10% reduction of chemical fertilizer with fungus dregs, at 121,192.2 kg/ha. Under the experimental conditions, the pattern of 20% reduction of nitrogen fertilizer with fungus dregs was observed to be the best. Osorio et al. (2012) hypothesized that vanilla plant nutrient uptake and growth depend on the substrate composition, fertilization dose and biofertilization. The results indicated that there was a significant ($P \leq 0.01$) interaction between substrate composition and fertilization on plant growth.

Juarez-Rosete et al. (2014a) conducted a greenhouse experiment with thyme to find out the mineral and organic nutrition in biomass production and quality of essential oils. The results showed that inorganic fertilization increased plant height by 36.8%, fresh shoot biomass by 72.19%, fresh root biomass by 59.27%, stem diameter by 12.15% and dry shoot and root biomass by 69.85 and 68.15%, respectively. Mineral fertilizing increased the concentration (μg/ml) of thymol (46%) and carvacrol (38.4%) at 90 days to harvest.

In a similar experiment, Juarez-Rosete et al. (2014b) measured the biomass production as well as qualitative and quantitative properties of the principal components of the essential oil of thyme as affected by the fertilizer source. The results showed that fertilizer source modified fresh and dry biomass production in thyme plants. The total yield of essential oils was not affected by days to harvest and fertilizer source. However, essential oil quality was higher in the mineral treatment at 90 days to harvest due to the content of thymol and carvacrol in the extract.

4.7 PLANT GROWTH REGULATORS

Shivakumar et al. (2008) studied the effect of plant growth regulators (PGRs) on growth and development of vanilla (*V. planifolia*) fruits. Growth regulators were applied to the pods after 10 days of completion of hand pollination in all the flowers in a bunch. The results revealed that application of GA3 (20 ppm) alone or in combination with NAA (2 ppm) and BAP (0.2 ppm) influenced length, girth and fresh weight of fruits compared with control treatment.

An investigation on the effect of mulching, irrigation, water spray, biostimulants, nutrients and growth regulators on growth and productivity was conducted by Subha et al. (2010) during winter season in comparison with monsoon season. It could be concluded that the fresh curry leaf yield and quality characters could be improved by foliar spray of 3% *panchagavya* at 30 days intervals for 2 times per season was found to be economical.

Another study showed that the best quantitative parameters of thyme raw material were observed after the application of the growth simulators. The root forecrop (carrot) also had a beneficial effect on thyme yield. Thyme crop protection without the application of the growth simulators resulted in the deterioration of the biometric traits and yields of this crop (Kwiatkowski, 2011).

4.8 WATER MANAGEMENT

In a microplot experiment, root celery was irrigated with simulated sea water, tap water or diluted sea water (1:1) by Rumasz et al. (1999) while studying the influence of saline water irrigation on celery yield. Irrigation increased root yield by an average of 38%. Diluted sea water was the most effective increasing yields by 52% compared with 35% for tap water and 25% for sea water. The effect of various salinity of irrigation water on the onion and celery yield was estimated by Rumasz-Rudnicka et al. (2005). The average yield increase caused by irrigation was 55.8% for celery and only 3.9% for onion. Tap water caused the highest increment (15.4%) of onion yields. Celery responded best to diluted saline water which increased the yield by 72.8%, tap water by 53% and saline water by 41% as compared with the nonirrigated control.

Eissa et al. (2004) investigated on the effect of saline water irrigation on the DM production, nutrition status and essential oil content of thyme plants (*T. vulgaris* L.) grown on a sandy soil. Saline water irrigation treatments were added at (0, 1500, 3000, 4500, 6000 ppm as NaCl) and phosphorus was applied at three levels (0, 50, 100 ppm as KH_2PO_4). The highest dry weight and N, P, K, Ca, Fe, Mn and Cu uptake were recorded at treatment with tap water irrigation and 50 ppm phosphorus application while the highest N, P, K, Zn and Cu content was recorded with tap water irrigation with 100 ppm phosphorus application. On the contrary, the highest Ca content was obtained at 3000 ppm NaCl with 50 ppm phosphorus application. The highest Fe content was recorded at treatment of 4500 ppm NaCl and 50 or 100 ppm phosphorus application. Mn content was recorded at treatment of 3000 ppm NaCl saline water irrigation and 50 ppm phosphorus application.

According to Rozek (2007), irrigation increased leaf yield of celery (*A. graveolens* L. var. Secalinum), on average, by 55.9%. Depending on the cultivar and planting density, leaf yield of nonirrigated plants was 5.29–6.34 kg/m^2, while for the irrigated plants, it ranged from 7.70 to 11.97 kg/m^2.

The effect of irrigation frequency on herbage biomass and oil production of thyme (*T. vulgaris*) and hyssop (*Hyssopus officinalis*) was determined by Khazaie et al. (2008) in Iran. Irrigation intervals did not change total harvested herbage biomass and oil production of both crops. However, there is a high potential for saving water through longer irrigation intervals (e.g. 14 days) using locally adapted plants in the semiarid conditions of Khorasan. These crops serve as alternative sources of income in dry years.

The effect of drip irrigation and cultivation methods on the yield and quality of parsley roots was studied by Kaniszewski and Dysko (2008).

Irrigation water was supplied through drip lines, which in subsurface irrigation were placed at a depth of 50 mm below the surface of the ridges, along the centre line between two rows of plants. In the case of surface irrigation, the drip lines were placed on the surface of the ridges between two rows of plants. Irrigation started when soil water potential was between −30 and −40 kPa. Both surface and subsurface irrigation used in the cultivation on ridges and on flat ground had a significant effect on the marketable yield of parsley roots. Parsley plants cultivated on ridges produced significantly longer, better-shaped storage roots compared with those cultivated on flat ground.

Hamed (2009) estimated the effect of date palm seeds with soil on the growth of parsley growing under drought stress. Parsley was cultivated in a soil incorporated with 3% (w/w) fine-crushed date palm seeds and irrigated either with 100 or 60% field capacity (FC). Incorporation at 100% FC led to increases in soluble sugars, soluble starch, nonstructural carbohydrates, protein, free amino acids, total nitrogen, photosynthetic pigments and photosynthetic efficiency, vitamin A, vitamin B complex (B_1, B_2, B_3, B_6), vitamin C as well as total oil, total lipids and the concentration of the estimated elements (N, P, K, Ca, Mg, Fe, Mn and S). Drought led to an accumulation in soluble carbohydrate, soluble starch, proline, free amino acids and the estimated elements but retarded the other investigated parameters.

To evaluate the effect of the irrigation system on fruit retention of *V. planifolia*, Castro-Bobadilla et al. (2011) conducted an experiment in Mexico. Irrigation was provided daily at 18:00 h during the dry season (May to June) in four treatments: (1) 1.0 l/m^2 of water, (2) 0.5 l/m^2, (3) 0.25 l/m^2 and (4) a control group without irrigation. The orchard with semi-technical management shows retention of 69.5% of fruits, attributable to irrigation and cropping system characteristics. The combination of soil moisture and management of fields of vanilla crops seem to have a strong influence on fruit retention, improving the condition of the crop.

Mameli et al. (2011) estimated the effects of different irrigation management on biomass and essential oil production of *T. vulgaris* L., *Salvia officinalis* L. and *Rosmarinus officinalis* L., cultivated in the Southern Sardinian climate (Italy). The results showed that different water-saving irrigation strategies could have a strong significant influence and improve the fresh and dry biomass of the three crops. The total oil content production of all the three crops was affected by irrigation treatments.

Keser and Buyuk (2012) carried out a research work to assess the effects of wastewater irrigation on chemical and physical properties of *P. crispum*. The application of waste water significantly decreased DM, while photosynthetic pigment content increased in parsley. Overall, the results indicated that municipal wastewater is not suitable for irrigation of parsley because it has negative effects on plant and causes heavy metal accumulation.

Najla et al. (2012) quantified the morphological and biochemical changes in two parsley varieties upon water stress. Besides the control, three irrigation treatments (10, 30 and 50% of control water amount) were applied. The responses of plant yield and quality varied according to the hybrid and water treatment. The high quality was observed in curly-leafed parsley if compared with plain-leafed parsley, which had higher yield (stem length and diameter, leaf area). Moreover, water stress improved the quality but decreased yield parameters in both hybrids.

Periodic irrigation and fertilization are important factors that influence the quantity and quality for aromatic plants. Qualitative and quantitative changes in the essence of thyme have been studied by Saffari et al. (2013) under different levels of irrigation and phosphorus level. The irrigation level of 5 days produced highest fresh and dry substance yield, essence and thyme and highest essence percentage in irrigation level of three, while phosphorus had a significant effect on dry and caroacrol yield in level three.

Harizanova-Petrova and Ovcharova (2013) determined the influence of irrigation regime on the yield of root celery by drip irrigation. The experiment was carried out in six variants: (1) irrigation with 130% of the optimal irrigation application, (2) optimal irrigation/100% of the calculated irrigation application by soil moisture before irrigation up to 80% of FC in the depth 0–0.40 m, (3) irrigation with 70% of the optimal irrigation application, (4) irrigation with 50% of the optimal irrigation application, (5) irrigation with 30% of the optimal irrigation application and (6) without irrigation. It was found that reducing irrigation rate leads to reduced yields in third, fourth, fifth and sixth variants, but in variant 1 (irrigation with 130% of the optimal irrigation application), the yield was greater than the optimal irrigation.

Phytochemical screening of *S. officinalis* and *T. vulgaris* irrigated with different levels of saline water (fresh water, electrical conductivity [EC] = 1.04 dS/m; saline water, EC = 7.3 dS/m; 50% saline water) was done by

El-Sakka and Tubail (2003). Photochemical and spectral analysis showed increased sterols and volatile oils in sage and saponins in thyme with an increase in salinity levels. No significant differences were observed in the other tested substances with the different treatments.

4.9 WEEDS AND OTHER MANAGEMENT

Weeds are one of the major limiting factors for economically viable spice production. Ogbuchiekwe and McGiffen (2001) analysed the efficacy and economic value of weed control for drip- and sprinkler-irrigated celery in California, USA. The nine treatments included an untreated control, cultivation as needed for weed control, a pre-emergent herbicide (trifluralin) and post-emergent herbicides (glyphosate, prometryn, linuron and sethoxydim). The treatments that reduced weed populations under drip and sprinkler irrigation also increased yield, net returns and rate of returns.

A field experiment was carried out in Poland by Kwiatkowski (2007) to study on weed infestation and yielding of garden thyme (*T. vulgaris* L.) in relation to protection method and forecrop. The complex mechanical–chemical protective operations resulted in the greatest weed reduction. Herbicide Goltix 700 SC used after emergence of thyme showed good selectivity in the crops of this plant. Spring barley turned out to be the worst forecrop, limiting the herbage yield and increasing canopy weed infestation.

Efficacy of weed control practices was also analysed by Kaur and Gill (2011) in celery crop production. All the herbicides and hand-weeding treatments significantly reduced total weed population and their dry weight when compared with unweeded/control. An integration of one hand weeding (40 days after transplanting) with the lowest dose of each herbicide, that is, pendimethalin (0.5 kg/ha, trifluralin (0.75 kg/ha) and oxyfluorfen (0.15 kg/ha) proved very effective.

The critical period for weed competition in parsley (*P. crispum* (Mill.) Nyman ex A. W. Hill) in Mediterranean areas was determined by Karkanis et al. (2012). *Amaranthus retroflexus* L., *Datura stramonium* L. and *Solanum nigrum* L. contributed most of the total weed dry weight because of their large size and density. Total dry biomass of weeds increased as the duration of weed infestation increased. Moreover, weed competition reduced parsley dry weight by 93 and 95% in 2 years, respectively. There was a significant negative correlation between the dry weight of parsley and weed dry weight. Results indicated that early weed removal is necessary to prevent yield loss.

Rahimi et al. (2014) presented a floristic evaluation of saffron field's weed and ways to improve their control. Weed problem is very severe to saffron, because the crop is perennial and having very low vegetative growth in the first year. More than 25% of saffron production is special to Iran, so it seems that dominant weed identification in these communities has a special importance and in the article ways to weeds control were evaluated.

4.10 INTERCROPPING

Korikanthimath et al. (1999) investigated on the feasibility of vanilla cultivation with coconut at Karnataka, India. The financial feasibility measures indicated the viability of mixed cropping of vanilla in the coconut gardens in lower elevations (plains) and low rainfall areas under irrigation. The studies on intercropping of parsley and demsisa with tomato under different rates of nitrogen fertilization were conducted by Awad et al. (2001) in Egypt. Intercropping system reduced net assimilation rate, total yield and N, P and K uptake by tomato and companion crops, while earliness index of tomato was not significantly affected. Intercropping demsisa on tomato plants increased N, P and K requirements for the mixture crops, and vice versa for parsley. Demsisa was more aggressive for tomato plants than parsley. Demsisa was more aggressive for Castle-Rock than the hybrid E-448.

According to Ruf et al. (2004) vanilla was adopted in an already-existing coffee farming system in Indonesia. The coffee was shaded by *Gliricidia*, which is being used by vanilla creepers as stakes. Vanilla development on the island of Bali was a direct and easy source of information for Balinese migrants based in Sumatra. Vanilla adoption illustrates how agroforestry can make tropical agriculture more sustainable by enabling its rapid adaptation to ecological and market changes through diversification and replanting.

Santosa et al. (2005) opined on the profitability of vanilla intercropping in pine forests in West Java, Indonesia. The profitability of intercropping with vanilla was slightly higher than that of intercropping with either cash crops or banana. The income level of the farmers was positively correlated with the size of the vanilla-intercropped areas, but negatively correlated with the size of paddy fields. The present study showed that vanilla cultivation is effective for increasing the income level of the farmers who own few or do not own paddy fields.

Response of vanilla (*V. planifolia* A.) intercropped in areca nut to irrigation and nutrition in humid tropics of India was assessed by Sujatha and Bhat (2010) on a laterite soil. They concluded that vanilla responded well to irrigation and nutrition in areca nut-based cropping system with a better economic output and improved soil fertility.

4.11 PLANT PROTECTION

4.11.1 DISEASES

For some years in the Ligurian coastal area of Italy and in Southern France, cultivation of aromatic pot plants (rosemary, salvia, thyme, mint, etc.) has intensified. Among these species, rosemary in pots represents 40% of the entire production. Attacks of rosemary by a powdery mildew (*Oidium* sp.) were repeatedly observed in the Albenga area (Savona). The symptoms of damage are described. Severe attacks led to the death of stems. Plants grown in the shade were particularly susceptible to attack. The perfect stage of the pathogen was never observed. Control measures are suggested (Minuto and Garibaldi, 1997).

Thomas and Bhai (2000) first reported about sclerotium rot as a new disease of vanilla (*V. planifolia*) in India. The disease was characterized by rotting of bean bunches and subsequent development of thick fungal mat over the bean surface. The causal organism was identified as *Sclerotium rolfsii*. The fungus was brought into pure culture and its pathogenicity was proved. This is thought to be the first report of the fungus on vanilla from India.

Venugopal and Mathew (2000) worked on integrated management of viral diseases of cardamom. Cardamom (*E. cardamomum*) is affected by three major viral diseases in India: cardamom mosaic virus, cardamom vein-clearing virus and cardamom necrosis virus. Disease spread, the production of virus-free planting material, vector management, resistant sources, removal of virus sources and early detection are also discussed by them.

Bhai and Thomas (2000) first reported on phytophthora rot as a new disease of vanilla in India, which affected beans, leaves and stems of vanilla during the south-west monsoon season in Kerala, India. The causal organism was identified as *Phytophthora meadii* and its pathogenicity was proved.

Doshi and Thakore (2002) demonstrated the efficacy of systemic and non-systemic fungicides (Aliette, Dithane M45 [mancozeb], Dithane Z-78 [zineb], Ridomil MZ [mancozeb + metalaxyl], phytoalexin, Blitox [copper oxychloride] and copper sulphate) for the control of downy mildew of opium poppy. Non-systemic infection was controlled best by Ridomil-ZM (6%) followed by Aliette (0.25%). Spraying of these fungicides three times at the seedling, rosette and elongation stages resulted in maximum increase in latex and seed yields.

Bhai et al. (2003) also reported on mosaic disease of vanilla (*V. planifolia*) characterized by symptoms of mosaic or mottling, leaf distortion, chlorotic streaks and leaf curling as the first report from India.

Nawrocki and Mazur (2004) estimated the effectiveness of some means using against root rot on parsley seedling root. The results of a 2-year study concern the effectiveness of four substances—biopreparate Chitosol (β-1,4-D-glucosamine polymer) and fungicides: Rovral Flo 255 SC (BAS iprodione 255 g/dm^3), Sportak Alpha 380 EC (BAS prochloraz 300 g/dm^3 and carbendazim 80 g/dm^3) and Zaprawa Funaben T (carbendazim 20% + tiuram 45%)—used against root rot and plant rot (*Alternaria*, *Fusarium*, *Phoma*, *Rhizoctonia* and *Sclerotinia*) was described. Among substances used for spring dressing of seedling roots, the best efficacy was exhibited by Zaprawa Funaben T and Sportak Alpha 380 EC.

The occurrence and distribution of viral diseases on vanilla (*V. planifolia*) in India was reported by Bhat et al. (2004). Two viral diseases, namely mosaic and stem necrosis, with an average incidence ranging from 0 to 5% and 0 to 10%, respectively, in various locations were noticed. Electron microscopy of leaf dip preparations of the diseased plants showed three kinds of flexuous particles resembling Potexvirus, Potyvirus and Closterovirus, and an isometric particle. Isoenzyme electrophoresis was used by Li et al. (2005) to analyse nine *Fusarium oxysporum* f.sp. vanillae isolates and three *F. solani* isolates causing root rot in vanilla.

Cucumber mosaic virus (CMV) on vanilla in India was reported by Madhubala et al. (2005) characterized on the basis of biological and coat protein (CP) nucleotide sequence properties. Double-antibody sandwich-enzyme linked immunosorbent assay method was standardized for the detection of CMV infection in vanilla plants. CP gene of the virus was amplified using reverse transcriptase-polymerase chain reaction (RT-PCR), cloned and sequenced. Sequence analyses with other CMV isolates revealed the greatest identity with black pepper isolate of CMV (99%) and the phylogram clearly showed that CMV infecting vanilla belongs to subgroup 1B or one B. This is the first report of the occurrence of CMV on *V. planifolia* from India.

Another new disease characterized by the appearance of small water-soaked spots, which later developed into characteristic brownish sunken lesions on vanilla beans (*V. planifolia*) was noticed in Kozhikode District (Kerala, India) where vanilla is grown as an intercrop with coconut, areca nut and clove. Studies on the causal organism and pathogenicity tests showed that the disease is caused by the fungus *Cylindrocladium quinqueseptatum* (Bhai et al., 2006).

In vitro and *in planta* assays for biological control of *Fusarium* root rot disease of vanilla was confirmed by Bhai et al. (2009). Two isolates of *Trichoderma* sp., one isolate of *Paecilomyces* sp. and two isolates of *P. fluorescens* were found antagonistic to the pathogen. *Paecilomyces* sp. provided 100% protection against root rot. *T. harzianum* and *P. fluorescens* provided 40% protection. Thus, the present study indicated the possibility of using *Paecilomyces* sp. as a potential antagonist for *F. oxysporum* f. sp. vanillae.

Ashoka et al. (2006) presented a new report of root rot of vanilla caused by *Rhizoctonia bataticola*. The symptoms initially appeared in the collar region of the plant and later moved towards the root portion resulting in dry rot. The later stages resulted in shrivelling and shredding of the affected portion. The pathogen was isolated and identified as *R. bataticola*.

Dhanapal et al. (2006) studied the improvement of seedling quality and management of rot diseases of *E. Cardamomum* Maton. (small cardamom) through microbial inoculants. It was observed that combined use of microbial inoculants followed by copper oxychloride spray and soil application resulted in better establishment and growth of cardamom seedlings. The incidence of rot diseases was also reduced in these seedlings.

According to Marthe and Scholze (2006), Alternaria leaf blight is a serious crop disease of parsley. *Alternaria radicina* causes leaf necrosis, which results immediately in a loss of quality. The diagnosis is not easy in the early stages of the disease because of confusion with the important leaf-blight-causing pathogen, *Septoria petroselini*. However, with the progression of the disease, characteristic lesions will develop. They first published the method of *A. radicina* resistance testing in a climate-controlled chamber, which offers the opportunity to commence resistance breeding in parsley.

Fungal diseases infecting *V. planifolia* were studied by Ashoka and Hegde (2006) in Karnataka. *Colletotrichum gloeosporioides* and *S. rolfsii* infected leaves, stems and beans. *F. oxysporum* caused stem, root and shoot tip rot. *R. bataticola* infected only the roots of *V. planifolia.*

Justin et al. (2008) reported on seasonal incidence and recommended for the management of Giant African snail, *Achatina fulica* on vanilla. It was observed that the snail incidence remained low between January and March but increased later to reach a peak at 12.2 snails/m^2 area by July. The population remained steady till October but declined thereafter. The control trial against the snails suggested that spray of garlic extract (2%) or Neemazal (2 ml/l) was found as effective as soil application of phorate (Phorate 5G; 5 g/vine) granules in controlling the snails on vanilla vines.

Bhai and Jithya (2008) reported on the occurrence of fungal diseases in vanilla in Kerala. The disease was found associated with *Colletotrichum vanillae* Massae, predisposed by high temperature and low relative humidity. *Phytophthora*-incited stem and bean infection were noticed in isolated areas in Kozhikode and Idukki districts. The survey at Kozhikode revealed the recurrence of flower shedding and yellowing and premature bean shedding, besides root rot caused by *F. oxysporum* f. sp. vanillae. Survey in Wayanad area revealed the prevalence of stem and root rot incited by *Fusarium* sp. along with leaf axil rot incited by *Colletotrichum* sp. In addition, *R. solani, C. quinqueseptatum* and *Mucor racemosus* were also isolated as pathogens from different symptomatic parts. Other minor infections noticed were leaf rot, bean rot and brown rot. *Fusarium* and *Colletotrichum* spp. were the predominant pathogens found to be associated with most of the infections on vanilla.

Jayasekhar et al. (2008) developed an integrated biocontrol strategy for the management of stem rot disease (*F. oxysporum* esp. vanillae) of vanilla. Under field conditions soil application of Pf (NI) followed by carbendazim spray (0.2%) after 30 days of *Pseudomonas* application recorded the lowest disease incidence of 3.77, 4.03 and 3.67% in 3 years and exerted 79.95, 80.50, 80.48% reduction in disease over the control plot, respectively. Among the bioagents, the soil application of Pf (NI), followed *T. harzianum* after 30 days and another dose of Pf (NI) after

30 days recorded a significant disease reduction of 69.84, 68.24 and 70.21% in 3 consecutive years.

Richard et al. (2009) suggested control of virus diseases in intensively cultivated vanilla plots of French Polynesia. The data confirmed a potential for high incidence of aphid-borne viruses in particular CMV and Watermelon mosaic virus as well as the non-vectored Cymbidium mosaic virus (CymMV). CMV had a particularly high prevalence (over 30% of the plots) and could severely damage up to 50% of the vines before blossom. Severe outbreaks of CMV were correlated to the presence of the weed *Commelina diffusa* as a reservoir of virus and aphid vectors.

Vijayan et al. (2009) worked on management of rot diseases of vanilla using bioagents. The results indicated that *T. harzianum* (cfu × 109/ml) as basal application + *P. fluorescens* as foliar spray and basal drenching (cfu × 109/ml) significantly decreased the incidence of rot diseases in the field. Siddiqui and Meon (2009) isolated and identified the fungi associated with leaf and stem blight of vanilla. Result showed that leaf blight was mainly associated with *C. gloeosporioides*, *Phytophthora* sp., *Curvularia lunata* and stem rot with *F. oxysporum*. Pathogenicity testing on detached leaves revealed that *C. gloeosporioides* and *F. oxysporum* were pathogenic, causing deep sunken spots covered with dense mycelial mat and fruiting bodies.

Talubnak and Soytong (2010) threw light on biological control of vanilla anthracnose using *Emericella nidulans*. *E. nidulans* was isolated from fallen leaves of vanilla and tested against mycelial growth and sporulation of *C. gloeosporioides*. Bi-culture test showed that *E. nidulans* could inhibit mycelial growth and sporulation at 49.44 and 75.31%, respectively. Methanol crude extract inhibited the mycelial growth and sporulation at the concentration of 1000 μg/ml and the effective dose (ED_{50}) were 2910 μg/ml and 0.0001 μg/ml, respectively.

Shahida et al. (2010) studied on the efficacy of native bioagents against *P. meadii* causing phytophthora rot in vanilla and its compatibility with fungicides. In vitro screening of native antagonists against *P. meadii* revealed that *T. harzianum* (VKA) and *Pseudomonas* spp. (MVR) were effective against the pathogen. Among the fungicides tested, *T. harzianum* (VKA) was compatible with mancozeb and potassium phosphonate but sensitive to copper fungicides. However, fluorescent pseudomonad (MVR) was compatible with potassium phosphonate alone. Hence, potassium phosphonate may be combined with both biocontrol agents for the management of phytophthora rot of vanilla.

Pinaria et al. (2010) investigated on vanilla stem rot in Indonesia and found out its association with *Fusarium* spp. Previous reports of vanilla stem rots in the Asia-Pacific region include those caused by *Fusarium, Colletotrichum* and *Phytophthora* spp. In the paper, they reported *Fusarium* sp. (more specifically *F. oxysporum* f. sp. Vanilla) associated with the disease. A total of 542 *Fusarium* isolates were recovered, comprising 12 species, namely *F. decemcellulare, F. fujikuroi, F. graminearum, F. mangiferae, F. napiforme, F. oxysporum, F. polyphialidicum, F. proliferatum, F. pseudocircinatum, F. semitectum, F. solani and F. subglutinans. F. oxysporum* was the most commonly isolated species from all areas surveyed, followed *by F. solani and F. semitectum. F. oxysporum, F. solani and F. semitectum* were tested for pathogenicity to vanilla but only *F. oxysporum* was shown to be pathogenic.

Partial CP sequence analysis of CMV isolates from *V. tahitensis* crops in French Polynesia and *V. planifolia* crops on Reunion Island detected several different sequence variants of CMV isolates that were related to their geographic origin. Experimental inoculation with a New Zealand isolate (NZ100) belonging to the CMV-II subgroup, demonstrated that vanilla is able to be infected by a very diverse range of CMV isolates (Farreyrol, 2010).

A study was carried out on biological control of phytopathogenic fungi of vanilla through lytic action of *Trichoderma* sp. and *P. fluorescens*. The lytic enzymes degrade the cell walls of the pathogenic fungi, enabling *Trichoderma* to utilize both their cell walls and cellular contents for nutrition. The antagonistic effect of *Trichoderma* was studied towards a range of phytopathogenic fungi; *F. oxysporum* f.sp. vanillae, *P. meadii* and *C. vanillae* in vanilla. In vanilla field evaluation with microbial consortia involving *P. fluorescens* strains and *Trichoderma* isolates, the combinations, namely, Pf1 + FP7 + chitin (soil and foliar), Pf1 + Tricho1 + chitin (soil) + Pf1 + chitin (foliar) and Pf1 + Tricho17 + chitin (soil) + Pf1 + chitin (foliar) performed well in reducing the disease incidence (Radjacommare et al., 2010).

A chapter in a book by Hernandez-Hernandez (2011) describes the main diseases of vanilla such as root and stem rot (*F. oxysporum* f.sp. vanillae), black rot (*Phytophthora* sp.), anthracnose (*Colletotrichum* sp.), rust (*Uromyces* sp.), rotting of recently planted cuttings (*F. oxysporum* and *R. solani*), yellowing and shedding of young fruits (*Fusarium incarnatum*-equiseti and *Colletotrichum* sp.), CymMV, Vanilla mosaic virus,

Vanilla Necrosis Potyvirus, Odontoglossum ringspot virus, their damage and control. Damage due to adverse climatic factors, sunburn and hurricanes are also discussed (Hernandez-Hernandez, 2011).

The low productivity of nutmeg (only 0.25 t/ha) at North Maluku in Central Indonesia is caused by high infestation of fruit dry blight (*Stigmina myristicae*) with disease intensity up to 50%. Lala et al. (2011) reported on its control. Technological package introduced was mechanical + cultural + chemical control technique which was then compared with farmers' technology. The results showed that the technology was able to decrease fruit dry blight intensity by 28.7%, increased fruit yield by 76.1%, and enhance farmers' income by 76.1%. Farmers gave positive response and satisfied to the technology introduced.

Bayman et al. (2011) found out the mycorrhizal relationships of vanilla and prospects for biocontrol of root rots. According to them, orchid mycorrhizas are distinct from those of other plants, in two ways: first, the fungi appear to receive little or nothing from the orchids, and second, their choice of partners. Molecular systematics studies have shown that the fungi formerly known as *Rhizoctonia* comprise several sexual genera, of which three have been found in vanilla: *Thanatephorus*, *Ceratobasidium* and *Tulasnella*. The application of these mycorrhizal fungi and Rhizoctonia (i.e. *Ceratobasidium*) as biocontrols for a variety of plant pathogens, including *Fusarium*, is highlighted.

According to Naik et al. (2011), management of stem rot disease of vanilla is possible to a limited extent, through biocontrol agents and fungicides. Among the treatment combinations, *T. harzianum* (2 g/plant) + carbendazim (0.1%), *T. harzianum* + cymoxanil and mancozeb (0.2%) and *T. harzianum* + hexaconazole (0.2%) showed 100% survival of plants up to 120 days after treatment imposition followed by Hexzol and *T. harzianum* combination with 77.77% survival of plants was observed both in pot and field conditions.

Bhadramurthy et al. (2011) highlighted the occurrence and CP sequence-based characterization of Bean yellow mosaic virus (BYMV) associated with vanilla in India. In RT-PCR, the BYMV-specific primers amplified a 950-bp expected product that was cloned and sequenced. The sequenced region contained 957 bases spanning nuclear inclusion b and CP genes. Sequence analyses and phylogenetic studies based on the CP region confirmed the identity of the virus as a strain of BYMV. This is the first report on the occurrence of BYMV infecting vanilla in India.

He-Zhen et al. (2011) determined the complete sequence of CymMV from *V. fragrans* in Hainan, China. It comprised 6224 nucleotides; sequence analysis suggested that the isolate they obtained was a member of the genus Potexvirus, and its sequence shared 86.67–96.61% identities with previously reported sequences. Phylogenetic analysis suggested that CymMV from *V. fragrans* was clustered into subgroup A and the isolates in this subgroup displayed little regional difference.

The molecular diversity of *F. oxysporum* causing rot diseases of vanilla in South India was explored. A total of 60 isolates of *F. oxysporum* were obtained from diseased samples, and 9 morphologically different isolates were taken for molecular characterization using randomly amplified polymorphic DNA (RAPD) markers to study the genetic variability if any, among them. PCR amplification of total genomic DNA with random oligonucleotide primers generated unique banding patterns depending upon primers and isolates. Nine oligonucleotide primers were selected for the RAPD assays, which resulted in 384 bands for 9 isolates of *F. oxysporum*. It is inferred that *F. oxysporum* infecting vanilla in South India consists of a single clonal lineage with a moderate level of genetic diversification (Vijayan et al., 2012).

Athul et al. (2012) worked on biocontrol of Fusarium wilt of vanilla (*V. planifolia*) using combined inoculation of *Trichoderma* sp. and *Pseudomonas* sp. These results suggested that using of *T. harzianum* with *P. fluorescens* through soil-mixing plus root-dipping treatment could provide not only additional protection against crop loss due to Fusarium diseases but also significantly increase vegetative growth of vanilla. The mechanism of biocontrol also facilitates the production of volatile and non-volatile organic acids, siderophore, chitinase, peroxidase and salicylic acid (SA).

Glen (2012) reported on the effect of non-chemical and chemical protection on healthiness and yielding of horseradish. Applied protection methods modified the yield, the structure of horseradish root fraction and leaf infection by *Albugo candida*, *Alternaria* spp. and *Pyrenopeziza brassicae*. In comparison with biological protection, chemical protection gave better yield-forming effects. The value of the total yield of horseradish roots per hectare was on average higher by 1.23 t and the marketable yield by 1.24 t. During all vegetation seasons, chemical protection proved to be more efficient in limiting horseradish leaves infection by *Alternaria* spp. and *A. candida*, whereas biological preparations produced a better effect in light leaf spot control.

'Chirke' is a viral disease of large cardamom (*A. subulatum*) characterized by light and dark green streaks on the leaf lamina. Sucker, the commonly used planting material, is the major source for spread of the disease. The virus is sap transmissible to the popular large cardamom cultivars Golsey, Ramsey, Swaney and Varlangey and vectored by *Rhopalosiphum maidis* and *Myzus persicae* in a non-persistent manner. The results suggested that the virus associated with the chirke disease of large cardamom is a new species under the genus *Macluravirus* in the family Potyviridae for which the name large cardamom chirke virus is proposed (Mandal et al., 2012).

Palama etal. (2012) had also an organized research study on metabolome of vanilla and related species under CymMV infection. Out of four accessions grown in intensive cultivation systems under shade house, *Vanilla pompona* had qualitatively more phenolic compounds in leaves and a virus titre that diminished over time. No differences in the metabolomic profiles of the shade house samples obtained by NMR were observed between the virus infected versus healthy plants. However, using in vitro *V. planifolia* plants, the metabolomic profiles were affected by virus infection. Under these controlled conditions, the levels of amino acids and sugars present in the leaves were increased in CymMV-infected plants, compared with uninfected ones, whereas the levels of phenolic compounds and malic acid were decreased. The metabolism, growth and viral status of *V. pompona* accession contrasted from that of the other species suggesting the existence of partial resistance to CymMV in the vanilla germ plasm.

Saju et al. (2013) conducted a research work on Colletotrichum blight control in large cardamom (*A. subulatum* Roxb.) nursery using bioagents and chemicals to produce healthy planting materials free from blight. The results indicated that selection of apparently disease free mother plants and treating the rhizome and pseudostem with carbendazim + mancozeb (0.3%) or copper oxychloride (0.3%) would help in reducing the disease incidence in the nursery. Since Sikkim is an organic state, use of copper oxychloride can be recommended.

Sandheep et al. (2013b) performed a colonization study of antagonistic *Pseudomonas* sp. in *V. planifolia* using green fluorescent protein (GFP) as a marker. *P. fluorescens* P7 and *Pseudomonas putida* P4 were isolated from vanilla rhizosphere soil and checked for their capability to control fungal pathogens of vanilla both under in vitro and in vivo conditions. The endophytic colonization ability of the selected rhizosphere bacteria was evaluated after genetically tagging them with a constitutively expressing green fluorescent protein gene (*gfp*). The green fluorescent endophytic bacteria were observed within the plant tissue when cross sections of the petiole were viewed under the confocal laser scanning microscope. The bacterial isolates were effectual in controlling the selected fungal pathogens of vanilla. The *gfp*-tagged *Pseudomonas* sp. was populated within the intercellular spaces of the vanilla leaves 1 week after its foliar spraying.

Beauveria bassiana strain, obtained from a larva of the opium poppy stem gall wasp (*Iraella luteipes*), endophytically colonizes opium poppy (*P. somniferum* L.) plants and protects them against this pest. The goal of a study was to monitor the dynamics of endophytic colonization of opium poppy by *B. bassiana* after the fungus was applied to the seed and to ascertain whether the fungus is transmitted vertically through seeds. These results demonstrate for the first time the vertical transmission of an entomopathogenic fungus from endophytically colonized maternal plants. This information is crucial to better understand the ecological role of entomopathogenic fungi as plant endophytes and may allow development of a sustainable and cost-effective strategy for *I. luteipes* management in *P. somniferum* (Quesada Moraga et al., 2014).

Becka et al. (2014) studied the poppy root weevils (*Stenocarus ruficornis*, Stephens 1831) control in opium poppy (*P. somniferum* L.). The levels of root damage caused by the insect pest larvae (expressed as number of boreholes per root) and yield got from individual treatments were compared. Sprays applied for 18 days after the first record of poppy root weevils in trials showed the highest effects on a decrease of the levels of root damage (40% of untreated control).

Sandheep et al. (2014) exploited the biocontrol potential of indigenous *Trichoderma* sp. against major phytopathogens of vanilla (*V. planifolia*). The pathogens, namely *F. oxysporum*, *R. solani* and *S. rolfsii* causing wilt and rot diseases of vanilla were isolated from naturally infected vanilla plants. Among the ten isolates, mainly two species *T. harzianum* and *Trichoderma virens,* which showed 80–90% inhibitions against the selected fungal pathogens were selected. *T. harzianum* inoculation resulted into 90–95% disease suppression under greenhouse experiment and was superior to the inoculation of *T. virens* and the standard *T. harzianum*.

4.11.2 INSECT–PESTS

Vercruysse et al. (2000) evaluated the insecticides for control of *Cavariella aegopodii* and carrot motley dwarf (CMD) disease in parsley. Toxicity values of different insecticides against *C. aegopodii* were determined using a biological test procedure. Lambda-cyhalothrin proved to possess the highest acute toxicity against *C. aegopodii* of the insecticides tested. Imidacloprid acted considerably slower than the other insecticides. Control of CMD disease and of its aphid vector *C. aegopodii* was studied in field experiments during two consecutive growing seasons in two different plantings of parsley. Foliar application of both imidacloprid (Confidor) and of a mixture of lambda-cyhalothrin and pirimicarb (Okapi) gave sufficient control of the aphid vectors on the parsley plants in both plantings. The percentage of plants showing CMD symptoms was lower in the insecticide-treated plots compared to the untreated controls, except in the late planting, treated with Confidor. Incidence of CMD was considerably lower in the Okapi-treated plots than in the Confidor-treated plots. In contrast to the untreated control plots, the incidence of CMD in both plantings of parsley was comparable when insecticides were applied. At the end of the growing season, a reduction in the number of parsley plants showing specific CMD symptoms was observed.

Results of investigations in Germany on the effects of nutrient deficiencies (nitrogen, potassium and phosphorus) on parsley grown under greenhouse conditions were presented by Wiethaler and Fischer (2001). Data on germination delay, leaf-edge necrosis through excess boron, chlorosis through iron deficiency, necrosis through phosphate deficiency, changes in leaf colour, changes in leaf size, root problems, and yield reduction were given.

A study on insect pests of vanilla, a popular intercrop in areca nut and coffee plantations, was carried out in Karnataka. During the survey conducted over the vanilla growing tracts of Karnataka, nine insect pest species belonging to four orders have been recorded. One of the pests, Vanilla bug—*Halyomorpha* sp. nr. *picus* (Fabricius) has been reported to cause severe economic damage. The insect pests are confined to the aerial parts of the crop. Detailed studies on biology and management aspects are in progress (Prakash et al., 2002).

Varadarasan et al. (2002) reported about vanilla vine weevil, a new insect pest on vanilla (*V. planifolia*). Vanilla vine weevil, *Sipalus* sp.

(Dryophthoridae: Coleoptera: Insecta) is recorded for the first time in India, damaging vine and shoot of vanilla (*V. planifolia*). Both the adult weevil and grub feed on vanilla, resulting in necrosis and rotting of the affected portion. While the weevil feed on vine and leaves, the grubs feed on the inner tissue of the vine by making tunnel. The symptom of attack by the pest and information on the pest biology are discussed.

Torres and Hoy (2005) assessed the relationship between carrot weevil (*Listronotus oregonensis*) infestation and parsley yield. Fresh weight was measured in one to two cuttings of parsley planted on two planting dates. The average weight declined with increasing numbers of oviposition scars in the later planting. Plant mortality increased as number of oviposition scars per plant increased in the second planting and in the first cutting of the first planting. One oviposition scar per plant is sufficient to result in significant reduction in fresh weight per plant. In commercial parsley fields, the relationship between fresh weight of parsley per 30-cm row section of parsley was best described as a linear function of the proportion of plants with root feeding. Economic damage to parsley that is equivalent to the cost of controlling carrot weevil was estimated to result from ~1% of plants with root damage. Based upon this estimated economic injury level, we suggest an action threshold of 1% of plants containing carrot weevil oviposition scars earlier in the growing season when controls could be applied to prevent the damage.

Another study was conducted to understand the most preferred feeding site on vanilla and the most preferred host for the Giant African snail, *A. fulica*. On vanilla, terminal leaves were most preferred followed by bottom leaves. Among six different hosts tested, cauliflower and cabbage leaves were highly preferred by the snails, which can be used to attract snails from vanilla field in future. Besides, the present investigation also indicated that barrier substances can also be used to arrest the movement of snails from one place to another. Among the four substances tested, copper sulphate even at 3 cm thickness and common salt at 6 cm thickness were found to be very effective barriers against these snails (Vanitha et al., 2011a).

In another study, Vanitha et al. (2011b) determined the occurrence and management of white grubs (*Holotrichia serrata*) in major vanilla-growing areas of Tamil Nadu. The treatments comprised *Beauveria brongniartii* (10 g at 1×109 spores/l/plant), *Metarhizium anisopliae* (10 g at 1×109 spores/l/plant), chlorpyrifos at 5 ml/l/plant, chlorpyrifos at 7.5 ml/l/plant, castor-oil cake at 300 g/plant and untreated control. Observations were recorded on the root damage and coiled shoot damage at weekly intervals. Results showed that among the treatments, chlorpyrifos at 7.5 ml/plant significantly reduced the per cent damage and recorded 26.6 and 28.3% root and shoot damage, respectively, at 7 days after treatment. The fungi and castor-oil cake treatments were not effective in reducing the damage as the grubs were at the late instar stage and were at par with the untreated control.

Vanitha et al. (2011c, 2012) investigated on pests of vanilla and their natural enemies in Tamil Nadu, India. A total of seven arthropods, seven gastropods and two invertebrates were recorded as pests of vanilla. Out of 60 farms surveyed, only 9 had the incidence of pest attack. Almost all parts of vanilla plant, that is, stem, leaf, flower, bud, roots, pods, and so forth. were found to be attacked. Among the pests, white grubs were found to cause considerable damage followed by vanilla bug and shoot and leaf webber, while others were not at the economic level. Among the pests, white grubs and Giant African Snail were found to cause considerable damage to the vanilla plants, while others were not at economic levels. Among the natural enemies, parasitoids such as *Euplectrus* sp., *Glyptapanteles* sp., *Aprostocetus* sp., *Chelonus* sp. and Uropoda mites were found to be associated with the pests of vanilla.

4.12 BREEDING, BIOTECHNOLOGY AND TISSUE CULTURE

Reghunath and Priyadarshan (1993) studied on the somaclonal variation in cardamom derived from axenic culture of juvenile shoot primordia. Since the variation in panicle characters may affect yield, the results indicated that reliable methods for early detection of somatic variants have to be introduced in the micropropagation protocol of cardamom, especially when they are produced on a massive scale for distribution among farmers. Jamwal and Kaul (1995) reported about colchicine-induced morphological variations in celery (*A. graveolens*). Later, Jamwal and Kaul (1997) detailed on cytomorphological studies in colchicine-induced autotetraploids of *A. graveolens* var. dulce.

Ovule culture of vanilla and its potential in crop improvement was explored by Minoo-Divakaran et al. (1997). Seeds were cultured on MS medium supplemented with combinations of 0.5 or 1.0 mg of kinetin, BA, IBA and NAA, respectively. Individual shoots produced multiple shoots and conversion of root meristem into shoots was observed on MS medium supplemented with 1 mg BA and 0.5 mg IBA/l. Plantlets could be transferred to soil with 80% success. Studies on leaf size, internode length and growth rate and isozyme analysis of leaf tissue by native polyacrylamide gel electrophoresis indicated genetic variability among progenies.

Sawant et al. (1999) screened some promising kokum genotypes. Among them, S8 exhibited consistently high yields over 7 years with short harvesting period (78 days) and minimum number of harvests (3). The fruits of S8 had the highest average width (4.15 cm), average circumference (13.15), average weight (34.45 g), average rind thickness (4.45 mm). Fruits also had the longest shelf life (15 days).

Nielsen and Siegismund (1999) detailed on interspecific differentiation and hybridization in *Vanilla* sp. The vegetative parts of *Vanilla claviculata*, *Vanilla barbellata* and *Vanilla dilloniana* are similar; however, their conspicuous flowers easily distinguish them. Electrophoresis of seven polymorphic enzymes revealed that the genetic composition of the three species is also very similar: they deviate mainly from each other in allele frequencies rather than by specific alleles. Nevertheless, they are efficiently recognized by their genotypic compositions.

Pollen viability and stigma receptivity period before and after complete flower opening was studied by hand pollinating the flowers/flower buds with its own pollen and fresh viable pollen by Shadakshari et al. (2003). Based on the results, the pollen viability periods were 23 h before complete flower opening and 16 h and 30 min after complete flower opening. The stigma receptivity periods were 41 h before complete flower opening and 17 h after complete flower opening under hill zone conditions.

Pank and Kruger (2003) detailed the sources of variability of thyme populations (*T. vulgaris* L.) and conclusions for breeding. Agroscope Changins-Wadenswil (ACW) carried out a selection program to optimize quality and yield of thyme varieties, obtaining 56 new hybrids by crossing male sterile with male fertile clones in Switzerland. The most interesting one, named Varico 3, is characterized by high homogeneity, thymol chemotype, high essential oil content (>5%) and the variety is currently cultivated in Switzerland.

The occurrence of different introductions (e.g. *V. planifolia*, *V. tahitensis* and *V. pompona*) from which modern vanilla cultivars have evolved in Reunion Island (Indian Ocean) and other humid tropical areas was demonstrated by Besse et al. (2004) by using RAPD markers. It was also suggested that *V. tahitensis* is probably not a species of direct hybrid origin (*V. planifolia* × *V. pompona*) but rather a species related to *V. planifolia.* These results are essential to guide further genetic analysis of cultivated vanilla specimens in introduction areas.

Karamian (2004) carried out plantlet regeneration through somatic embryogenesis in four species of *Crocus*, namely, *C. sativus*, *C. cancellatus*, *C. michelsonii* and *C. caspius*, using shoot meristem culture on LS medium containing 4 mg NAA/l and 4 mg BA/l or 1 mg 2,4-D/l and 4 mg kinetin/l. Complete plantlets were obtained by transferring germinated embryos into half-strength MS medium supplemented with 1 mg NAA/l and 1 mg BA/l at 20°C under a 16/8 h (light/dark) cycle.

Kuruvilla et al. (2005) studied on the performance evaluation of tissue culture vis-à-vis open-pollinated seedlings of cardamom. Results indicated that tissue culture plants performed better compared to open-pollinated seedlings with regard to growth attributes and yield irrespective of the seasons and zones. Further, the findings also support the concept that tissue culture plants are 'true to type' unlike open-pollinated seedlings of cardamom.

Paul et al. (2006), however, quantitatively evaluated the induced macromutants in celery (*A. graveolens* L.) and ajowan (*Trachyospermum amni* L.). Nissar et al. (2006) studied on pollination, interspecific hybridization and fruit development in vanilla. The study indicates that pollination can be done with much success during the time between 6 a.m. and 12 p.m. The present study showed that cross-fertilization has superiority over the existing self-fertilization on fruit setting and hence suitable techniques have to be evolved to carry out the same for commercial production of beans. Hybridization between *V. planifolia* and *Vanilla aphylla* resulted in 77.7% fruit set and its reciprocal cross gave 88.8%, fruit set. *V. aphylla* also offers better chance to undertake inter specific hybridization due to its synchronized flowering with the cultivated species.

Sreedhar et al. (2007) highlighted on the genetic fidelity of long-term micropropagated shoot cultures of vanilla as assessed by molecular markers. One thousand microplants were established in soil of which 95 plantlets (consisting of 4 phenotypes) along with the mother plant were subjected to genetic analyses using RAPD and inter-simple sequence repeat (ISSR) markers. Out of the 45 RAPD and 20 ISSR primers screened, 30 RAPD and 7 ISSR primers showed 317 clear, distinct and reproducible band classes resulting in a total of 30,115 bands. However, no difference was observed in banding patterns of any of the samples for a particular primer, indicating the absence of variation among the micropropagated plants. It was concluded that the micropropagation protocol used for in vitro proliferation of vanilla plantlets for the last 10 years might be applicable for the production of clonal plants over a considerable period of time.

Bory et al. (2008) reported on natural polyploidy in *V. planifolia.* It exhibited an important variation in somatic chromosome number in root cells, and endoreplication as revealed by flow cytometry. Given that 2C-values, mean chromosome numbers, and stomatal lengths were positively correlated and that Sterile and Grosse Vanille accessions were indistinguishable from Classique accessions using molecular markers, the occurrence of recent autotriploid and autotetraploid types in Reunion Island is supported. This is thought to be the first report showing evidence of a recent autopolyploidy in *V. planifolia* contributing to the phenotypic variation observed in this species.

Mewes et al. (2008) researched on physiological, morphological, chemical and genomic diversities of different origins of thyme. The criteria investigated were winter hardiness, beginning of flowering, growth height, yield of the dry herb, content of essential oil, composition of the essential oil, DNA content of cell nuclei and number of chromosomes.

Loskutov et al. (2008) evaluated the callus selection (CS) and flamingo-bill explant (FB) methods for celery transformation using *Agrobacterium tumefaciens* and the bar gene as selectable marker. Using CS, a total of 34 GR callus clones were selected, and shoots were regenerated from over 50% of them on Gamborg B5 medium + 6-(γ, γ-dimethylallylamino) purine 2ip (4.9 μM) + NAA (1.6 μM) and rooted on MS in 5–6 mo total time. Conversely, using FB with inoculation by GV3101/pDHB321.1 on cultivar XP166 yielded putative transgenic celery plants confirmed by polymerase chain reaction (PCR) in just 6 weeks. Transformation of the bar gene into celery was confirmed by PCR for 5 and 6 CS and FB lines, respectively. Herbicide assays on whole plants with 100 and 300 mg/l glufosinate indicated a range of low to high tolerance for lines derived by both methods. The bar gene was found to be Mendelian inherited in one self-fertile CS-derived line.

Agayev et al. (2009) reported the results of a unique, large-scale research project that focused on the problem of clonal selection of saffron (*C. sativus* L.). A comparison of many hundreds and thousands of clones, each grown from one corm of the same weight, resulted in the identification of 'superior' clones in terms of exceptionally large numbers of flowers and large (≥10 g) corms. Those clones would also be very suitable for facilitating the mechanization of saffron agriculture in terms of the lifting, sorting and planting of corms, weeding, softening ground and harvesting flowers.

Verma et al. (2009) investigated on the genetic diversity among nine leafy and leaf-less *Vanilla* sp. employing 30 decamer RAPD primers and 10 ISSR primers. A total of 154 RAPD polymorphic markers (83.24%, h =0.378) and 93 ISSR polymorphic markers (86.11%, h=0.363) were used to generate a genetic similarity matrix followed by the cluster analysis. Among the nine species studied, *V. planifolia, V. aphylla* and *V. tahitensis* revealed very low level of variation within their collections, thus indicating a narrow genetic base.

Shal-Chandran and Puthur (2009) worked out the assorted response of mutated variants of *V. planifolia* towards drought. The results of this study denoted the different level of drought tolerance of mutated variants of *V. planifolia.* The mutated variants of *V. planifolia* exhibited much variability in their morphology as well as their growth rate. Tuttolomondo et al. (2009) undertook a study of thyme germ plasm in Sicily. Thyme germ plasm constitutes a large part of the wild flora in Sicily and is a potentially useful source of genetic variability. On the basis of this, a research was carried out on wild thyme accessions. These were used to evaluate both its productivity and individual biotypes with the functional characteristics most sought-after by the various sectors which use this species.

Chaterjee et al. (2010) studied on the prospects of in vitro production of thebaine in opium poppy. A variant plant of opium poppy (*P. somniferum* L.) having high thebaine was obtained from the dose of 10 kR + 0.4% EMS during an extensive mutation breeding experiment. Alkaloid profile of variant showed higher thebaine in stem followed by leaf callus, stem callus and cotyledons. Such a high content of thebaine from tissue raised plant material opens a new vista for the extraction of thebaine at commercial level. A comparison was also made between in vivo and in vitro cost of production and extraction of thebaine.

A research work was done by simultaneous elimination of CMV and CymMV infecting *V. planifolia* through meristem culture. Apical meristem measuring 0.1–0.25 mm were isolated and cultured in MS medium supplemented with 0.45 μM thidiazuron for 40–45 days to initiate the growth. Following the enlargement of meristem, it was transferred to MS medium supplemented with 4.43 μM 6-benzylaminopurine (BAP) and 2.68 μM α-NAA (Retheesh and Bhat, 2010).

Ozudogru et al. (2011) developed a protocol on in vitro propagation from young and mature explants of thyme resulting in genetically stable shoots. In vitro-grown shoot tips of thymes were exposed to cytokinins alone or in combination with auxins, GA3 and/or silver nitrate in order to optimize in vitro shoot proliferation. Optimum shoot proliferation (97% regeneration rate, with 8.6 shoots produced per explant) was obtained when semi-solid MS medium was supplemented with 1 mg/l kinetin and 0.3 mg/l GA3. Rooting of the shoots was easily obtained on semi-solid MS medium that was either hormone-free or supplemented with auxins. However, the best root apparatus (92.5% rooting rate, with 19 adventitious roots per shoot) developed on MS medium supplemented with 0.05 mg/l 2,4-dichlorophenoxyacetic acid.

General and specific combining ability (SCA) and heterosis were studied by Singh and Pandey (2011) for morphine content, opium yield/plant, seed yield/plant, capsule weight/plant, capsule size, capsule/plant, plant height and days to 50% flowering involving 8 parents and 28 F1s derived from them in diallel excluding reciprocals in opium poppy. The hybrid BR 233 × BR 242 exhibited the highest magnitude of positive significant SCA effects with high positive heterosis over mid-parent for morphine content.

Yadav and Singh (2011) studied on heterosis and inbreeding depression for seed and opium yield in opium poppy in set of crosses derived from partial diallel breeding design. The crosses ND1002 × NBRI-11 and ND1002 × BR241 showed higher better parent heterosis both for seed and opium yield. A considerable amount of inbreeding depression in F2 generation was also recorded.

Umamaheswari and Mohanan (2011a) studied on the association of agronomic characters in vanilla. Most of the agronomic characters of vanilla are polygenic in nature and they show different levels of interrelationship due to sharing of common genes. This phenomenon leads to different levels of interrelationship and association of these characters. The most important agronomic characters of this plant have been analysed presently for their association by factor analysis so as to group them into groups with maximum gene sharing and also to identify the lead characters.

In a separate study, Umamaheswari and Mohanan (2011b) compared the vegetative and floral morphology of *V. planifolia* and *V. tahitensis*. Leaves of *V. tahitensis* are significantly narrow and leaf area is significantly low when compared to *V. planifolia*. Inflorescence length is significantly higher in *V. tahitensis*. Average duration from flower initiation to first flower opening is lower in *V. tahitensis*. Among flower characters, significant reduction is found in *V. tahitensis* in the case of bract length, labellum length, labellum breadth and gynostemium length.

The peak production of essential oil takes place in the morning and maximal concentration in essential oil for the second harvest is between July 20 and August 20 for the studied year (Vouillamoz et al., 2011).

Two families (VG26 × VG20 and SG35II × VE01) of opium poppy were analysed to study the gene actions involved in the inheritance of yield and component traits (plant height, leaves per plant, capsules per plant, peduncle length, capsule index, seed and straw yield per plant and morphine content). Simple additive, dominance and epistatic genetic components were found to be significant. Biparental mating followed by recurrent selection involving desired recombinants may be utilized to improve the component traits (Kumar and Patra, 2012).

Radhakrishnan and Mohanan (2012) described cardamom breeding in India. This overview elaborated the germ plasm; genetic improvement of planting materials through selection, hybridization, polyploid breeding, mutation breeding and tissue culture; breeding for resistance to drought, pests and diseases; and some cultivars and high yielding varieties of cardamom.

Rao et al. (2012) research was based on fruit set and seed germination of vanilla and found some interesting features like natural fruit set for hybridization between two species of vanilla followed by embryo rescue and subsequent raising of four suspected vanilla hybrid progeny. However, these suspected hybrids could not be verified by phenotype as the plantlets resembled the female parent. ISSR primers were employed to assess the hybrid nature of the four putative hybrids. The genetic distance between the two parents was 0.48, whereas the genetic distances between hybrids HY1, HY2, HY3 and HY4 parents ranged from 0.48 to 0.80. These data demonstrate that hybrids HY1, HY2, HY3 and HY4 are crossbreeds between *Vanilla* sp., *V. planifolia* and *V. wightiana* and that hybridization between *Vanilla* spp. are resourceful.

Osinska et al. (2012) evaluated the yield and quality of three cultivars of leafy parsley, namely, Amphia, Festival and Verta. The yield of leaves per 1 m^2 was on average from 0.55 to 3.57 kg with Amphia produced higher yields than others (2.61 kg/m^2). The content of biologically active compounds in the leaves of parsley significantly depended on the variety, date of harvest and method of conservation. The content of assimilated pigments and vitamin C in leaves of parsley clearly decreased under the influence of freezing and drying. Season dependence of available in vitro protocols of saffron limits their use in commercialization. Making use of in vitro corm cultures as ex-plant not only made the protocol seasonal independent but also increased the proliferation ratio from the single cycle of 1:120–1:1800. MS media supplemented with PGRs (BAP/NAA) and growth retardant (CCC) and sucrose has been found stimulating agents for inducing sprouting/shooting/multiple shooting and corm development (Salwee and Nehvi, 2013).

Ahmad et al. (2014) reviewed saffron in the light of biotechnological approaches. Saffron (*C. sativus* L.) is a triploid plant belonging to the Iridaceae. It is propagated by corms as the flowers are sterile and fail to produce viable seeds. Biotechnological approaches offer the capability to produce large quantities of propagating material in short time as well as the production of commercially important chemical constituents such as crocin, picrocrocin, crocetin and safranal under in vitro conditions. However, the protocols available so far need further refinement for their commercial utilization. The review described the progress made in genus *Crocus*, and highlighted the potential for future expansion in this field through biotechnological interventions.

Azofeifa-Bolanos et al. (2014) described the importance and conservation challenges of *Vanilla* spp. in Costa Rica taking into account the vulnerability of *V. planifolia* to extinction and the genetic potential in the country with respect to the concentration of vanillin. The comprehensive conservation strategy proposed in the paper aims to perpetuate vanilla plant genetic resources in gene banks through a formal system of seed production, for further use for breeding, research and commercial production. Agroecological conditions of Costa Rica as well as the genetic material of both wild and cultivated vanilla open the possibility of developing production models based on organic agriculture.

Rao et al. (2014) reported on a deep transcriptomic analysis of pod development in the vanilla orchid. Using next-generation sequencing technologies, they have generated very large gene sequence data sets from vanilla pods at different times of development, and representing different tissue types, including the seeds, hairs, placental and mesocarp tissues. This developmental series was chosen as being the most informative for interrogation of pathways of vanillin and C-lignin biosynthesis in the pod and seed, respectively. This database provides a general resource for further studies on this important flavouring species.

4.13 HARVEST, POSTHARVEST, QUALITY AND VALUE ADDITION

Maletic and Jevovic (1998) conducted trials to estimate planting and harvesting period on the yield and quality of mint leaves and its essential oil. Leaves were harvested twice each year and their essential oils were analysed. Menthofurane, menthyl acetate, limonene and 1,8-cineole contents were higher in the first harvest and menthone content higher in the second harvest. Harvesting in July is recommended. Rema and Krishnamoorthy (1998) evaluated the packing materials and storage of scions on graft success in nutmeg. The research results indicated that storage of scions in sealed polythene bags and in moist coir dust in sealed polythene bags were the best and 63.7 and 62.2% graft success, respectively, was obtained after up to 10 days of storage.

Leaf yields from Hamburg (cultivar Berlinska) and leafy (cultivar Paramount) types of parsley were compared in trials in Poland. For all the parameters, the yield of the leafy type was significantly higher than that of the Hamburg type. In both types, the proportion of the leaf blade tended to increase as the growing season progressed (Kmiecik and Lisiewska, 1999).

Ram and Kumar (1999) optimized the interplant space and harvesting time for high-essential oil yield in seven mint cultivars (MAH-3, Himalaya, Hariti, SS-15, Kalka, Gomti and Shivalik). A row spacing of 50 cm was optimum for the short cultivars, such as Kalka, and 70 cm was best for those which regenerated well after the first harvest, for example, Himalaya, with respect to essential oil production. Kalka produced essential oil yields 23 and 20% higher than Himalaya and Shivalik at the first harvest; Himalaya gave the highest yield at the second harvest. Kalka, irrespective of plant spacings and harvesting periods, produced a better-quality essential oil (higher menthol content, 81.5–87.1%) than other cultivars. Himalaya grown at a wider row spacing was the second best choice in terms of essential oil quality (81.2–83.2% menthol content).

The quality of this spice is dependent on methods of harvesting and drying. Sun-drying and kitchen fire-drying are the methods employed for the drying of nutmeg in Kerala. Nutmeg samples collected from different locations were dried in an imported tunnel drier and quality of the solar tunnel-dried spice was compared with commercial samples. Quality enhancement and reduction in drying time were evident for tunnel-dried samples. Oleoresin content of the experimental samples was high with respect to market samples. Overall quality of nutmeg increased upon tunnel drying (Joy et al., 2000).

Studies were conducted by Kulathooran et al. (2000) to determine the effect of drying method on the quality of curry leaf with respect to physico-chemical parameters such as colour, texture, and volatile oil, chlorophyll and ascorbic acid contents. Tabulated data on the 39 volatile curry oil components identified and the relative levels of flavouring constituents of fresh, sun- and shade-dried curry leaf oils are presented. Pinene hydrate, pinocarvone, borneol, dihydrocarveol, piperitol (*cis*), piperitol (*trans*), isodihydrocarveol, pulegone, α-terpinyl acetate and germacrene are constituents reported for the first time in curry leaf oil.

Rao et al. (2001) estimated the drying characteristics of large cardamom. For long-duration storage of cardamom and in order to bring out its aroma, the fresh cardamom capsules (with 80–85% moisture) has to be dried immediately after harvesting to bring down its moisture content to less than 10% (w.b.) through a curing (drying) process. Still a primitive and inefficient (operating efficiency level of about 5–15%) smoking method (using traditional *bhatti*) is being used for drying of large cardamom resulting in huge (estimated 20,000 MT/year) wastage of fuelwood and poor-quality (charred and blackened) product. The paper gave a brief description of efforts made to obtain basic drying parameters of large cardamom (which is a prerequisite for dryer design) under different operating conditions.

Kattimani and Reddy (2001) investigated on the effect of time of harvest on growth, biomass, oil yield and nutrient uptake by Japanese mint under semi-arid tropical climate of Andhra Pradesh. Harvesting of Japanese mint at 100 days after planting in the first and 60 days after the first harvest recorded the highest growth attributes, biomass yield and oil yield.

A study was undertaken by Menon et al. (2002) to develop a model for forecasting the yield in the vanilla plantations using biometric characters. A multiple regression equation was derived using yield as dependent variable and it exhibited a precision of about 95%. The model was refined using the most significant attributes, which are also easy to record in the plantations such as number of inflorescence, beans and bean length. The truncated model can be employed for forecasting the yield in vanilla plantations with a precision of about 93%.

Molina et al. (2004) worked on extending the harvest period of saffron. An earlier flowering was obtained when the corms were lifted before leaf withering (late May) and/or when the corms were heated at 30°C during 20 days before incubation at 25°C. On the other hand, flowering was delayed when corms were stored at low temperatures. A deviation from the optimum conditions resulted in flower abortion and/or a reduction in flower size.

Corm weight (13 and 15 g) had a significant effect on the quality of stigma of saffron flowers. The colour, bitter compounds and aroma of stigma produced from 15 g corms were more suitable than those produced from 13 g corms (Omidbaigi, 2005). Le et al. (2005) evaluated the seed storage methods of star anise (*Illicium verum*) and cinnamon (*Cinnamomum cassia*). Studies on seed storage of these species were undertaken by the Forest Science Institute of Vietnam in order to improve their storage longevity. The results showed that seeds of these species differ in morphological characters such as fruit shape, initial moisture content and initial germination rate. Seeds of star anise and cinnamon with moisture contents of 25–40% and 30–40%, respectively, had the highest germination percentage after storage at 5°C for 9 and 12 months.

Tuteja et al. (2006) found out the effect of harvesting intervals on herbage, oil yield and economics of different varieties of Japanese mint. Himalaya cultivar gave significantly higher mean herbage, oil yield and net returns when it was harvested at 120 days after planting and 75 days after the first harvest. In the case of Koshi cultivar, the maximum herbage and oil yields and net returns were obtained when it was harvested at 110 days after planting and 75 days after the first harvest.

Havkin-Frenkel and Frenkel (2006) studied on postharvest handling and storage of cured vanilla beans. This insight suggested all the four stages of curing of vanilla beans, namely killing (carried out by hot water scalding, freezing or by other methods), sweating (entailing high temperatures and high humidity), drying and conditioning (to preserve the formed flavour compounds). The postharvest handling of cured vanilla beans is a continuation of the curing process, aimed at preserving quality attributes achieved in the production and curing of vanilla beans. Temperature, humidity, gas composition and type of packaging are some important factors that determine bean quality in storage.

With the aim of reducing the curing period, effects of pretreatments on flavour formation in vanilla beans during accelerated curing at 38°C for 40 days were studied by Sreedhar et al. (2007). The use of NAA (5 mg/l) or Ethrel (1%) with blanching pretreatment resulted in threefold higher vanillin on the 10th day. Scarification of beans resulted in nearly 4- and 3.6-fold higher vanillin formations on the 10th day in NAA- and Ethrel-treated beans, respectively. The pretreatment methods of the present study may find importance for realizing higher flavour formation in a shorter period because the major quality parameters were found to be comparable to those of a commercial sample.

Arslan et al. (2007) investigated on corm size and different harvesting times on saffron. In this study, the effects of different harvesting times (every year and every 2 years) and corm size on the number of corms obtained was investigated. Harvesting time and corm size affected flowering of saffron and new corm regeneration.

Alvares et al. (2007, 2010) determined the effect of pre-cooling on the postharvest of parsley leaves. The objective of this work was to determine the effects of a hydrocooling procedure, followed by storage at 5°C, on the shelf life of parsley cultivar Grauda Portuguesa leaves. Bunches of parsley leaves were pre-cooled for 15 min in a water and ice mixture kept at 5°C. During this period of pre-cooling, approximately 43% of the initial field heat was removed from the leaves. After the pre-cooling treatment, the leaf bunches were placed at 5°C for 7 days. Based on the data, the authors recommend pre-cooling as an efficient method to retard the beginning of wilt for stored parsley leaves.

Waliszewski (2007) described a simple and rapid high-performance liquid chromatography (HPLC) technique for vanillin determination in alcohol vanilla extract. Vanillin was separated on a Nucleosil C18 column by using water and methanol (40:60) as the mobile phase and retention time was only 2.2 min. The measurements were made by using a photodiode array detector of the most adequate maximum wavelength absorbance at 231 nm. This method has been successfully applied for the determination of vanillin in some commercial extracts.

Five sample preparation methods (steam distillation, extraction in the Soxhlet apparatus, supercritical fluid extraction, solid-phase microextraction and pressurized liquid extraction [PLE]) used for the isolation of aroma-active components from thyme are compared. As per the obtained data, PLE is the most efficient sample preparation method in determining the essential oil from the thyme herb. Moreover, the relative peak amounts of essential components revealed by PLE are comparable with those obtained by steam distillation, which is recognized as standard sample preparation method for the analysis of essential oils in aromatic plants (Dawidowicz et al., 2008).

Patil et al. (2009) evaluated the variability studies in physico-chemical parameters in kokum (*G. indica* Choicy) for syrup preparation. Fifteen kokum genotypes from the Western Ghat region of Karnataka, Goa and Maharashtra, India, were for the quality of fruits intended for syrup production was studied. The highest scores for overall acceptability (3.83, 3.71 and 3.67 at 0, 3 and 6 months after storage, respectively) were recorded for S14. These parameters should be taken into account along with rind weight and thickness when selecting kokum genotypes for the improvement of syrup quality.

The diversity of yeast from rhizosphere soil in curing and storing of vanilla bean has been studied by General et al. (2009). A total of 4655 isolates with 86 morphological different colonies were screened, of which 35 forms belonging to 17 genera were found to be β-glycosidase producers. Further testing for their ability to convert ferulic acid to vanillic acid proved all the 35 forms utilized ferulic acid. *Buttera* sp. (MVY22) was found to be the most promising, which was further investigated for its conversion ability. HPLC analysis proved the conversion of ferulic acid to vanillic acid by this isolate.

The study by Jadhav et al. (2009) dealt with extraction of vanillin from cured vanilla pods using conventional Soxhlet extraction and ultrasound-assisted extraction. For Soxhlet extraction, an increase in operating temperature from 90 to 100°C was found to increase extent of extraction by 30%, whereas for ultrasound-assisted extraction, pre-leaching stage for 30 min duration was found to be beneficial in enhancing the extent of extraction by about 20%.

Brunschwig et al. (2009) evaluated the chemical variability of cured vanilla beans. In order to establish a chemical fingerprint of vanilla diversity, 30 samples of *V. planifolia*- and *V. tahitensis*-cured beans from seven producing countries were examined for their aroma and fatty acid contents. Both fatty acid and aroma compositions were found to vary between *Vanilla* spp. and origins. Vanillin was found in higher amounts in *V. planifolia* (1.7–3.6% of DM) than in *V. tahitensis* (1.0–2.0%), and anisyl compounds were found in lower amounts in *V. planifolia* (0.05%) than in *V. tahitensis* (1.4%-2.1%). The study highlighted the role of the curing method as vanilla-cured beans of two different species cultivated in the same country were found to have quite similar fatty acid compositions.

Saidj et al. (2009) conducted the kinetic study and optimized the operating conditions of the extraction by steam distillation of the essential oil of *Thymus numidicus* of Algeria. The results revealed that the simple effect of the bed porosity is preponderant in comparison to the simple effects of the mass and the steam water flow rate. Concerning the combined effects of two parameters, the steam water flow rate and the bed porosity, is the more important. The yield is also affected by the interaction of the three parameters.

Edris et al. (2009) studied the effect of organic agriculture practices on the volatile flavour components of thyme essential oil in Egypt. The obtained results indicated that the oil content of thyme at the first cut was slightly higher for chemically fertilized herb compared with the organic one. However, in the second cut, there was no difference in the oil content. Organic fertilization at high level (20 m^3/acre) increased the percentage of oxygenated components in thyme oil, especially thymol (51.1%), compared with any other treatments in which thymol content ranged between 31.4 and 43.2%. High levels of organic fertilization also shifted the major components of thyme oil; thymol, carvacrol, and γ-terpinene to the high margins specified by the European Pharmacopoeia.

Links between phenology, yield and composition of the essential oil of common thyme, *T. vulgaris* L., grown in Central Spain were determined by Arraiza et al. (2009) in the different phases of the biological cycle during 1 year. Data showed an average yield of about 2%. The analysis of the oil components was carried out by gas chromatography–flame ionization detector (GC-FID) and GC-MS. The main oil constituents were thymol (36.3–47.5%), p-cymene (13.0–27.8%) and delta-terpinene (5.3–16.2%), which is in accord with a thymol chemotype. The highest yield of oil was obtained in the period of full flowering and the highest concentration of thymol in the period of initial flowering.

Borges et al. (2012) determined the essential oils of basil (*Ocimum gratissimum* L.), Cured vanilla beans were irradiated (5, 10, 15, 20 and 30 kGy) for a possible enhancement of vanillin content by radiolysis of vanillin glucoside. Results obtained revealed that the more stable one absorbing at 360 nm is aldehydic radical. Hence, the highly stable oxygen–carbon linkage between vanillin and glucose limits the possible enhancement of aroma quality of irradiated beans (Kumar et al., 2010).

The effect of ultra-high-pressure treatment on the extraction efficiency of vanillin in vanilla was studied by Wang et al. (2011). The effect of the different pressure and retention time on the extraction efficiency of vanillin was studied by using 80% ethanol as extractant. The results demonstrated that ultra-high-pressure treatment has a great influence on the extraction efficiency of vanillin. The per cent content of vanillin can be up to 72% in the optimized conditions of 200 MPa and 20 min. The extraction efficiency was increased by 17.5% compared with traditional methods.

A book chapter by Toth et al. (2011) presented a lexicon of aroma and flavour descriptors that sensory evaluators often use to grade vanilla. It also included a list of volatile and semi-volatile compounds reported to be naturally present in vanilla. For each entry, the individual compounds are listed using International Union of Pure and Applied Chemistry nomenclature, common name or synonym. Volatile compounds reported by the authors for the first time are highlighted in bold.

There is limited information on how harvest time and drying influence peppermint and spearmint yield, oil composition, and bioactivity. In a 2-year study by Zheljazkov et al. (2010), the effects of harvest time and drying on the yield, oil composition and bioactivity of peppermint, Scotch spearmint, and Native spearmint were evaluated.

Krasaekoopt et al. (2010) described the processing of vanilla pods grown in Thailand and its application. The methods include soaking of green vanilla pods in hot water, sun-drying, slow-drying in shed and finally conditioning. Another study was undertaken by Dyk et al. (2010) to investigate methods of curing that could greatly reduce the time to complete the process and yield cured beans that retain high concentrations of vanillin and other flavour compounds with high sensory quality rating. The study revealed that a mild hot water blanching treatment followed by sweating at 35–45°C and rapid drying is required to produce cured beans with excellent appearance and attractive aroma.

A book chapter by Frenkel et al. (2011) focused on the curing process and outlines various aspects that might influence the flavour quality of cured vanilla beans, including botany of the vanilla bean, the nature and purpose of the curing process, as well as effects of curing practices in various production regions (Addendum).

Another study was focused on equations describing drying behaviours of vanilla in soaked and non-soaked drying methods. Equations with high confidence intervals were found for describing the drying process. Reducing weight loses while conserving the quality of vanilla will have positive effects for farmers and producers. Results of this study showed that the soaked drying method for vanilla was much better than the non-soaked drying in terms of drying period, energy saving and product quality (Gurdil et al., 2011).

Golparvar and Bahari (2011) determined the effects of phenological stages on herbage yield and quality/quantity of oil in garden thyme. Plants were harvested in five phenological stages, that is, in the vegetative stage, beginning of blooming, 50% blooming, full blooming and fruit set stages. On the whole, it is recommended to harvest this plant in 50% blooming to gain the highest essence and thymol yield as well as fresh and dry herbage.

Chagas et al. (2011) evaluated the biomass yield and essential oil content in function of the age and harvest period of mint plants. Two experiments were set: the first evaluated the effect of three harvest ages (80, 100 and 120 days) after transplanting, and three ages at the second harvest (60, 75 and 90 days). In the second experiment, three harvest periods were evaluated—early January, late March and early June. Plant age at first harvest did influence essential oil content and dry biomass of the aerial parts in response to plant age at the second harvest. The harvest period in which the essential oil content was higher was in early June.

Dambrauskiene et al. (2011) determined the biochemical value of leaf parsley. Parsley with even leaves accumulates more sugars, dry soluble solids and DM. In raw matter of all the investigated parsley, very high amount of ascorbic acid was established: from 170.0 mg/100 g in Gigant D'Italia up to 241.6 mg/100 g in Moss Curled. The higher amount of chlorophylls was found in parsley cultivars, which are dark green —Astra and Festival. Curly parsley has more essential oils in comparison with even-leaved parsley. In a raw matter of cultivar Astra, there was 0.13% of essential oil, in Moss Curled—0.09%, in other cultivars—0.03%.

Jabbari et al. (2011) evaluated the nitrogen and iron fertilization methods affecting essential oil and chemical composition of thyme. Results showed that nitrogen foliar application increased the vegetative yield, amount and percentage of essential oil and chemical compositions of thyme. On the other hand, application of iron had a suppressing effect on the studied traits.

Chemat et al. (2012) characterized the composition and microbial activity of thyme essential oil. The essential oil obtained by hydrodistillation is characterized by a rich content of phenolic compounds such as thymol (71%) and carvacrol (4%) versus a minor presence of monoterpene hydrocarbons like p-cymene (3%) and γ-terpinene (0.5%). The extracted essential oil has presented a high inhibiting activity against eight bacterial strains up to 75 mm and two species of fungus up to 45 mm.

Kamble and Pillai (2012) evaluated the efficacy of kokum health drink in obesity. The rinds of garcinia fruits are the richest source of hydroxycitric acid, which has a nutraceutical potential for obesity. By taking into account the nutraceutical properties of kokum, health drink was prepared as a therapeutic food supplementation. The result of serum cholesterol found reduced from 300 to 215 mg/100 ml, triglyceride levels noted decreased from 200 to 116 mg/100 ml of blood among selected obese adolescents.

El-Nakhlawy et al. (2013) evaluated the response of essential oil and terpene contents of two mint genotypes to different drying temperatures before distillation. Menthone and menthol concentrations significantly decreased as plant leaves were dried >60°C/14 days. As drying temperature increased by 1°C essential oil, menthone and menthol decreased by 0.005, 0.013 and 0.071%, respectively. Correlation coefficients were 0.96 and 0.94 between oil (%) and menthone and menthol, respectively.

Singh and Pothula (2013) reviewed the postharvest processing of large cardamom in the Eastern Himalaya and recommended for increasing the sustainability of a niche crop. The article reviewed the crop's postharvest processing, quality issues and trade patterns, and identifies research topics that could contribute to increasing its quality and value and thereby to protecting and promoting the livelihoods of several thousands of people in the value chain.

Mechanical devices for supporting human labour in harvesting flowers and separating stigmas are presented and described by Gambella et al. (2013). Results from the field indicated the number of successes in terms of picked flowers with respect to the number of semi-oscillations required. The data revealed a 35% success rate in tests with two cam semi-oscillations, 31% with three cam semi-oscillations, 16% with one cam semi-oscillation, and 18% with more than three cam semi-oscillations. The separation of the stigmas was tested in three airflow systems. Two of the three separation systems produced damage to filaments of the stigmas on the order of 65% with weight equal to 0.05 g.

Jaramillo-Villanueva et al. (2013) estimated the economic efficiency of vanilla curing in Mexico. The methodology uses the policy analysis matrix, which generates the private cost ratio and the domestic resources cost ratio, which indicates the level of economic efficiency of the systems. The traditional curing system (sun wilting) shows an efficient use of domestic resources and has a comparative advantage over other systems; it is profitable, while the heating furnace system resulted in low efficiency and showed no comparative advantage.

Abbaszadeh and Haghighi (2013) determined the effect of nutrition and harvest time on growth and essential oil content of *T. vulgaris* L. in Iran. Results indicated that fertilizer, harvest date and their interactions significantly affected most of the traits. It can be concluded from the results that *T. vulgaris* responds well to fertilization, and selecting the best treatment depends on the objective of production.

Dyk et al. (2014) determined the harvest maturity of vanilla beans. Current methods for determining the maturity of vanilla (*V. planifolia*) beans are unreliable. Yellowing at the blossom end, the current index, occurs before beans accumulate maximum glucovanillin concentrations. Beans left on the vine until they turn brown have higher glucovanillin concentrations but may split and have low quality. To find a better index, changes in bean dimensions, DM and glucovanillin accumulation were followed over four seasons. Therefore, optimum harvest time occurs when DM accumulation slows down and should be measured in the central portion of beans. Two near-infrared spectrometers using interactance geometry were trialled for non-invasive assessment of DM. Cross-validation r and root mean square error of cross-validation values of 0.87 and 1.76, respectively, for a unit using wavelengths between 1100 and 2300 nm, and 0.82 and 1.05 for a portable unit using wavelengths between 800 and 1050 nm were obtained from the second derivative of absorbance spectra. The latter unit allows infield monitoring of maturation.

Salehi et al. (2014) assessed the effect of harvest time on yield and quality of *T. vulgaris* L. essential oil in Iran. The results showed that the phenological stages had very significant effects ($P<0.01$) on essential oil yield and percentage as well as thymol percentage and yield. The highest essential oil content of thyme (2.42%) was extracted at the beginning of blooming stage. The highest thymol content of thyme (74.8%) was extracted at the full-blooming stage. According to the results of this research, harvesting the thyme at 50% blooming stage have maximum essential oil quality and quantity in Isfahan Province.

Safaii et al. (2014) investigated the effect of harvesting stages on quantitative and qualitative characters of essential oil and phenolic yield composition in two thyme species—*T. daenensis* produced the highest essential oil yield and phenolic yield at the full-flowering stage (82.5 and 73.5 kg/ha, respectively), however in *T. vulgaris,* these factors increased at fruit set (31.7 and 27.6 kg/ha, respectively). Mid of flowering and fruit set stages had the maximum essential oil percentage in *T. daenensis* (2.8%) and *T. vulgaris* (1.2%), respectively. *T. daenensis* is better than *T. vulgaris* in view of essential oil yield, phenolic yield and essential oil percentage; therefore, it can be recommended for extensive cultivation and use in medicine industries.

Gallage et al. (2014) opined that vanillin formation from ferulic acid in *V. planifolia* is catalysed by a single enzyme. More precisely, a single hydratase/lyase-type enzyme designated vanillin synthase (VpVAN) catalyses direct conversion of ferulic acid and its glucoside into vanillin and its glucoside, respectively. VpVAN localizes to the inner part of the vanilla pod and high transcript levels are found in single cells located a few cell layers from the inner epidermis. A gene encoding an enzyme showing 71% sequence identity to VpVAN was identified in another vanillin-producing plant species *Glechoma hederacea* and was also shown to be a vanillin synthase as demonstrated by transient expression in tobacco.

Liang et al. (2014) analysed the volatile components in the fruits of *V. planifolia* by headspace–solid-phase microextraction (HS-SPME) combined with GC-MS. HS-SPME combined with GC-MS was used to analyse the volatile components from *V. planifolia* fruits for the first time, which provided scientific basis for further research and development of volatile components in *V. planifolia* fruits. A total of 28 components were separated and identified, accounted for 98.17% of total volatile components. Among them, vanillin had the maximum content (48.28%), followed by guaiacol (15.54%).

The electrochemical behaviour of vanillin was studied using the AuPd–graphene hybrid-based electrode. It presented high electrocatalytic activity and vanillin could produce a sensitive oxidation peak at it. Under the optimal conditions, the peak current was linear to the concentration of vanillin in the ranges of 0.1–7 and 10–40 μM. The sensitivities were 1.60 and 0.170 mA/mM/cm, respectively; the detection limit was 20 nM. The electrode was successfully applied to the detection of vanillin in vanilla bean, vanilla tea and biscuit samples (Lei et al., 2014).

The effect of foliar application of SA and reduced irrigation on growth, oil yield, chemical components and antibacterial and antioxidant activities of thyme (*T. daenensis*) in field condition were investigated by Pirbalouti et al. (2014). Foliar application of SA influenced growing degree days from early growing stage to 50% and full flowering, minimum radius and canopy diameter. The results showed that foliar application of SA reduced the negative effect of water deficit on thymol content in the essential oil of *T. daenensis*. The essential oils of *T. daenensis* exhibited antioxidant and antibacterial activities when plants were sprayed with 1.5 and 3.0 M SA, respectively.

A domestic microwave oven was modified and Clevenger apparatus was attached to it to make it an extraction unit. A multivariate study based on a Box–Behnken design was used to evaluate the influence of three major variables (soaking time, temperature and power density) affecting the performance of microwave-assisted extraction on celery seed. It was found that microwave-assisted process gave approximately same oil yield (1.90%) in less time (93.5 min) and with low energy consumption (58,191.78 MJ/kg oil) compared with the traditional hydrodistillation (Talwari and Ghuman, 2014).

Petrovic et al. (2014) investigated on the chemical composition and antioxidative activity of essential oil of *Thymus serpyllum.* Wild thyme essential oil was isolated from the dried herb *T. serpyllum* by hydrodistillation. The isolated essential oil is a liquid of light yellow colour and the odour characteristic of the genus *Thymus*. The study showed that the essential oil of *T. serpyllum* can be an important source for the production and application in the food industry as nutritional supplements, functional food components or natural food antioxidants.

KEYWORDS

- **morphology**
- **propagation**
- **intercropping**
- **breeding**
- **germination**
- **naphthalene acetic acid**

CHAPTER 5

UNDEREXPLOITED SPICE CROPS: FUTURE THRUST AREAS AND RESEARCH DIRECTIONS

CONTENTS

5.1 INTRODUCTION

Once upon a time spices were often equated with gold and precious stones. Now, they are no longer luxury items, but have high demand with growing importance. With increasing health consciousness in people around the world this trend continues to be consistent. The communication revolution and demand-driven market dynamics recreate the situation favourable to the innovators with greater competitiveness in certain areas like organic spices, encapsulated spices, spice blends and different other forms of food supplements. Technology in many spheres integrating biology, physics, chemistry, physiology, nuclear science, biotechnology, nanotechnology, geographic information system (GIS) and bioinformatics paves the way for the scientists, researchers, growers, investors and policymakers to go hand in hand towards sustainability with a deterministic and futuristic vision. However, certain obvious changes such as change in physical and intrinsic qualities of the inputs, change in climate, change in market components and even change in human preferences over space and time are creating issues and taking us into situations which offer stiff challenges. All these necessitate formulation of future strategies to bring underutilized spices out of their niche role.

The main researchable areas on underutilized spices to accomplish the objectives are discussed below.

5.2 PROPER CONSERVATION AND IMPROVEMENT OF EXISTING GENETIC RESOURCES

Large arrays of already existing genetic resources with significant variation levels are not only native to the primary centre of origin but also contribute for future improvement. As far as the underutilized spice crops are concerned, these vast genetic resources are often uncared for and having with untapped potential. Having complete information of this precious gene pool is only the primary step and is extremely essential towards the development of varieties with high yield, superior traits and disease/pest resistance through the most appropriate breeding methods such as selection, hybridization, mutation breeding, polyploidy breeding and biotechnological methods. Identifying and exploring the resistance/ tolerance mechanism to biotic and abiotic stresses as well as to combat the ill effects of climate change is necessary.

5.3 INCREASING PRODUCTIVITY OF SPICES

5.3.1 QUALITY PLANTING MATERIAL PRODUCTION

Good quality planting material, as the key to any crop development programme, should be the major priority area. For common spice crops such as black pepper, small cardamom, ginger, turmeric, chilli, and so forth, already a good amount of varieties are available. However, the area is still gloomy for underutilized spice crops. Along with the common practices, necessary protocols for micropropagation of several lesser-known spice crops, wherever applicable, are to be standardized in mint, saffron, horseradish, cinnamon, cassia, kokum, parsley, bay leaf, asafoetida, rosemary, thyme, basil, curry leaf, ajowan, dill, and so forth, and their supply is to be ensured.

5.3.2 EFFICIENT INPUT MANAGEMENT

Efficient use of inputs taking into account the region-specific climatic, physical and physiological needs for stable yield, enhanced quality and increased income from the underutilized spice crops is the need of the hour. This will express the potential of the crops as maximum productivity with the available natural resources, soil fertility and water. Research directions may be oriented towards varietal responses to climate change and enhance the usage efficiency of all the inputs.

5.3.3 INTEGRATED PLANT PROTECTION RESEARCH

This is one of the most important areas demanding due attention. The transformation and evolution of protection mechanism and methods should be more concerted and organized towards raising the productivity without impairing quality. Biotic stress is the major factor influencing the yield and quality in underexploited spices. Besides conducting surveys, emphasis is also to be laid on development of better crop protection models taking cultural, chemical and biological formulae. The existing tools such as integrated disease management (IDM) and integrated pest management (IPM) are to be redefined for specific underutilized spice crops, at any specific region, in the context of emerging diseases and pests day by day.

A holistic approach is desired and innovative diagnostic techniques are to be encouraged.

5.3.4 POSTHARVEST MANAGEMENT

Postharvest management is the key to recover major economic losses happening in spice produce management. The future research should address the issues of postharvest loss minimization. For spices and herbs, it is virtually impossible to recondition for the following potential contaminants, namely, mycotoxins, heavy metals, pesticide residues, allergens, undeclared colours, and so forth. In such a scenario, the only option is to prevent these potential contaminants from either getting into the product or being formed during postharvest handling.

Lack of proper sanitation in certain developing countries compared to the western world makes food more vulnerable. Many researchers reported increased incidents of food-borne infections and intoxications due to spices during the last decades of the 20th century (Buckenhuüskes and Rendlen, 2004; Jackson et al., 1995; Burow and Pudich, 1996; Lehmacher et al., 1995; Weber, 2003a, 2003b). Microbial decontamination generally includes fumigation with ethylene oxide (though some countries banned it for carcinogenic and mutagenic interaction), irradiation with gamma rays, electron beam and X-rays in permissible doses by preventing the chances of any sort of secondary contamination, steam treatment especially on whole spices (as due to high temperature of the steam there are chances of discolouration and volatile oil reduction in spice powders), high hydrostatic pressure ranging from 100–1000 MPa, and so forth. The future research in this direction also includes extrusion technology (through reduction of microbial loads, prevention of mycotoxin contamination and inhibition of endogenous enzymes), alternative spice powder production process characterized by low microbial loads, bright colour and high flavour compounds (by integrating safe production and processing that includes good agricultural practices (GAP), good manufacturing practices (GMP) and hazard analysis critical control point (HACCP)), enzyme-assisted liquefaction of spice materials to improve yields, enhance the release of secondary metabolites and better recovery of functional food ingredients.

5.3.5 BIOTECHNOLOGY—THE EMERGING AREA OF INTERVENTION

Biotechnological tools have supplemented various conventional approaches in conservation, characterization, improvement and utilization for boosting production and productivity of not only the major and important spices but also the so-called underutilized spices. In many such spices, viable micropropagation technologies are available for commercial production and generation of disease-free planting material. Somaclonal variation is important in crops in which natural variability is low and a few useful somaclonal variants have been identified in a few rhizomatous crops and vanilla. Protoplast technology is also available for fennel, fenugreek, garlic, saffron and peppermint. In vitro cryopreservation, synseed and micro-rhizome technologies are available for safe propagation, conservation, movement and the exchange of spices germplasm. Studies are in progress for in vitro production of flavour and colouring compounds such as capsaicin, vanillin, anethole, crocin, picrocrocin, safranal, and so forth, using immobilized and transformed cell cultures. Use of molecular markers for crop profiling, fingerprinting, molecular taxonomy, identification of duplicate hybrids, estimation of genetic fidelity and tagging of genes for marker aided selection (MAS) is gaining importance. Isolation of important and useful genes and development of transgenics is also a promising proposition (Peter et al., 2006).

The biotechnological aspects are unexplored with respect to underutilized spice crops, in depth studies with respect to genomic constitution, genes responsible for aroma and other quality products need to be tagged. A detailed scanning of the transcriptome and proteome will deliver genes that are of high significance for idiotype breeding. Artificial seed production has been reported from explants like horseradish (Rao, 1977), cardamom (Nirmal Babu et al., 1998), vanilla, cinnamon, anise (Nirmal Babu et al., 1998), celery, *Carum carvi* (Rao, 1977) and coriander (Stephen and Jayabalan, 1998). Somatic embryo formation or plantlet regeneration from several species such as coriander (Kim et al., 1996; Stephen and Jayabalan, 1998), caraway (Ammirato, 1974; Furamanova et al., 1991), cumin (Dave et al., 1996), fennel (Hunault and Maatar, 1995; Umetsu et al., 1995), saffron (Ahuja et al., 1994) and celery (Choi and Soh, 1997) have also been reported.

5.3 ENHANCING QUALITY AND VALUE ADDITION

Quality issues are gaining topmost priority whenever marketing of spice crops are viewed globally. This has an obvious and inevitable consequence for the underutilized spice crops too. The global trade is no longer concentrated on the popular demand-supply theory of economics. Rather the value of spices in general and lesser-known spices, in particular, depends on consumer preferences and acceptability in the developing new global markets. In this very context, value addition offers great scope. The research directions should be oriented by focussing on this particular issue. Both pre- and postharvest technologies are essential to add value in these kinds of spice crops. The new methods of extraction, identification and quantification exploit chemo-profiling and omic tools that fully utilize the bioactive principles to develop new products and thus help in putting them into an attractive and powerful value chain.

5.4 TRANSFER OF TECHNOLOGY TO THE TARGET GROUPS

Technology dissemination, adoption and further refinement effectively ensure a product to be sustainable. This concept is true for the kind of spice crops which are not so popular and still do not have a strong footing to stand and survive. Many approaches are thought of to fulfil this very objective. Participatory planning, refinement and monitoring along with evaluation of the tools to perfectly fit in the immediate target group maybe the featured areas of research in future. Public–private partnership approach may also be given importance.

5.5 CLIMATE CHANGE AND UNDERUTILIZED SPICES

Global industrial development and technological advancements lead to the emission of a huge amount of greenhouse gases resulting in rise in temperature and associated problems. Climate change is one of the most important global challenges for human beings with implications on agriculture production, natural ecosystems, freshwater supply, health, and so forth. The anticipated rising CO_2 level appears to favour the seed spices production.

With the rise in temperature, population of natural enemies of insect pests of a wide array of underutilized spices will possibly increase and this will be beneficial in natural control. Salinity, alkalinity and water stress problems need to be managed in an effective manner. Introduction of short duration varieties, management of biotic and abiotic stresses, introduction of climate smart cultivation technique, evolving climate resilient varieties, comprehensive disease and pest surveillance and coping mechanisms against the aberrant weather conditions maybe some of the immediate steps to be taken.

Coordinated approaches on the local, regional and international levels are essential to maximize the potential of underutilized spice crops. This attempt can only be brought into reality under the umbrella of the global sustainable development goals and converging the ideas and agreements of several global organizations and associations, namely, United Nation (UN), Food and Agriculture Organization (FAO), Gesellschaft für Internationale Zusammenarbeit (GIZ), International Fund for Agricultural Development (IFAD), Technical Centre for Agricultural and Rural Cooperation (CTA), Crops for the Future (CFF), and so forth, to name a few. If possible, a mark of geographical indication (GI) should be given to the product for imparting trade value in association with the region of cultivation for harnessing high price by a value tag to these products. There is still much to explore in these crops with respect to compounds having high industrial or medicinal importance. Concerted efforts of scientists have delivered numerous research findings in various aspects which will definitely help in increasing the productivity and quality of spices.

A successful strategy to promote and mainstream all minor, neglected, underutilized spice crops is to identify and analyse the crops as one of the highest priority criteria. The next job will be to find out the root cause of all the negative perceptions and distracting attitudes of the growers and the end users as well. The leverage points within the entire food supply chain must be identified thereafter, which include production, storage, processing and consumer's expectations. The promotion and mainstreaming through removal of barriers must address the following key issues:

- There should not be any conflict or competition with the staple and major crops.
- The leverage points within the supply chain should be efficiently designed.

- A sound baseline database should be present with nutritional and protective properties.
- Concerted research directions with a view to human welfare
- Organized management of information and communication
- Policy-supported extension services
- Removal or simplification of trade and regulatory barriers confronting regional, national and international markets.

KEYWORDS

- **integrated disease management**
- **biotechnology**
- **target groups**
- **climate change**
- **Food and Agriculture Organization**

REFERENCES

Abbaszadeh, B.; Haghighi, M. L. Effect of Nutrition and Harvest Time on Growth and Essential Oil Content of *Thymus vulgaris* L. *J. Med. Plants By-Prod.* **2013,** *2*(2), 143–151.

Abbaszadeh, B.; Sharifi, A. E.; Ardakani, M. R. and Aliabadi, F. H. Effect of Drought Stress on Quantitative and Qualitative of Mint. Abstracts Book of 5th International Crop Science Congress & Exhibition. Korea, 2008, 23.

Abirami, K.; Rema, J.; Mathew, P. A.; Srinivasan, V.; Hamza, S. Effect of Different Propagation Media on Seed Germination, Seedling Growth and Vigour of Nutmeg (*Myristica fragrans* Houtt.). *J. Med. Plants Res.* **2010,** *4*, 2054–2058.

Abirami, K.; Rema, J.; Mathew, P. A. Evaluation of Nutmeg Genotypes for Seed Germination, Seedling Growth and Epicotyl Graft Take. *J. Med. Aromat. Plant Sci.* **2012,** *34*, 91–94.

Aga, F. A.; Wani, G. M.; Hassan, B., Wani, M. A. Cultivating Saffron Scientifically in Kashmir. *Indian Hortic.* 2006, *51*(1), 21–24.

Agayev, Y. M.; Fernandez, J. A.; Zarifi, E. Clonal Selection of Saffron (*Crocus sativus* L.) the First Optimistic Experimental Results. *Euphytica* **2009,** *169*(1), 81–99.

Ahmad, M.; Zaffar, G.; Mehfuza, H.; Dar, N. A.; Dar, Z. A. Saffron (*Crocus sativus* L.) in the Light of Biotechnological Approaches: A Review. *Sci. Res. Essays.* **2014,** *9*, 13–18.

Ahmed, B.; Alam, T.; Varshney, M.; Khan, S. A. Hepatoprotective Activity of Two Plants Belonging to the Apiaceae and the Euphorbiaceae Family. *J. Ethnopharmacol.* **2002,** *79*(3), 313–316.

Ahuja A.; Koul, S.; Ram, G.; Kaul, B. L. Somatic Embryogenesis and Regeneration of Plantlets in Saffron, *Crocus sativus* L. *Indian J. Exp. Biol.* **1994,** *32*, 135–140.

Ait-Oubahou, A.; El-Otmani, M. Saffron Cultivation in Morocco. In *Saffron: Crocus sativus L.*; Negbi M., Ed.; Harwood Academic Publishers: Australia, 1999; pp 87–94.

Akgul, A. *Spice Science and Technology*. Turkish Association Food Technologists Publ. No. 15, Ankara, Turkey, 1993.

Akgül, A. Volatile Oil Composition of Sweet Basil (*Ocimum basilicum* L.) Cultivating in Turkey. *Nahrung* **1989,** *33*, 87–88.

Al Ramamneh, E. A. M. Plant Growth Strategies of *Thymus vulgaris* L. in Response to Population Density. *Ind. Crops Prod.* **2009,** *30*(3), 389–394.

Alam, M.; Samad, A.; Khaliq, A.; Ajayakumar, P. V.; Dhawan, O. P.; Singh, H. N. Disease Incidence and its Management on Opium Poppy: A Global Perspective. International Symposium on Papaver. ISHS *Acta Hortic.* **2014,** *1036,* 14.

Aliabadi F. H; Valadabadi S. A. R.; Daneshian J.; Khalvati M. A. Evaluation Changing of Essential Oil of Balm (*Melissa officinalis* L.) Under Water Deficit Stress Conditions. *J. Med. Plant Res.* **2009,** *3*, 329–333.

Alonso, G. L.; Salinas, M. R.; Garijo, J.; Sanchez-Fernadez, M. A. Composition of Crocins and Picrocrocin from Spanish Saffron (*Crocus sativus* L). *J. Food Qual.* **2001,** *24*(3), 219–233.

Alqasoumi, S. Anxiolytic Effect of *Ferula assafoetida* L. in Rodents. *J. Pharmacogn. Phytother.* **2012,** *4*(6), 86–90.

Al-Snafi, A. E. The Pharmacological Importance of *Anethumgraveolens*: A Review. *Int. J. Pharm. Pharm. Sci.* **2014**, *6*(4), 11–13.

Alvares,V. de S.; Negreiros, J. R. da S.; Ramos, P. A. S.; Mapeli, A. M.; Finger, F. L. Pre-Cooling and Packing in the Conservation of Parsley Leaves *Braz. J. Food Technol.* **2010,** *13*(1/4), 107–111.

Alvares, V. S.; Finger, F. L.; Santos, R. C. de A.; Negreiros, J. R. da S.; Casali, V. W. D. Effect of Pre-Cooling on the Postharvest of Parsley Leaves. *J. Food Agric. Environ.* **2007,** *5*(2), 31–34.

Ammirato, P. V. The Effects of Abscisic Acid on the Development of Somatic Embryos from Cells of Caraway (*Carum carvi* L.). *Bot. Gaz.* **1974,** *135*, 328–337.

Anandaraj, M.; Devasahayam,T.; Zachariah, J. T.; Mathew, P. A.; Rema, J. Nutmeg Extension Pamphlet, IISR, 2005, 7.

Annonymous. Technical Bulletin 8. Cultivation and Processing of Cinnamon. Department of Minor Export Crops. Colombia, Sri Lanka, 1973.

Anonymous. *The Wealth of India, Publication and Information Directorate;* Council of Scientific and Industrial Research: New Delhi, India, 1988; vol. 7, pp 79–89.

Anonymous. Star anise. *J. Indian Spices* **1991,** *4*(28).

Anonymous. *The Wealth of India Vol IV*, 1st ed.; CSIR: New Delhi, 2002, pp 20–21.

Antonopoulos, A.; Karapanos, I. C.; Petropoulos, S. A.; Passam, H. C. The Effect of Two Levels of Ammonium Nitrate Application on the Yield of Plain-Leaf, Curly-Leaf and Turnip-Rooted Parsley and the Quality and Essential Oil Composition of the Leaves Before and After Storage in a Partially Dehydrated Form. *An. Univ. Oradea, Fasc. Biol.* **2014,** *21*(2), 65–69.

Anuradha, M. N.; Farooqui, A. A.;Vasundhara, M.; Kathiresan, C.; Srinivasappa, K. N. In *Effect of Biofertilizers on the Growth, Yield and Essential Oil Content in Rosemary (Rosmarinus officinalis L.)*, Proceedings of National Seminar on the Strategies for Production and Export of Spices, Calicut, India, Oct 24–26, 2002.

Arraiza, M. P.; Andres, M. R.; Arrabal, C; Lopez, J. V. Seasonal Variation of Essential Oil Yield and Composition of Thyme (*Thymus vulgaris* L.) Grown in Castilla—La Mancha (Central Spain). *J. Essent. Oil Res.* **2009,** *21*(4), 360–362.

Arslan, N.; Gurbuz, B.; Ipek, A.; Ozcan, S.; Sarhan, E.; Daeshian, A. M. The Effect of Corm Size and Different Harvesting Times on Saffron (*Crocus sativus* L.) Regeneration. *Acta Hortic.* **2007,** *739*, 113–117.

Arulselvan, P.; Subramanian. S. P. Beneficial Effects of *Murraya koenigii* Leaves on Antioxidant Defense System and Ultra Structural Changes of Pancreatic Beta-Cells in Experimental Diabetes in Rats. *Chem. Biol. Interact.* **2007,** *165*(2), 155–164.

Ashoka, S.; Hegde, Y. R. Scenario of Fungal Diseases of *Vanilla planifolia* in Karnataka. *Int. J. Plant Sci.* **2006,** *1*(2), 269–272.

Ashoka, S.; Hegde, Y. R.; Srikant, K. New Report of Root Rot of Vanilla Caused by *Rhizoctonia bataticola*. *J. Plant Dis. Sci.* **2006,** *1*(2), 223–224.

Atanda, O. O.; Akpan, I.; Oluwafemi, F. The Potential of Some Spice Essential Oils in the Control of *A. parasiticus* CFR 223 and Aflatoxin Production. *Food Control* **2007,** *18*, 601–607.

Ateia, E. M.; Osman, Y. A. H.; Meawad, A. E. A. H. Effect of Organic Fertilization on Yield and Active Constituents of *Thymus vulgaris* L. Under North Sinai Conditions. *Res. J. Agric. Biol. Sci.* **2009,** *5*(4), 555–565.

Athul, S. R.; Asok, A. K.; Jisha, M. S. Biocontrol of Fusarium wilt of Vanilla (*Vanilla planifolia*) Using Combined Inoculation of *Trichoderma* sp. and *Pseudomonas* sp. *Int. J. Pharm. Bio Sci.* **2012,** *3*(3), B-706–B-716.

Awad, S. S.; Hassan, H. M. F.; Shahien, A. H.; Zayed, A. A. Studies on Intercropping of Parsley and Demsisa with Tomato Under Different Rates of Nitrogen Fertilization. *Alexandria J. Agric. Res.* **2001,** *46*(2), 97–112.

Azofeifa Bolanos, J. B.; Paniagua, V. A.; Garcia Garcia, J. A. Importance and Conservation Challenges of *Vanilla* spp. (*Orchidaceae*) in Costa Rica. *Agron. Mesoamericana* **2014,** *25*(1), 189–202.

Babu, K. N.; Anu, A.; Remashree, A. B.; Praveen, K. Micropropagation of Curry Leaf Tree. *Plant Cell Tissue Organ Cult.* **2000,** *61*, 199–203.

Bailer, J.; Aichinger, T.; Hackl, G.; Hueber, K. D.; Dachler, M. Essential Oil Content and Composition in Commercially Available Dill Cultivars in Comparison To Caraway. *Ind. Crops Prod.* **2001,** *14*, 229–239.

Bailey, L. H.; Bailey, E. Z. *Hortus Third. A Concise Dictionary of Plants Cultivated in the United States and Canada;* Macmillan Publishing Co. Inc.: New York, 1976.

Balakumbahan, R.; Rajamani, K.; Kumanan, K. *Acoruscalamus*: An Overview. *J. Med. Plants Res.* **2010,** *4*, 2740–2045.

Baldermann, S.; Blagojevic, L.; Frede, K.; Klopsch, R.; Neugart, S.; Neumann, A.; Ngwene, J.; Norkeweit, J.; Schröter, D.; Schröter, A.; Schweigert, F. J.; Wiesner, M.; Schreiner, M. Are Neglected Plants the Food For the Future? *Crit. Rev. Plant Sci.* **2016,** *35*(2), 106–119.

Bali, A. S.; Sagwal, S. S. Saffron—A Cash Crop of Kashmir. In *Agricultural Situation India;* Lichtfouse, E., Navarrete, M, Debaeke, P, Véronique, S, Alberola, C. Eds.; Spriger Publishers, 1987; pp 965–968.

Balyan, S. S.; Chowdhary, D. K.; Kaul, B. L. Response of Celery to Different Row Spacings. *Indian Perfum.* **1990,** *34*(2), 168–170.

Banerjee, N. S., Manoj, P.; Das, M. R. Male Sex Associated RAPD Markers in *Piper longum* L. *Curr. Sci. India* **1999,** *77*, 693–695.

Barceloux, D. G. Nutmeg *Myristica fragrans* Houtt. *Res. J. Spices* **2009,** *55*(6), 373–379.

Baritaux, O.; Richard, H.; Touche, J.; Derbesy, M. Effects of Drying and Storage of Herbs and Spices on the Essential Oil: Part I. Basil, *Ocimum basilicum* L. *Flavour Fragrance J.* **1992,** *7*, 267–271.

Barrera Rodriguez, A. I.; Herrera Cabrera, B. E.; Jaramillo Villanueva, J. L.; Escobedo-Garrido, J. S.; Bustamante Gonzalez, A. Characterization of Vanilla Production Systems (*Vanilla planifolia* A.) on Orange Tree and Mesh Shade in the Totonacapan Region. *Trop. Subtrop. Agroecosyst.* **2010,** *10*(2), 199–212.

Barrera Rodriguez, A. I.; Jaramillo Villanueva, J. L.; Escobedo Garrido, J. S.; Herrera Cabrera, B. E. Profitability and Competitiveness of the Vanilla (*Vanilla planifolia* J.) Production Systems in the Totonacapan region, Mexico. *Agrociencia Montecillo* **2011,** *45*(5), 625–638.

Baruah, A.; Saikia, N. Vegetative Anatomy of the Orchid *Vanilla planifolia* Andr. *J. Econ. Taxon. Bot.* **2002,** *26*(1), 161–165.

Bauman, D.; Hadolin, M.; Rizner, H. A.; Knez, Z. Supercritical Fluid Extraction of Rosemary and Sage Antioxidants. *Acta Alimentaria Budapest* **1999,** *28*(1), 15–28.

Bayman, P.; Mosquera, E. A. T.; Porras, A. A. Mycorrhizal Relationships of Vanilla and Prospects for Biocontrol of Root Rots. *Handb. Vanilla Sci. Technol.* Blackwell Publishing Ltd.: UK, 2011; Vol. 1, pp 266–280.

Baytop, T. *Treatment with Plants in Turkey*; Publications of Istanbul University: Istanbul. Publ. No. 3255, 1984.

Baytop, T. *Therapy with Medicinal Plants in Turkey;* Publications of Istanbul University: Istanbul, 1984, p 194.

Becka, D.; Cihlar, P.; Vlazny, P.; Pazderu, K.; Vasak, J. Poppy Root Weevils (*Stenocarus ruficornis*, Stephens 1831) Control in Opium Poppy (*Papaver somniferum* L.). *Plant Soil Environ.* **2014,** *60*, 470–474.

Besharati-Seidani, A., Jabbari, A.; Yamini, Y. Headspace Solvent Microextraction: A Very Rapid Method for Identification of Volatile Components of Iranian *Pimpinella anisum* Seed. *Anal. Chim. Acta* **2005,** *530*(1), 155–161.

Besse, P.; Silva, D. da.; Bory, S.; Grisoni, M.; Bellec, F. le.; Duval, M. F. RAPD Genetic Diversity in Cultivated Vanilla: *Vanilla planifolia* and Relationships with *V. tahitensis* and *V. pompona. Plant Sci.* **2004,** *167*(2), 379–385.

Bettazzi, F.; Palchetti, I.; Sisalli, S.; Mascini, M. A. Disposable Electrochemical Sensor for Vanillin Detection. *Anal. Chim. Acta* **2006,** *555*(1), 134–138.

Bhadramurthy, V.; Bhat, A. I., Biju, C. N. Occurrence and Coat Protein Sequence-Based Characterization of Bean Yellow Mosaic Virus (BYMV) Associated with Vanilla (*Vanilla planifolia* Andrews) in India. *Arch. Phytopathol. Plant Prot.* **2011,** *44*(18), 1796–1801.

Bhagyalakshmi, N. Factors Influencing Direct Shoot Regeneration from Ovary Explants of Saffron. *Plant Cell Tissue Organ Cult.* **1999,** *58*, 205–211.

Bhai, R. S.; Anandaraj, M. Brown Spot—A New Disease of Vanilla (*Vanilla planifolia* Andrews) from India. *J. Spices Aromat. Crops* **2006,** *15*(2), 139–140.

Bhai, R. S.; Jithya, D. Occurrence of Fungal Diseases in Vanilla (*Vanilla planifolia* Andrews) in Kerala. *J. Spices Aromat. Crops* **2008,** *17*(2), 140–148.

Bhai, R. S.; Thomas, J. Phytophthora Rot—A New Disease of Vanilla (*Vanilla planifolia* Andrews) in India. *J. Spices Aromat. Crops* **2000,** *9*(1), 73–75.

Bhai, R. S.; Thomas, J.; Potty, S. N.; Geetha, L.; Solomon, J. J. Mosaic Disease of Vanilla (*Vanilla planifolia* Andrews)—the First Report from India. *J. Spices Aromat. Crops* **2003,** *12*(1), 80–82.

Bhai, R. S.; Remya, B.; Jithya, D.; Eapen, S. J. In vitro and In Planta Assays for Biological Control of Fusarium Root Rot Disease of Vanilla. *J. Biol. Control* **2009,** *23*(1), 83–86.

Bhasin, M. *Ocimum*—Taxonomy, Medicinal Potentialities and Economic Value of Essential Oil. *J. Biosphere* **2012,** *1,* 48–50.

Bhaskaran, S.; Mehta, S. Stabilized Anthocyanin Extract from *Garcinia indica*. US Patent 2006/0230983A1, 2006.

Bhat, S.S.; Sudharshan, M. R. Vanilla-Crop Improvement at Sakleshpur. *Spice India* **2004,** *17*(3), 42–45.

Bhuyan, A. K.; Pattnaik, S.; Chand, P. K. Micropropagation of Curry Leaf Tree *(Murraya koenigii* (L.) Spreng.) by Axillary Proliferation Using Intact Seedlings. *Plant Cell Rep.* **1997,** *16,* 779–782.

Bureau of Indian Standards (BIS). *Spices and Condiments—Large Cardamom—Capsules and Seeds—Specification;* Bureau of Indian Standards: New Delhi, India, 1999; IS 13446.

Blamey, M.; Fitter, R.; Fitter, A. Wild Flowers of Britain and Ireland: The Complete Guide to the British and Irish Flora; A & C Black: London, 2003.

Boelens, M. H. The Essential Oil of *Rosmarinus officinalis* L. *Perfum. Flavor* **1985,** *10,* 21–37.

Borges, A. M.; Pereira, J.; Cardoso, M. G.; Alves, J. A.; Lucena, E. M. P. Determination of Essential Oils of Basil (*Ocimum gratissimum* L.), Oregano (*Ocimum gratissimum* L.) and Thyme (*Thymus vulgaris* L.). *Rev. Bras. Plant. Med.* **2012,** *14*(4), 656–665.

Boriquet, G. *Le vanillier et la vanille dans le monde*; kditions Paul Lechevalier: Paris, 1954.

Bory, S.; Catrice, O.; Brown, S.; Leitch, I. J.; Gigant, R.; Chiroleu, F.; Grisoni, M.; Duval, M. F.; Besse, P. Natural Polyploidy in *Vanilla planifolia* (*Orchidaceae*). *Genome* **2008,** *51*(10), 816–826.

Boutekedjiret, C.; Belabbes, R.; Bentahar, F.; Bessiere, J. M. Study of *Rosmarinus officinalis* L. Essential Oil Yield and Composition as a Function of the Plant Life Cycle. *J. Essent. Oil Res.* **1999,** *11*, 238–240.

Boutekedjiret, C.; Bentahar, F.; Belabbes, R.; Bessiere, J. M. The Essential Oil from *Rosmarinus officinalis.* L. in Algeria. *J. Essent. Oil Res.* **1998,** *10*, 680–682.

Bozan, B.; Karakaplan, U. Antioxidants from Laurel (*Laurus nobilis* L.) Berries: Influences of Extraction Procedure on Yield and Antioxidant Activity of Extracts. *Acta Aliment.* **2007,** *36*, 321–328.

Brillouet, J. M.; Odoux, E.; Conejero, G. A Set of Data on Green, Ripening and Senescent Vanilla Pod (*Vanilla planifolia*; Orchidaceae): Anatomy, Enzymes, Phenolics and Lipids. *Fruits Paris* **2010,** *65*(4), 221–235.

Brown, R. W. *Composition of Scientific Words: A Manual of Methods and A Lexicon of Materials for the Practice of Logotechnics;* Smithsonian Institution Press: Washington, D.C., 1956.

Bruneton, J. Pharmacognosy, Phytochemistry, Medicinal Plants, 2nd ed.; Intercept Ltd: London, 1999; pp 519–520.

Brunschwig,C.; Collard, F. X.; Bianchini, J. P.; Raharivelomanana, P. Evaluation of Chemical Variability of Cured Vanilla Beans (*Vanilla tahitensis* and *Vanilla planifolia*). *Nat. Prod. Commun.* **2009,** *4*(10), 1393–1400.

Buckenhüskes, H. J.; Rendlen, M. Hygienic Problems of Phytogenic Raw Materials for Food Production with Special Emphasis to Herbs and Spices. *Food Sci. Biotechnol.* **2004,** *13*, 262–268.

Burow, H.; Pudich, U. Different *Salmonella* Serovars in Spices, Flavoured Potato Crisps and Other Food. *Fleischwirtschaft* **1996,** *76*, 640–643.

Carmona, M.; Alonzo, G. L. A New Look At Saffron: Mistakes Beliefs. *Proceedings of the First International Symposium on Saffron Biology and Biotechnology. Acta Hort.*, **2004,** *650*, 373–391.

Carmona, M.; Sanchez, A. M.; Ferreres, F.; Zalacain, A.; Tomas-Berberan, F.; Alonso, G. L. Identification of the Flavonoid Fraction in Saffron Spice by LC/DAD/MS/MS: Comparative Study of Samples from Different Geographical Origin. *Food Chem.* **2007,** *100,* 445–450.

Carmona, M.; Zalacain, A.; Pardo, J. E.; Lòpez, E.; Alvarruiz, A.; Alonso, G. L. Influence of Different Drying and Aging Conditions on Saffron Constituents. *J. Agric. Food Chem.* **2005,** *53,* 3974–3979.

Castro Bobadilla, G.; Martinez, A. J.; Martinez, M. L.; Garcia Franco, J. G. Application of Located Irrigation System to Increase the Retention of fruit of *Vanilla planifolia* in the Totonacapan, Veracruz, Mexico. *Agrociencia Montecillo* **2011,** *45*(3), 281–291.

Central Bank of Sri Lanka, Annual Report 2015. http://www.cbsl.gov.lk/pics_n_docs/10_pub/_docs/efr/annual_report/AR2015/English/9_Chapter_05.pdf, 137–140 (accessed April 29, 2017).

Ceylan, S.; Yoldas, F.; Cakici, H.; Mordogan, N.; Bayram, E. Effects of Nitrogen Rates On Nitrogen Accumulation, Ascorbic Acid and Essential Oil Contents in Parsley. *Asian J. Chem.* **2006,** *18*(3), 2113–2118.

Chagas, J. H.; Pinto, J. E. B. P.; Bertolucci, S. K. V.; Santos, F. M.; Botrel, P. P.; Pinto, L. B. B. Production of Mint Depending on Organic Manure at Planting and Dressing. *Hortic. Bras.* **2011,** *29*(3), 412–417.

Chagas, J. H.; Pinto, J. E. B. P.; Bertolucci, S. K. V.; Santos, F. M. do. Biomass Yield and Essential Oil Content in Function of the Age and Harvest Period of Mint Plants. *Acta Sci. Agron.* **2011,** *33*(2), 327–334.

Chandre, G. M.; Shiva, P. B L.; Vasundhara, M. Effect of Growth Regulators and Methods of Application on Rooting of Thyme (*Thymus vulgaris* L.) Cuttings. *Biomed* **2006,** *1*(3), 198–204.

Chaterjee, A.; Shukla, S.; Mishr, P.; Rastogi, A.; Singh, S. P. Prospects of In Vitro Production of Thebaine in Opium Poppy (*Papaver somniferum* L.). *Ind. Crops Prod.* **2010,** *32*(3), 668–670.

Chauhan, H. S. A. High Rhizome and Oil Yielding Variety "CIM-Balya" of Vach (Acoruscalamus). *J. Med. Arom. Plant. Sci.* **2006,** *28*, 632–634.

Chauhan, K.; Solanki, R.; Patel, A.; Macwan, C.; Patel, M. Phytochemical and Therapeutic Potential of *Piper longum* Linn. A Review. *Int. J. Res. Ayurveda Pharm.* **2011,** *2*(1), 157–161.

Chemat, S.; Cherfouh, R.; Meklati, B. Y.; Belanteur, K. Composition and Microbial Activity of Thyme (*Thymus algeriensis* genuinus) Essential Oil. *J. Essent. Oil Res.* **2012,** *24*(1), 5–11.

Chen, Q.Y.; Shi, H.; Ho, C. T. Effects of Rosemary Extracts and Major Constituents on Lipid Oxidation and Soybean Lipoxygenase Activity. *J. Am. Oil Chem. Soc.* **1992,** *69*, 999–1002.

Chen, X. B.; Yang, B.; Feng E, L.; Xu, Y. Z.; Yan, F. Quantitative Index of Water Fertilizer Coupling Effect on Greenhouse Celery in Desertified Areas. *Jiangsu J. Agric. Sci.* **2008,** *24*(5), 674–678.

Chenard, C. H.; Kopsell, D. A.; Kopsell, D. E. Nitrogen Concentration Affects Nutrient and Carotenoid Accumulation in Parsley. *J. Plant Nutr.* **2005,** *28*(2), 285–297.

Cheng-hua, X.; Yong-song, Z.; Xian-yong, Shao-ting, D.; Cheng-yan, Y.; Xing-hua, S. Effects of NDSA Fertilization on the N Uptake, Yield and Quality of Celery. *J. Plant Nutr. Fert.* **2006,** *12*(3), 388–393.

Choi, Y. E.; Soh, W. Y. Origin and Early Developmental Patterns of Somatic Embryos From Freely Suspended Single Cells in Cell Culture of Celery (*Apium graveolens* L.). *Phytomorphology* **1997,** *47*, 107–117.

Chopra, R. N.; Roy, D. N.; Ghosh, S. M. Insecticidal and Larvicidal Action of the Essential Oils of *Ocimum basilicum* Linn. and *Ocimum sanctum* Linn. *J. Malaria Ins. India* **1941,** *4,* 109.

Choudhary, D. K.; Kaul, B. L. Gamma Rays Induced Early Flowering Mutant of Celery (*Apium graveolens* L. var. dulce). *Mutat. Breed. Newsl.* **1993,** *40,* 8–9.

Coolbear, P.; Toledo, P. E.; Seetagoses, U. Effects of Temperature of Pre-Sowing Hydration Treatment and Subsequent Drying Rates on the Germination Performance of Celery Seed. *New Zealand J. Crop Hortic. Sci.* **1991,** *19*(1), 9–14.

Coppen, J. J. W. *Gums, Resins and Latexes of Plant Origin. Non Wood Forest Products;* Food and Agriculture Organisation of the United Nations: Rome, 1995, 1–141.

Cronquist, A. *The Evolution and Classification of Flowering Plants;* The New York Botanical Garden: New York, USADK, 1988.

Cu, J.Q.; Zhong, J. Z.; Par, P. GC-MS Analysis of the Essential Oil of Celery Seed. *Indian Perfumer* **1990,** *34*(1–4), 6–7.

D' Antuono, L. F.; Neri, R.; Angioloni, A.; Moretti, A. Nitrogen and Celery (Bologna). *Colture Protette* **2001,** *30*(4), 71–72.

Dalby, A. *Dangerous Tastes: The Story of Spices*; British Museum: London, 2000a.

Dalby, A. *Empire of Pleasures: Luxury and Indulgence in the Roman World*; Routledge: London, 2000b.

Dambrauskiene, E. Influence of Various Factors on the Amount of Essential Oils in Fresh Raw Matter of Medicinal Thyme. *Sodininkyste ir Darzininkyste* **2010,** *29*(2), 55–60.

Dave, A., Batra, A. Somatic Tissues Leading to Embryogenesis in Cumin. *Curr. Sci.* **1995,** *68*(7), 754–755.

Dawidowicz, A. L.; Rado, E.; Wianowska, D.; Mardarowicz, M.; Gawdzik, J.; Application of PLE for the Determination of Essential Oil Components from *Thymus vulgaris* L. *Talanta* **2008,** *76*(4), 878–884.

De Baggio, T.; Belsinger S. *Basil: An Herb Lover's Guide;* Interweave Press: *Loveland,* CO, 1996; p 62.

Deshpande, R. S.; Tipnis, H. P. Insecticidal Activity of *Ocimum basilicum* Linn. *Pesticides* **1977,** *11*(5), 11.

Devasagayam, T. P. A.; Tilak, J. C.; Boloor, K. K.; Sane, K. S.; Ghaskadbi, S.; Lele, R. D. Free Radicals and Antioxidants in Human Health: Current Status and Future Prospects. *J. Assoc. Phys. India* **2004,** *52,* 794–804.

Dhanapal, K.; Thomas, J.; Kannan, V. R.; Udayan, K.; Gopakumar, B. Studies on Improvement of Seedling Quality and Management of Rot Diseases of *Elettaria Cardamomum* Maton. (Small cardamom) Trough Microbial Inoculants. *J. Plant. Crops* **2006,** *34,* 461–466.

Dietrich, A.; Afawih. In *Encyclopaedia of Islam,* 2nd ed.; Bearman, P.; Bianquis, T., Bosworth, C. E., vanDonzel, E., Heinrichs, W. P., Eds.; Brill: Leiden, 2008.

Doshi, A.; Thakore, B. B. L. Efficacy of Systemic and Non-Systemic Fungicides for the Control of Downy Mildew of Opium Poppy. *Plant Dis. Res.* **2002,** *17*(1), 40–45.

Du Toit, R.; Volsteedt, Y.; Apostolides, Z. Comparison of the Antioxidant Content of Ruits Vegetables and Teas Measured As Vitamin C Equivalents. *Toxicology* **2001,** *166,* 63–69.

Dube, S.; Upadhyay P. D.; Tripathi, S. C. Antifungal, Physico–Chemical And Insect-Repelling Activity of the Essential Oil of *Ocimum basilicum. Can. J. Bot.* **1989,** *67,* 2058–2087.

Dubey, R. B. Correlation and Path Analysis in Opium Poppy (*Papaver somniferum*). *J. Med. Aromat. Plant Sci.* **2010,** *32,* 212–216.

Dyk, S. van; McGlasson, W. B.; Williams, M.; Gair, C. Influence of Curing Procedures on Sensory Quality of Vanilla Beans. *Fruits Paris* **2010,** *65*(6), 387–399.

Dyk, S. van; Holford, P.; Subedi, P.; Walsh, K.; Williams, M.; McGlasson, W. B. Determining the Harvest Maturity of Vanilla Beans. *Sci. Hortic.* **2014,** *168,* 249–257.

Economist.com/specialreports (www.economist.com.rights).

Edris, A. E.; Shalaby, A. S.; Fadel, H. M. Effect of Organic Agriculture Practices On the Volatile Flavor Components of Some Essential Oil Plants Growing in Egypt: III. *Thymus vulgaris* L. Essential Oil. *J. Essent. Oil Bear. Plants* **2009,** *12*(3), 319–326.

Eissa, A. M.; Omer, E. A.; Abou Hadid, A. F. Effect of Saline Water Irrigation On the Dry Matter Production, Nutrition Status and Essential Oil Content of Thyme Plants (*Thymus vulgaris* L.) Grown On A Sandy Soil. *Egyptian J. Hortic.* **2004,** *31*(1), 59–69.

El Nakhlawy, F. S.; Shaheen, M. A.; Al Shareef, A. R. Response of Essential Oil And Terpene Contents of Two Mint Genotypes to Different Drying Temperatures Before Distillation. *J. Agric. Sci. Toronto* **2013,** *5*(12), 126–131.

El Sakka, M. A. Tubail, K. M. Phytochemical Screening of *Salvia officinalis* and *Thymus vulgaris* Irrigated with Different Levels of Saline Water. *Alexandria J. Agric. Res.* **2003,** *48*(1), 159–165.

Elkner, K.; Kaniszewski, S. The Effect of Nitrogen Fertilization on Yield and Quality Factors of Celery (*Apium graveolens* L. var. Dulce Mill/Pers.). *Veg. Crops Res. Bull.* **2001,** *55,* 49–59.

Elumalai, A.; Chinna Eswaraiah, M. A. Pharmacological Review on *Garcinia indica. Int. J. Univ. Pharm. Life Sci.* **2011,** *1*(3), 57–60.

Embong, M. B.; Hadziyev, D.; Molnar, S. Essential Oils from Spices Grown in Alberta. Anise oil (*Pimpinella anisum*). *Can. J. Plant Sci.* **1997,** *57,* 681–688.

Escribano, J.; Alonso, G. L.; Coca-Prados, M.; Fernandez, J. A. Crocin, Safranal and Picrocrocin from Saffron (*Crocus sativus* L.) Inhibit The Growth of Human Cancer Cells in Vitro. *Cancer Lett.* **1996,** *100,* 23–30.

Esiyok, D.; Ilbi, H. Effects of Different Sowing Periods of Parsley (*Petroselinum crispum* Mill.) on Yield under Bornova Conditions. *Ege Universitesi Ziraat Fakultesi Dergisi* **2000,** *37*(2/3), 77–84.

Evans, W. C. *Volatile oils and Resins, Trease and Evans Pharmacognosy*, 15th ed.; W.B. Saunders: London, 2002, p 286.

Farooqi, A.A.; Sreeramulu, B. S. *Cultivation of Medicinal and Aromatic Plants*. University Press India Ltd: Hydrabad, India, **2001**, 43343.

Farooqi, A. A.; Sreeramu, B. S.; Srinivasappa, K. N. *Cultivation of Spice Crops*; Universities Press: Hyderabad, 2005, ISO 6754, 1996.

Farrel, K. T. *Spices, Condiments and Seasonings,* 2nd ed.; AVI Book, Van Nostrand Reinhold: New York, 1990.

Farreyrol, K.; Grisoni, M.; Pearson, M.; Richard, A.; Cohen, D.; Beck, D. Genetic Diversity of Cucumber Mosaic Virus Infecting Vanilla in French Polynesia and Reunion Island. *Aust. Plant Pathol.* **2010,** *39*(2), 132–140.

Fatemeh, H.; Bahram, M.; Farrokh, R. K.; Mehrdad, Y.; Alireza, T. Effect of Bio and Chemical Fertilizers on Seed Yield and Its Components of Dill (*Anethumgraveolens*). *J. Med. Plants Res.* **2013,** *7*(3), 111–117.

Fazal, S. S.; Singla, R. K. Review on the Pharmacognostical and Pharmacological Characterization of *Apium graveolens* Linn. *Indo Global Journal of Pharmaceutical Sciences* **2012,** *2*(1), 36–42.

Fernández, J. A. Biology, Biotechnology and Biomedicine of Saffron. *Recent Res. Dev. Plant Sci.* **2004,** *2,* 127–159.

Ferrao, J. E. M. Spices: Cultivation Technology—Trade. *Especiarias cultura Tecnologia Comercio.* Centro de Documentação e Informação, Instituto de Investigação Científica Tropical: Lisboa, Portugal, 1993; p 413.

Filipovic, V.; Radivojevic, S.; Ugrenovic, V.; Jacimovic, G.; Kuzevski, J.; Subic, J.; Grbic, J. Effects of a Certified Organic Fertilizer on the Yield and Market Quality of Root Parsley (*Petroselinum crispum* (Mill) Nym. ex A.W. *Hill* ssp. *tuberosum* (Bernh.) Crov.). *Afr. J. Biotechnol.* **2012,** *11*(38), 9182–9188.

Fochesato, M. L.; Martins, F. T.; Souza, P. V. D.; Schwarz, S. F.; Barros, I. B. I. Leaf Amounts and Indolebutyric Acid on *Laurus nobilis* l. Cuttings Propagation. *Rev. Bras. Plant. Med.* **2006,** *8*(3),72–77.

Frabboni, L.; de Simone, G.; Russo, V. The Influence of Different Nitrogen Treatments on the Growth and Yield of Basil (*Ocimum basilicum* L.). *J. Chem. Chem. Eng.* **2011,** *5,* 799–803.

Fraser, S.; Whish, J. P. M. A Commercial Herb Industry for NSW—An Infant Enterprise, RIRDC Research Paper, 1997, Series No. 97/18.

Freedman, P. *Out of the East: Spices and the Medieval Imagination;* Yale University Press: New Haven, 2008.

Frenkel, C.; Ranadive, A. S.; Tochihuitl-Vazquez, J.; Havkin-Frenkel, D. Curing of Vanilla. In: *Handbook of vanilla science and technology. Havkin-Frenkel,-D. & F. Belanger (Eds.)* Wiley Blackwell Publishing Ltd., **2011**, 79–106.

Furamanova, M.; Sowinska, D.; Pietrosiuk, A. In *Biotechnology in Agriculture and forestry*; Bajaj, Y. P. S., Ed.; Springer-Verlag: Berlin, 1991, p 176.

Gale, J. Nutmeg Market Largely Unmoved. The Public Ledger, Published: 21 December 2016. https://www.agra-net.com/agra/public-ledger/commodities/spices/nutmeg/ nutmeg-market-largely-unmoved-536574.htm. (accessed May 1, 2017).

Galigani, P.F.; Garbati Pegna, F. Mechanized Saffron Cultivation Including Harvesting. In *Saffron: Crocus sativus L.*; Negbi, M., Ed.; Harwood Academic Publishers: Australia, 1999; pp 115–126.

Gallage, N. J.; Hansen, E. H.; Kannangara, R.; Olsen, C. E.; Motawia, M. S.; Jorgensen, K.; Holme, I.; Hebelstrup, K.; Grisoni, M.; Moller, B. L. Vanillin Formation from Ferulic Acid in Vanilla Planifolia is Catalysed by A Single Enzyme. *Nature Commun.* **2014,** *5*(6), 4037.

Gambella, F.; Paschino, F.; Bertetto, A. M.; Ruggiu, M. Application of Mechanical Device and Airflow Systems in the Harvest and Separation of Saffron Flowers (*Crocus sativus* L.). *Trans. ASABE* **2013,** *56,* 1259–1265.

Ganesh, D. S.; Sreenath, H. L.; Jayashree, G. Micropropagation of Vanilla Through Node Culture. *J. Plant. Crops* **1996,** *24*(1), 16–22.

Gao, G. X.; Wang, W. T.; Wu, F.; Liu, H. J.; Xia, J F.; Hao, J. Quan. Effects of Foliar Fertilizer On Yield and Quality of Celery Seed. *China Vegetables* **2013,** *4,* 65–68.

Garlet, T. M. B.; Santos, O. S.; Medeiros, S. L. P.; Manfron, P. A.; Garcia, D. C.; Borcioni, E.; Fleig, V. Production and Quality of Essential Oil of Mint Under Hydroponics With Four Potassium Levels. *Ciencia Rural* **2007,** *37,* 956–962.

Garner-Wizard, M. Interest in Chinese Star Anise as Source of Drug for Avian Flu. *Washington Post* November 18, 2005, http://content.herbalgram.org/bodywise/HerbClip/review.asp? i=44500.

Gavazzi, C.; Tabaglio, V.; Sartori, D.; Nervo, G.; Longo, C. Courgette and Celery: Less Nitrates with Organic. *Informatore Agrario* **2009,** *65*(22), 52–55.

General, T.; Mamatha, V.; Divya, V.; Appaiah, K. A. A. Diversity of Yeast with Beta-Glycosidase Activity in Vanilla (*Vanilla planifolio*) Plant. *Curr. Sci.* **2009,** *96*(11), 1501–1505.

George, C. K.; Sandana, A. Report of the Visit to Vietnam under the Project INT/61/77 on Co-operative Programme on Quality Assurance of Spices, International Trade Centre, Geneva, 2000.

Ghorbani, M. The Efficiency of Saffron's Marketing Channel in Iran. *World Appl. Sci. J.* **2008,** *4*(4), 523–527.

Gilbert, J.; Nursten, H. E. Volatile Constituents of Horseradish Roots. *J. Sci. Food Agric.* **1972,** *23,* 527–539.

Girma, H.; Digafie, T.; Habtewold, K.; Haimanot, M. The Effect of Different Node Number Cuttings on Nursery Performance of Vanilla (*Vanilla planifolia* syn. *fragrans*) Cuttings in South Western Ethiopia. *Int. Res. J. Agric. Sci. Soil Sci.* **2012,** *2*(9), 408–412.

Gleń, K. Effect of Non-Chemical and Chemical Protection on Healthiness and Yielding of Horseradish. *Progress in Plant Protection/Postępy W Ochronie Roślin* **2012,** *52*(4), 1165–1169.

Golcz, A.; Bosiacki, M. Effect of Nitrogen Fertilization Doses And Mycorrhization On The Yield And Essential Oil Content In Thyme (*Thymus vulgaris* L.). *J. Res. Appl. Agric. Eng.* **2008,** *53*(3), 72–74.

Golcz, A.; Politycka, B. Contents of Chlorophyll and Selected Mineral Components and the Yield of Common Thyme (*Thymus vulgaris* L.) at Differentiated Nitrogen Fertilization. *Herba Polonica* **2009,** *55*(3), 69–75.

Goliaris, A. H. Saffron Cultivation in Greece. In *Saffron: Crocus sativus L.*; Negbi, M., Ed., Harwood Academic Publishers: Australia, 1999; pp 73–85.

Golparvar, A. R.; Bahari, B. Effects of Phenological Stages on Herbage Yield and Quality/Quantity of Oil in Garden Thyme (*Thymus vulgaris* L.). *Journal of Medicinal Plants Research* **2011,** *5*(25), 6085–6089.

Graifenberg, A.; Botrini, L.; Marchetti, L.; Carmassi, G. Effect of Increasing Doses of N, P and K in Curly Parsley (*Petroselinum crispum* Mill. spp. *crispum*). *Colture Protette* **2011,** *40*(3), 58–64.

Graifenberg, A.; Botrini, L.; Marchetti, L. Macro and Microelement Deficiencies in Curly Parsley (*Petroselinum crispum* Mill. spp. *crispum*). *Colture Protette* **2009,** *38*(5), 72–83.

Gregory, M. J.; Menary, R. C.; Davies, N.W. Effect of Drying Temperature and Air Flow on the Production and Retention of Secondary Metabolites in Saffron. *J. Agric. Food Chem.* **2005,** *53*(15), 5969–5975.

Gresta De Juan, J. A.; Moya, A.; Lopez, S.; Botella, O.; Lopez, H.; Muòoz, R. Influencia del tamaòo del cormo y la densidad de plantación de la producción de cormos de *Crocus sativus* L. *ITEA* **2003,** *99,* 169–180.

Grevsen, K.; Christensen, L. P. Common Thyme. *Gron Viden, Havebrug* **2004,** *161,* 12.

Grilli Caiola, M. Saffron Reproductive Biology. *Acta Hort.* **2004,** *650,* 25–37.

Gruszecki, R. Effect of Cultivar on Early Yield of Parsley Grown From the Late Summer Sowing. Folia Horticulturae **2004,** *16*(2), 27–32.

Gruszecki, R.; Salata, A. Effect of Sowing Date on Biometrical Features of Hamburg Parsley Plants. *Modern Phytomorphology* **2013,** *3,* 125–129.

Gudade, B. A.; Chhetri, P.; Gupta, U.; Deka, T. N.; Vijayan, A. K. Traditional Practices of Large Cardamom Cultivation in Sikkim and Darjeeling. *Life sciences Leaflets* **2013,** *9,* 62–68.

Gulcin, I. Antioxidant Activity of Caffeic Acid (3,4-dihydroxycinnamic acid). *Toxicology* **2006,** *217,* 213–220.

Gulcin, I.; Oktay, M.; Kirecci, E.; Kufrevioglu, O. I. Screening of Antioxidant and Antimicrobial Activities of Anise (*Pimpinella anisum* L.) Seed Extracts. *Food Chemistry* **2003,** *83*(3), 371–382.

Guo, J.; Kong, D; Hu, L. Comparative Analysis of Volatile Flavor Compounds of Poppy Seed Oil Extracted by Two Different Methods Via Gas Chromatography/Mass Spectrometry. *J. Pharm. Sci. Technol.* **2015,** *4*(2), 36–38.

Gupta, A., Gupta, R.; Lal, B. Effect of *Trigonella Foenum-Graecum* (fenugreek) Seeds on Glycaemic Control and Insulin Resistance in Type 2 Diabetes Mellitus: A Double Blind Placebo Controlled Study. *J.Assoc.Physicians India* **2001,** *49,* 1057.

Gurdil, G. A. K.; Demirel, B.; Herak,D.; Karansky,J.; Simanjuntak,S. Soaked and Non-Soaked Drying Methods for Vanilla (*Vanilla planofilia* L.). *AMA-Agricultural Mechanization in Asia Africa and Latin America* **2011,** *42*(3), 66–70.

Gyongyosi, R.; Pluhar, Z.; Sarosi, S.; Rajhart, P. Development of Cutting Technology of Garden Thyme (*Thymus vulgaris* L.). *Kertgazdasag Horticulture* **2008,** *40*(3), 66–75.

Habib U.; Athar M.; Honermeier, B. Essential Oil and Composition of Anise (*Pimpinella anisum* L.) with Varying Seed Rates and Row Spacing. *Pakistan Journal of Botany* **2014,** *46*(5), 1859–1864.

Habib, U.; Athar, M.; Masood, I. A.; Bernd, H. Effect of Row Spacing and Seed Rate on Fruit Yield, Essential Oil and Composition of Anise (*Pimpinella anisum* L.) *Pakistan Journal of Agricultural Sciences* **2015,** *52*(2), 349–357.

Haldankar, P. M.; Nagwekar, D. D.; Desai, A. G.; Rajput, J. C. Factors Influencing Epicotyl Grafting in Nutmeg. *Journal of Medicinal and Aromatic Plant Sciences.* **1999,** *21,* 940–944.

Haldankar, P. M.; Joshi, G. D.; Jamadagni, B. M.; Patil, B. P.; Kelaskar, A. J.; Sawant, V. S. Studies on Genotypic Response of Nutmeg to Softwood Grafting. *Journal of Spices and Aromatic Crops* **2003,** *12,* 139–145.

Haldankar, P. M.; Nagwekar, D. D.; Desai, A. G. Studies on Softwood Grafting in Nutmeg. *Indian Cocoa, Arecanut and Spices Journal* **1997,** *21,* 96–100.

Haldankar, P. M.; Nagwekar, D. D.; Desai, A. G.; Patil, J. L.; Gunjate, R.T. The Effect of Season and the Rootstocks on Approach Grafting in Nutmeg. *Indian Journal of Arecanut, Spices and Medicinal Plants* **1999,** *1,* 52–54.

Haldankar, P. M.; Salvi, M. J.; Joshi, G. D.; Patil, J. L. Effect of Season and Shade Provision on Softwood Grafting of Kokam. *Indian Cocoa, Arecanut and Spices Journal,* **1991,** *14*(4), 158–159.

Hamed, B. A. Effect of Incorporation of Date Palm Seeds with Soil on the Growth of Parsley Plant Growing Under Drought Stress. *American Eurasian Journal of Agricultural and Environmental Science* **2009,** *5*(6), 733–739.

Harizanova Petrova, B.; Ovcharova, A. Effect of the Irrigation Regime on the Productivity of Root Celery by Drip Irrigation in the Plovdiv region. *Agricultural Science and Technology* **2013,** *5*(1), 53–57.

Hartmann, H. D. Parsley Production. *Gemuse Munchen* **1996,** *32*(5), 359.

Havkin Frenkel, D.; Frenkel, C. Postharvest Handling and Storage of Cured Vanilla Beans. *Stewart Postharvest Rev.* **2006,** *2*(4), 1–9.

Heath, H. B. *Source book of flavors*. AVI. West port CT, **1981,** 222–223.

Hegde, N. K.; Siddappa, R.; Hanamashetti, S. I. Response of Curry Leaf (*Murraya koenigii* Spreng) 'Suvasini' for Foliar Spray of Vermiwash and Nutritional Treatments. *Acta Hortic.* **2012,** *933,* 279–284.

Hegnauer, R. The Content of Essential Oil and Carvone in Various Species of Mint. *Ber. Schweiz. Bot. Ges.* **1953,** *63,* 90–102.

Hemphill I. *Spice Notes—A Cook's Compendium of Herbs and Spices*; Pan MacmillanAustralia Pty Ltd: Sydney, **2000**.

Henriksen, K.; Bjorn, G. Cultivation of Horse Radish. *GronViden Havebrug* **2004,** *160,* 6.

Herb File. Global Garden. *global-garden.com.au.*

Hernandez, H. J. *Vanilla Diseases. Handbook of Vanilla Science and Technology*; 2011, pp 26–39.

He, Z.; Jiang, D. M.; Liu, A. Q.; Sang-Li, W.; Li-Wen, F.; Li-Shi, F. The Complete Sequence of Cymbidium Mosaic Virus from *Vanilla fragrans* in Hainan, China. *Virus Genes* **2011,** *42*(3), 440–443.

Hooker, J. D. *Flora of British India;* L. Reeve and Co. Ltd: Kent. IV, 1885; pp 607–609.

Hornok, L. *Cultivation and Processing of Medicinal Plants;* Academic Publication: Budapest, 1992; p 338.

http://www.cassiafromchina.com/ (accessed April 29, 2017).

http://www.foodmanufacturing.com/news/2014/03/consumer-trends-seasonings-spices-sector-grows-economy (accessed April 28, 2017).

http://www.persistencemarketresearch.com/mediarelease/seasoning-spices-market.asp (accessed April 28, 2017).

http://www.steirerkren.at/en/strong-demand/facts-and-figures.html (accessed April 29, 2017).

https://www.agra-net.com/agra/public-ledger/commodities/spices/cinnamon/cassia-and-cinnamon-markets-holding-firm-536569.htm (accessed April 29, 2017).

Hunault, G.; Maatar, A. Enhancement of Somatic Embryogenesis Frequency by Gibberellic Acid in Fennel. *Plant Cell Tissue Organ Cult.* **1995,** *41*(2), 171–176.

Hussein, A. H.; Said-Al Ahl Atef, M.; Sarhan Abou Dahab, M.; AbouDahab El-Shahat, N.; Abou-Zeid Mohamed, S.; Nabila, A. Y.; Naguib Magda, A.; El-Bendary. Essential Oils of *Anethum graveolens* L.: Chemical Composition and Their Antimicrobial Activities at Vegetative, Flowering and Fruiting Stages of Development. *Int. J. Plant Sci. Ecol.* **2015,** *1*(3), 98–102.

Ishikawa, T. M.; Kudo, M.; Kitajima, J. Water-Soluble Constituents of Dill. *Chem. Pharm. Bull.* **2002,** *55,* 501–507.

ISO. Dried Thyme (*Thymus vulgaris*)—Specification. ISO 6754 second edition, 1996, 04–16.

Jabbari, R.; Dehaghi, M. A.; Sanavi, A. M. M.; Agahi, K. Nitrogen and Iron Fertilization Methods Affecting Essential Oil and Chemical Composition of Thyme (*Thymus vulgaris* L.) Medical Plant. *Adv. Environ. Biol.* **2011,** *5*(2), 433–438.

Jackson, S. G.; Goodbrand, R. B.; Ahmed, R.; Kasatiya, S. *Bacillus cereus* and *Bacillus thuringiensis* Isolated in a Gastroenteritis Outbreak Investigation. *Lett. Appl. Microbiol.* **1995,** *21*, 103–105.

Jadhav, D.; Rekha, B. N.; Gogate, P. R.; Rathod, V. K. Extraction of Vanillin from Vanilla Pods: A Comparison Study of Conventional Soxhlet and Ultrasound Assisted Extraction. *J. Food Eng.* **2009,** *93*, 421–426.

Jaenicke, H.; Hoeschle-Zeledon, I. Strategic Framework for Underutilized Plant Species Research and Development, with Special Reference to Asia and the Pacific, and to Sub-Saharan Africa. ICUC, Colombo and GFU, Rome. 2006, 33.

Jaenicke, H. Nutzung vernachlässigter Pflanzenarten und ihres genetischen Potentials zur Verbesserung der Welternährungslage—Status und Aussichten eines nachhaltigen Beitrags deutscher Wissenschaftseinrichtungen. Gutachten erstellt für das Büro für Technikfolgenabschätzung beim Deutschen Bundestag im Rahmen des TA-Projekts, Welchen Beitrag kann die Forschung zur Lösung des Welternährungsproblems leisten? 2009, 39.

Jamwal, M.; Kaul, B. L. Studies on Floral-Biology and Breeding System of *Apium graveolens* L. (Celery). *J. Non Timber Forest Prod.* **1997,** *4*, 73–77.

Jaramillo-Villanueva, J. L.; Escobedo-Garrido, J. S.; Barrera-Rodriguez, A.; Herrera-Cabrera, B. E. Economic Efficiency of Vanilla Curing (*Vanilla planifolia* J.) in the Totonacapan Region, Mexico. *Rev. Mexicana Ciencias Agricolas* **2013,** *4*(3), 477–483.

Javadzadeh, S. M. Prospects and Problems for Enhancing Yield of Saffron (*Crocus sativus*) in Iran. *Int. J. Agric.* **2011,** *1*, 21–25.

Jayaprakasha, G. K.; Sakariah, K. K. Determination of Organic Acids in Leaves and Rinds of *Garcinia indica* (Desr.) by LC. *J. Pharm. Biomed. Anal.* **2002,** *28*, 379–384.

Jayasekhar, M.; Manonmani, K.; Justin, C. G. L. Development of Integrated Biocontrol Strategy for the Management of Stem Rot Disease (*Fusarium oxysporum* esp. vanillae) of Vanilla. Agric. Sci. Digest **2008,** *28*(2), 109–111.

Jiang, Z.; Tao Li, Rang, Y. Jimmy, C. Pungent Components from Thioglucosides in *Armoracia rusticana* Grown in China, Obtained by Enzymatic Hydrolysis. *Food Technol. Biotechnol.* **2006,** *44*(1), 41–45.

Johnykallupurackal, A.; Ravindran, P. N. High Yielding Spices Varieties Developed in India. In Advances in Spices Research; Ravindran, P. N., Nirmalbabu, K., Shiva, K. N., Johnykallupurackal, A. Eds.; Agrobios (India): Jodhpur, 2006, pp 93–139.

Joshi, M.; Dhar, U. Effect of Various Presowing Treatments on Seed Germination of *Heracleum candicans* Wall. Ex DC: A High Value Medicinal Plant. *Seed Sci. Technol.* **2003,** *31*, 737–743.

Joy, C. M.; Pittappillil, G. P.; Jose, K. P. Quality Improvement of Nutmeg Using Solar Tunnel Dryer. *J. Plant. Crops* **2000,** *28*, 138–143.

Joy, P. P.; Thomas, J.; Samuel, M.; Baby, P. S. Medicinal Plants. Kerala Agricultural University. Aromatic and Medicinal Plants Research Station, 1998, 1–211.

Juarez Rosete, C. R.; Aguilar Castillo, J. A.; Rodriguez Mendoza, M. N. Fertilizer Source in Biomass Production and Quality of Essential Oils of Thyme (*Thymus vulgaris* L.). Eur. J. Med. Plants **2014,** *4*(7), 865–871.

Juarez Rosete, C. R.; Aguilar Castillo, J. A.; Rodriguez Mendoza, M. N.; Trejo Tellez, L. I. Mineral and Organic Nutrition in Biomass Production and Quality of Essential Oils of Thyme (*Thymus vulgaris* L.). *Acta Hortic.* **2014,** *1030*, 119–123.

Justin, C. G. L.; Leelamathi, M.; Johnson, S. B. N.; Thangaselvabai, T. Seasonal Incidence and Management of the Giant African Snail, *Achatina fulica* (Bowdich) (Gastropoda: Achatinidae) on Vanilla. *Pest Manage. Econ. Zool.* **2008,** *16*(2), 235–238.

Kamble, R. M.; Pillai, P. Efficacy of Kokam Health Drink in Obesity. *Asian J. Home Sci.* **2012,** *7*(2), 508–513.

Kanakis, C. D.; Tarantilis, P. A.; Tajmir-Riahi, H. A.; Polissiou, M. G. Crocetin, Dimethylcrocetin, and Safranal Bind Human Serum Albumin: Stability and Antoxidative Properties. *J. Agric. Food Chem.* **2007,** *55*(3), 970–977.

Kaniszewski, S.; Dysko, J. Effect of Drip Irrigation and Cultivation Methods on the Yield and Quality of Parsley Roots. *J. Elementol.* **2008,** *13*(2), 235–244.

Karamian, R. Plantlet Regeneration via Somatic Embryogenesis in Four Species of Crocus. *Acta Hortic.* **2004,** *650*, 253–259.

Kareparamban, J. A.; Nikam, P. H.; Jadhav, A. P.; Kadam, V. J. *Ferula foetida* "Hing": A Review. *Res. J. Pharm. Biol. Chem. Sci.* **2012,** *3*(2),775–786.

Karkanis, A.; Bilalis, D.; Efthimiadou, A.; Katsenios, N. The Critical Period for Weed Competition in Parsley (*Petroselinum crispum* (Mill.) Nyman ex A.W. Hill) in Mediterranean Areas. *Crop Prot.* **2012,** *42*, 268–272.

Karkleliene, R.; Dambrauskiene, E.; Juskeviciene, D.; Radzevicius, A.; Rubinskiene, M.; Viskelis, P. Productivity and Nutritional Value of Dill and Parsley. *Hortic. Sci.* **2014,** *41*(3),131–137.

Kattimani, K. N.; Reddy, Y. N. Effect of Time of Harvest on Growth, Biomass, Oil Yield and Nutrient Uptake by Japanese Mint Under Semi-Arid Tropical Climate of Andhra Pradesh, Karnataka. *J. Agric. Sci.* **2001,** *14*, 704–707.

Kattimani, K. N.; Reddy, Y. N. Effect of Planting Material and Spacing on Growth, Biomass, Oil Yield and Quality of Japanese Mint (*Mentha arvensis* L.). *J. Essent. Oil Bear. Plants* **2000,** *3*, 33–43.

Kattimani, K. N.; Reddy, Y. N.; Pasha, S. N.; Sarma, P. S.; Rao, B. R. R.; Sathe, A. Standardization of Production Technology of Menthol Mint *Mentha arvensis* Under Semi-Arid Climate of Andhra Pradesh in Southern India. *J. Med. Aromat. Plant Sci.* **2001,** *22/23*, 417–425.

Kaul, V. K.; Nigam, S. S. Antibacterial and Antifungal Studies of Some Essential Oils. *J. Res. Indian Med. Yoga Homoeo.* **1977,** *12*(3), 132.

Kaur, G. J.; Arora, D. S. Bioactive Potential of *Anethum graveolens, Foeniculum vulgare* and *Trachyspermum ammi* Belonging to the Family Umbelliferae—Current Status. *J. Med. Plants Res.* **2010,** *4*(2), 87–94.

Kaur, G. J.; Arora, D. S. Antibacterial and Phytochemical Screening of *Anethum graveolens, Foeniculum vulgare* and *Trachyspermum ammi. BMC Complement Altern. Med.* **2009,** *9*, 30.

Kaur, R.; Gill, B. S. Efficacy of Weed Control Practices in Celery Crop Production. Environ. Ecol. **2011,** *29*(2a), 896–899.

Kavvadias, V.; Paschalidis, C.; Vavoulidou, E.; Petropoulos, D.; Koriki A Effects of Soil Amended with Cadmium and Lead on Growth, Yield, and Metal Accumulation and Distribution in Parsley. *Commun. Soil Sci. Plant Anal.* **2012,** *43*(1/2), 161–175.

Keita, S. M.; Vincent, C.; Schmit, J. P.; Belanger, A. Essential Oil Composition of *Ocimum basilicum* L., *O. gratissimum* L. and *O. suave* L. in the Republic of Guinea. *Flavour Fragr. J.* **2000,** *15*, 339–341.

Keser, G.; Buyuk, G. Effects of Wastewater Irrigation on Chemical and Physical Properties of *Petroselinum crispum*. *Biol. Trace Elem. Res.* **2012,** *146*(3), 369–375.

Kewalanand, P.; Bisht, C. S.; Singh, L. D., S. Influence of Stage of Umbel Harvesting and Nitrogen Levels on European Dill (*Anethum graveolens*). *J. Med. Aromat. Plant Sci.* **2001,** *23*(3), 361–364.

Khalid, K. A. Effect of NP and Foliar Spray on Growth and Chemical Compositions of Some medicinal Apiaceae Plants Grow in Arid Regions in Egypt. *J. Soil Sci. Plant Nutr.* **2012,** *12*(3), 581–596.

Khan, A.; Safdar, M. Role of Diet, Nutrients, Spices and Natural Products in Diabetes Mellitus. *Pak. J. Nutr.* **2003,** *2*(1), 1–12.

Khan, I. A.; Ahmad, Z. Studies on Cultivation, Seed Production and Nutrition Quality of Parsley (*Petroselinum hortense*) in Mid Hills of Uttarakhand. *Bioved* **2000,** *11*(1/2), 33–35.

Khandekar, R. G.; Joshi, G. D.; Daghoral, L. K.; Manjarekar, R. G.; Haldankar, P. M. Effect of Time of Softwood Grafting on Sprouting, Survival and Growth of Nutmeg (*Myristica fragrans* Houtt.) Grafts. *J. Plant. Crops* **2006,** *34*, 226–228.

Khare, C. P. *Indian Medicinal Plants: An Illustrated Dictionary;* Springer: New York, 2008; p 900. ISBN:9780387706375.

Khatoon, J.; Verma, A.; Chacko, N.; Sheikh, S. Utilization of Dehydrated Curry Leaves in Different Food Products. Indian *J. Nat. Prod. Resour.* **2011,** *2*(4), 508–511.

Khatri, I. M.; Nasir, M. K. A.; Saleem, R.; Noor, F. Evaluation of Pakistani Sweet Basil Oil for Commercial Exploition. *Pak. J. Sci. Ind. Res.* **1995,** *38*, 281–282.

Khazaie, H. R.; Nadjafi, F.; Bannayan, M. Effect of Irrigation Frequency and Planting Density on Herbage Biomass and Oil Production of Thyme (*Thymus vulgaris*) and Hyssop (*Hyssopus officinalis*). *Ind. Crops Prod.* **2008,** *27*(3), 315–321.

Khorana, M. L.; Vangikar, M. B. *Ocimum basilicum*. Part II. Antibacterial Properties. *Indian J. Pharm.* **1950,** *12*, 134.

Khorshidi, M.; Bahadori, F.; Behnamnia, M. The Effects of Arbuscular Mycorrhizal Fungi (*Glomus intraradices*) and Vermicompost Application on Yield and Nutrient Uptake in Garden Thyme (*Thymus vulgaris* L.) Under Field Conditions. *Int. J. Agric. Crop Sci.* **2013,** *5*(11), 1191–1194.

Khunsri, S.; Poeaim, A. *In Vitro Propagation of Vanilla planifolia Andr. from Axillary Bud Explants*. In Proceedings of the 47th Kasetsart University Annual Conference, Kasetsart, March 17–20, 2009, pp 630–635.

Kilic, A.; Hafizoglu, H.; Kollmannsberger, H.; Nitz, S. Volatile Constituents and Key Odorants in Leaves, Buds, Flowers, and Fruits of *Laurus nobilis* L. *J. Agric. Food Chem.* **2004,** *52*, 1601–1606.

Kim, S. W.; Park, M. K.; Liu, J. R. High Frequency Plant Regeneration via Somatic Embryogenesis in Cell Suspension Cultures of Coriander. *Plant Cell Rep.* **1996b,** *15*, 751–754.

Kincses, I.; Filep, T.; Kremper, R.; Sipos, M. Effect of Nitrogen Fertilization and Biofertilization on Element Content of Parsley. *Cereal Res. Commun.* **2008,** *36*(Suppl. (5)), 571–574.

Kirtikar, K. R.; Basu, B. D. *Indian Medicinal Plants*, 2nd ed.; Allahabad, 1991; vol. I.

Kirtikar, K. R.; Basu, B. D. *Indian Medicinal Plants;* Orients Longman: Mumbai, India, 1980; pp 21–28.

Klaus, A.; Beatovic, D.; Niksic, M.; Jelacic, S.; Petrovic, T. Antibacterial Activity Oils from Serbia Against the *Listeria monocytogenes*. *J. Agric. Sci.* (Belgrade) **2009,** *54*(2), 95–104.

Klc, C. C.; Anac, D.; Eryuce, N.; Klc, O. G. Effect of Potassium and Phosphorus Fertilization on Green Herb Yield and some Quality Traits of *Thymus vulgaris* L. Afr. J. Agric. Res. **2012,** *7*(48), 6427–6431.

Kmiecik, W.; Lisiewska, Z. Comparison of Leaf Yields and Chemical Composition of Hamburg and Leafy Types of Parsley. I. Leaf Yields and Their Structure. *Folia Hortic.* **1999,** *11*(1), 53.

Kolasinska, K.; Dabrowska, B. Effect of Seed Conditioning on Germination Ability, Vigour and Emergence in Carrot and Parsley. *Biuletyn Instytutu Hodowli I Aklimatyzacji Roslin* 1996, 197, 261–271.

Kolodziej, B. The Effect of Plantation Establishment Method and Foliar Fertilization on the Yields and Quality of Thyme. Ann. Univ. Mariae Curie Sklodowska Sec. E. Agric. **2009,** *64*(2), 1–7.

Kolota, E. Yield and Quality of Leafy Parsley as Affected by the Nitrogen Fertilization. *Acta Scientiarum Polonorum Hortorum Cultus* **2011,** *10*(3), 145–154.

Konstantinidou, E.; Takos, I.;Merou, T. Desiccation and Storage Behavior of Bay Laurel (*Laurus nobilis* L.) Seeds. Eur. J. Forest Res. **2008,** *127*(2),125–131.

Koota, E.; Winiarska, S. The Influence of the Method of Transplanting and Plant Covers on Yielding of Leafy Type of Parsley. *Veg. Crops Res. Bull.* **2004,** *61*, 61–67.

Korikanthimath, V. S.; Ravindra, M. Pre Sowing Seed Treatment to Enhance Germination in Cardamom Karnataka. *J. Agric. Sci.* **1998,** *11*, 540–542.

Korikanthimath, V. S. High Density Planting in Cardamom (*Elettaria cardamomum* Maton.). *South Indian Hortic.* **2001,** *49*, 27–28.

Korikanthimath, V. S. Rapid Clonal Multiplication of Elite Cardamom Selections for Generating Planting Material, Yield Upgradation and its Economics. J. Plant. Crops **1999,** *27*(1), 45–53.

Korikanthimath, V. S.; Rao, G.; Hiremath, G. M. Prospects of Conversion of Marshy Areas a Techno-economic Feasibility for Cardamom Cultivation. *Indian J. Hortic.* **1999,** *56*, 360–367.

Korikanthimath, V. S.; Rajendra, H.; Gaddi, A. V.; Parashuram, C. Nutrient Management in Cardamom. *Fert. News* **2001**, *46*, 37–38, 41–48, 51–53.

Krasaekoopt, W.; Abusali, S. B.; Chayasana, M. Processing of Vanilla Pods Grown in Thailand and its Application. *AU J. Technol.* **2010,** *13*(3), 135–142.

Krishnamoorthy, B. Sex Conversion in Nutmeg, ICAR. *Spice India* **2013,** *13*(5), 11–12.

Krishnamurthy, N.; Lewis, Y. S.; Ravindranath, B. Chemical Constituents of Kokam Fruit Rind. *J. Food Sci. Technol.* **1982,** *19*(3), 97–100.

Krishnamurthy, R.; Avani, K. Impact of Planting Space, Irrigation Pattern and Different Concentration of Fertilizers on the Yield and Yield Related Parameters of Sweet Flag (*Acorus calamus* Linn) in South Gujarat Region of India. *Int. J. Agric.* **2015,** *126*, 405–411.

Kshirsagar, P. J.; Pawar, C. D.; Haldankar, P. M.; Shirodkar, A.; Parulekar, Y. R. Kokam (*Garcinia indica*)—Prospects. SYMSAC VIII Towards 2050—Strategies for Sustainable Spices Production. Indian Society for Spices, Dec 2015, pp 15–23.

Kulathooran, R.; Rao, L. J. M.; Raghvan, B. Physico-chemical Changes on Processing of Curry Leaf (*Murraya koenigii* Spreng.). *J. Med. Aromat. Plant Sci.* **2000,** *22*, 510–516.

Kumar, B.; Patra, N. K. Inheritance Pattern and Genetics of Yield and Component Traits in Opium Poppy (*Papaver somniferum* L.). *Ind. Crops Prod.* **2012,** *36*, 445–448.

Kumar, N.; Lohani, H.; Lehari, A.; Dwivedi, K. N. Composition of Oil from Rosemary (*Rosmarinus officinalis* L.) Grown in Kumaon Hills of Uttaranchal. *Indian Perfumer* 2004, 48, 411–414.

Kumar, R.; Singh, V.; Devi, K.; Sharma, M.; Singh, M. K.; Ahuja, P. S. State of Art of Saffron (*Crocus sativus* L.) Agronomy: A Comprehensive Review. Food Rev. Int. **2009,** *25*, 44–85.

Kumar, S.; Yadav, R. P.; Singh, A. K. Potential and Business Opportunities in Essential Oil Sector. *J. Med. Aromat. Plant Sci.* **2008,** *30*, 336–339.

Kumar, K. K.; Ananthakumar, A. A.; Ahmad, R.; Adhikari, S.; Variyar, P. S.; Sharma, A. Effect of Gamma-Radiation on Major Aroma Compounds and Vanillin Glucoside of Cured Vanilla Beans (*Vanilla planifolia*). *Food Chem.* **2010,** *122*(3), 841–845.

Kuruvilla, K. M.; Madhusoodanan, K. J.; Sudharshan, M. R.; Natarajan, P.; Thomas, J. *Performance Evaluation of Tissue Culture vis-a-vis Open Pollinated Seedlings of Cardamom. Recent Trends in Horticultural Biotechnology,* vol. I and II ICAE National Symposium on Biotechnological Interventions for Improvement of Horticultural Crops: Issues and Strategies, Vellanikkara, Kerala, India, 2005, pp 163–169.

Kwiatkowski, C. A. The Effect of Growth Stimulators and Forecrop on Raw Material Quality and Yield of Garden Thyme (*Thymus vulgaris* L.). *Herba Pol.* **2011,** *57*(1), 5–15.

Kwiatkowski, C. Weed Infestation and Yielding of Garden Thyme (*Thymus vulgaris* L.) in Relation to Protection Method and Forecrop. Prog. Plant Prot. **2007,** *47*(3), 187–190.

Kybal, J.; Kaplicka, J. *Herbs and Spices;* Harveys Bookshop Ltd.; Wingston, Leicester, 1995.

Lahariya, A. K.; Rao, J. T. In Vitro Anti Microbial Studies of the Essential Oil of *Cyperus scariosus* and *Ocimum basilicum* L. *Ind. Drugs* **1979,** *16*, 150–152.

Lala, F.; Assagaf, M.; Mejaya, M. J. The Control of Fruit Dry Blight on Nutmeg Caused by *Stigmina myristicae* (Stein.) Mand.-Sum. et Rifai in Tidore Islands. *Indones. J. Agric.* **2011,** *4*, 52–57.

Lalitha, S.; Thamburaj, S.; Thangaraj, T. A Note on the Effect of Spacing and Nitrogen in Curry Leaf (*Murraya koenigii* Spreng). South Indian Hortic. 1997, 45, 303–304.

Lawrence, B. M. Rosemary Oil. Progress in Essential Oils. *Perfumes Flavors* **1977,** *2*, 34–35.

Lawrence, B. M. A Review of the World Production of Essential Oil. *Perfumes Flavors* **1985,** *10*, 2–16.

Lawrence, B. M. Rosemary Oil. Progress in Essential Oils. *Perfumes Flavors* **1986,** *11*, 81–82.

Lawrence, B. M. Rosemary Oil. Progress in Essential Oils. *Perfumes Flavors* **1989,** *14*, 49–54.

Lawrence, B. M. Rosemary Oil. Progress in Essential Oils. *Perfumes Flavors* **1991,** *16*, 59–60.

Lawrence, B. M. Rosemary Oil. Progress in Essential Oils. *Perfumes Flavors* **1995,** *20*, 47–50.

Lawrence, B. M. Rosemary Oil. Progress in Essential Oils. *Perfumes Flavors* **1997,** *2*, 71–73.

Lawrence, B. M. Rosemary Oil. Progress in Essential Oils. *Perfumes Flavors* **2007,** *32*, 48–54.

Le, D. K.; Nguyen, H. S.; Tran, H. Q.; Nguyen, T. H. Seed Storage Methods of Star Anise (*Illicium verum*), Cinnamon (*Cinnamomum cassia*) and Michelia (*Michelia mediocris*). In *Comparative Storage Biology of Tropical Tree Seeds;* Renouf Publishing Company Limited, 2005; pp 238–248.

Lehmacher, A.; Bockemühl, J.; Aleksic, S. Nationwide Outbreak of Human Salmonellosis in Germany due to Contaminated Paprika and Paprika-Powdered Potato Chips. *Epidemiol. Infect.* **1995,** *115*, 501–511.

Lei, S.; Zhao, F. Q.; Zeng, B. Sensitive Voltammetric Determination of Vanillin with an AuPd Nanoparticles Graphene Composite Modified Electrode. *Food Chem.* **2014,** *151*, 53–57.

Li, X.; Chen, L. S.; Wang, Y. Y. Isozyme Analysis of the Causal Reagents of Vanilla Root-Rot. *J. Yunnan Agric. Univ.* **2005,** *20*(1), 27–30.

Liang, H.; Lu, J. Q.; Dai, Y.; Li, X. S.; Guo, S. N.; Li, Q. Analysis of the Volatile Components in the Fruits of *Vanilla planifoli* Andrews by HS-SPME Combined with GC-MS. *Med. Plant* **2014,** *5*(4), 23–26.

Loi, D. T.; Thu, N. V. *Medicinal Trees and Pharmaceutical Material und Herbs in Vietnam;* Pharmacy College Publisher: Hanoi, 1970; pp 209–211.

Lokesh, M. S.; Suryanarayana, V.; Patil, S. V.; Gurumurthy, S. B.; Palakshappa, M. G.; Manjunath, G. O. Documentation and Management of Anthracnose—A New Nursery Disease of Garciniaindica Choice in Karnataka. *Int. J. Plant Prot.* **2012,** *5*(2), 275–277.

Lopez, M. G.; Sanchez-Mendoza, I. R.; Ochoa-Alejo N Compartive Study of Volatile Components and Fatty Acids of Plants and In Vitro Cultures of Parsley (*Petroselinum crispum* (Mill) nym ex hill). *J. Agric. Food Chem.* **1999,** *47*(8), 3292–3296.

Loskutov, A. V.; Song, G. Q.; Sink, K. C. Evaluation of Methods for Celery (*Apium graveolens* L.) Transformation Using *Agrobacterium tumefaciens* and the Bar Gene as Selectable Marker. *In Vitro Cell. Dev. Biol. Plant* **2008,** *44*(4), 239–245.

Low, T.; Rodd, T.; Beresford, R. *Magic and Medicine of Plants;* Reader's Digest Publications: New South Wales, Australia, 1994.

Lubinsky, P.; Romero Gonzalez, G. A.; Heredia, S. M.; Zabel, S. Origins and Patterns of Vanilla Cultivation in Tropical America (1500–1900): No Support for an Independent Domestication of Vanilla in South America. Handbook of Vanilla Science and Technology, 2011; pp 117–138.

MacLeod, A. J.; Snyder, C. H.; Subramanian, G. Volatile Aroma Constituents of Parsley Leaves. *Phytochemistry* **1985,** *24*(11), 2623–2627.

Madaiah, D. D. K. M.; Shadakshari, Y. G.; Murthy, P. V. Effect of Coffee Pulp Compost and Bio Fertilizers on Bean Yield of Vanilla (*Vanilla planifolia* Andrews). *J. Asian Hortic.* **2006,** *2*(1/2), 68–71.

Madhaiyan, M.; Santhanakrishnan, P.; Pragatheswari, D. Effect of Orchid Mycorrhizal Fungi on the Growth and Nutrient Status of *Vanilla planifolia* Andr. *South Indian Hortic.* **2001,** *49*(Special), 265–267.

Madhubala, R.; Bhadramurthy, V.; Bhat, A. I.; Hareesh, P. S.; Retheesh, S. T.; Bhai, R. S. Occurrence of Cucumber Mosaic Virus on Vanilla (*Vanilla planifolia* Andrews) in India. *J. Biosci.* **2005,** *30*(3), 339–350.

Maggi, L.; Carmona, M.; del Campo, C. P.; Kanakis, C. D.; Anastasaki, E.; Tarantilis, P. A. Worldwide Market Screening of Saffron Volatile Composition. *J. Sci. Food Agric.* **2009,** *89*(11), 1950–1954.

Mahdi, F.; Behjat, S.; Mostafa, H.; Mohammad, T. K.; Barat, A. G. Effects of Fertilizer and Plant Density on Yield and Quality of Anise (*Pimpinella anisum* L.). *J. Agric. Sci.* **2013,** *58*(3), 209–215.

Mahendra, P.; Bisht, S. *Ferula asafoetida*: Traditional Uses and Pharmacological Activity. *Pharmacogn. Rev.* **2012,** *6*(12), 141–146.

Mahesh, S. S.; Saju, K. A.; Vadiraj, B. A.; Thomas, J.; Deka, T. N. Efficacy of Microbial Inoculants and Organic Fertilizer for Establishing Large Cardamom (*Amomum subulatum* Roxb.) Sucker Nursery. *J. Spices Aromat. Crops* **2014,** *23*, 138–142.

Majid, P.; Ali, R.; Khalil, J. The Effects of Irrigation Intervals and Harvesting Time on Grain Yield and Essential Oil of Anise (*Pimpinella anisum* L.). *Iran. J. Field Crop Sci. (Iran. J. Agric. Sci.)* **2014,** *45*(3), 453–460.

Maletic, R.; Jevovic, R. Influence of Locality and Planting and Harvesting Period on the Yield and Quality of Mint Leaves (*Mentha piperita* L., Micham) and its Essential Oil. Review of Research Work at the Faculty of Agriculture. Belgrade 1998, 43, 75–82.

Mallavarapu, G. R.; Singh, M.; Chandra Shekhar, R. S.; Ramesh, S.; Kumar, S. Rosemary Oil: Prospects of its Production in India. *J. Med. Aromat. Plant Sci.* **2000,** *22*, 298–301.

Mameli, M. G.; Zucca, L.; Maxia, M.; Manca, G.; Satta, M. Effects of Different Irrigation Management on Biomass and Essential Oil Production of *Thymus vulgaris* L., *Salvia officinalis* L. and *Rosmarinus officinalis* L., Cultivated in the Southern Sardinian Climate (Italy). *Acta Hortic.* **2011,** *889*, 469–474.

Mandal, B.; Vijayanandraj, S.; Shilpi, S.; Pun, K. B.; Singh, V.; Pant, R. P.; Jain, R. K.; Varadarasan, S.; Varma, A. Disease Distribution and Characterisation of a New Macluravirus Associated with Chirke Disease of Large Cardamom. *Ann. Appl. Biol.* **2012,** *160*, 225–236.

Marthe, F.; Scholze, P. Alternaria Leaf Blight (*Alternaria radicina* Meier, Drechsler& Eddy)—A Serious Crop Disease of Parsley (*Petroselinum crispum* [Mill.] Nym.). *Zeitschrift-fur-Arznei–and-Gewurzpflanzen* **2006,** *11*(3), 145–149.

Martins, A. P. Composition of the Essential Oils of *Ocimum basilicum*. *Plants* **1999,** *65*, 187–189.

Martins, A. P.; Salgueiro, L. R.; Vila, R.; Tomi, F.; Cani-Gueral, S.; Casanova, J.; Proença da Cunha, A.; Adzet, T. Composition of the Essential Oils of *Ocimum canum*, *O. gratissimum* and *O. minimum*. *Planta Med.* **1999,** *65*, 187–189.

Mathew, K. M.; Rao, Y. S.; Kumar, K. P.; Sallykutty, J.; Lakshmanan, R.; Madhusoodanan, K. J. Micropropagation of Curry Leaf (*Murraya koenigii* L.). *J. Spices Aromat. Crops* **1999,** *8*, 77–79.

McGimpsey, J. A.; Douglas, M. H.; Wallace, A. R. Evaluation of Saffron (*Crocus sativus* L.) Production in New Zealand, N.Z. *J. Crop Hortic. Sci.* **1997,** *25*, 159–168.

Meena, M. K.; Kameriya, P. R.; Meena, A. K. Effect of Nitrogen and Phosphorus on Uptake at Different Stages and Harvest Stages of Dill (*Anethum sowa* Roxb). *J. Prog. Agric.* **2012,** *3*(2), 106–108.

Meena, S. S.; Mehta, R. S.; Lal, G.; Sharma, Y. K.; Meena, R. D.; Kant, K. Effect of Sowing Dates and Crop Geometry on Growth and Seed Yield of Dill (*Anethum sowa* L.). *Int. J. Seed Spices.* **2015,** *5*(1), 79–82.

Meena, S. S.; Mehta, R. S.; Meena, R. D.; Meena, N. K.; Singh, B. Effect of Sowing Time and Crop Geometry on Growth and Seed Yield of Dill (*Anethum sowa* L.). *Int. J. Seed Spices.* **2013,** *3*(2), 81–84.

Mehta, R. S.; Meena, S. S.; Vishal, M. K. Yield and Economic Feasibility of Ajwain (*Trochispermum ammi*) Production Under Varying Irrigation Interval, Nutrient Levels and Crop Geometry. *Agric. Sci. Digest.* **2013,** *33*(1), 56–59.

Menon, P. P.; Kuruvilla, K. M.; Madhusoodanan, K. J. *Yield Prediction in Vanilla* (*Vanilla planifolia* Andrews), In Proceedings of the 15th Plantation Crops Symposium Placrosym-XV, Mysore, India, December 10–13, 2002, 414–415.

Mewes, S.; Kruger, H.; Pank, F. Physiological, Morphological, Chemical and Genomic Diversities of Different Origins of Thyme (*Thymus vulgaris* L.). *Genet. Resour. Crop Evol.* **2008,** *55*(8), 1303–1311.

Milind, P.; Isha, D. Golden Benefits of Drinking Kokam-Cola. *Int. Res. J. Pharm.* **2013,** *4*(5), 5–9.

Mini, A.; Devadas, V. S.; Bridjit, T. K. Effect of Organic Nutrition on Disease Tolerance in Vanilla (*Vanilla planifolia* Andrews). *Indian J. Arecanut Spices Med. Plants* **2010,** *12*(2), 12–14.

Mini, A.; Kumar, P. S.; Devadas, V. S. Aerial Roots of Vanilla—The Most Nutrient Absorbing Plant Part. *Indian J. Arecanut Spices Med. Plants* **2010,** *12*(1), 11–12.

Minoo, D.; Sajina, A.; Babu, K. N.; Ravindran, P. N. *Ovule Culture of Vanilla and its Potential in Crop Improvement. Biotechnolog of Spices, Medicinal and Aromatic Plants,* In Proceedings of the National Seminar on Biotechnology of Spices and Aromatic Plants, Calicut, India, April 24–25, 1996, 1997, 112–118.

Minuto, A.; Garibaldi, A. Powdery Mildew of Rosemary Caused by Oidium sp. *Colture-Protette* **1997,** *26*(3), 83–85.

Mirdad, Z. M. Effect of Irrigation Intervals, Nitrogen Sources and Nitrogen Levels on Some Characters of Parsley (*Petroselinum crispum* Mill). *J. King Abdulaziz Univ. Meteorol. Environ. Arid Land Agric. Sci.* **2011,** *22*(1), 3–17.

Misra, A.; Srivastava, N. K. Influence of Water Stress on Japanese Mint. *J. Herbs Spices Med. Plants* **2000,** *7*, 51–58.

Mitchell, A. *Fertilizer Guide: Peppermint and Spearmint;* Oregon State University: Corvallis, 1998.

Moaveni, P.; Farahani, H. A.; Maroufi, K. Effects of Sowing Date and Planting Density on Quantity and Quality Features in Thyme (*Thymus vulgaris* L.). *Adv. Environ. Biol.* **2011,** *5*(7), 1706–1710.

Mohammadjavad, S.; Gholamreza, M.; Hamidreza, N. Effect of Irrigation and Planting Date on the Selected Morphophenological and Quality Traits of Ajowan (*Carum copticum* Benth. & Hook.f.). *J. Med. Plants By-prod.* **2014,** *2*, 97–106.

Mohan Rao, L. J. Quality of Essential Oils and Processed Materials of Selected Spices and Herbs. *J. Med. Aromat. Plant Sci.* **2000,** *22*, 808–816.

Mohanalakshmi, M.; Thangaraj, T.; Shanmugasundaram, K. A. Effect of Maturity and Season on Rooting of Curry Leaf Cuttings. Spices and Aromatic Plants Challenges and Opportunities in the New Century Contributory Papers Centennial Conference on Spices and Aromatic Plants, Calicut, Kerala, India, 2000, p 221.

Molina, R. V.; Garcia Luis, A.; Valero, M.; Navarro, Y.; Guardiola, J. L. Extending the Harvest Period of Saffron. *Acta Hortic.* **2004,** *650*, 219–225.

Mollafilabi, A. Experimental Findings of Production and Echo Physiological Aspects of Saffron (*Crocus sativus* L.). *Acta Hortic.* **2004,** *650*, 195–200.

Mollafilabi, A.; Aslami, M. H.; Shoorideh, H. Replacement of Saffron (*Crocus sativus* L.) with Poppy (*Papaver somniferum* L.) and its Socio-economic Results in Afghanistan. *Acta Hortic.* **2010,** *850,* 299–202.

Montvale, N. J. *PDR for Herbal Medicines;* Medical Economics Company: St. Louis, MO, 1998; pp 1184–1185.

Moorthy, V. K.; Moorthy, A. K. Organic Cultivation of Vanilla—Varanashi method. 6th IFOAM Asia Scientific Conference, Yangpyung, Korea, September 7–11, 2004, Benign Environment and Safe Food, 2004, 469–475.

Moorthy, V. K.; Moorthy, A. K.; Rao, K. B. *Effect of Compost on Growth of Cardamom. Water and Nutrient Management for Sustainable Production and Quality of Spices,* In Proceedings of the National Seminar, Madikeri, Karnataka, India, 1997, pp, 85–88.

Morton, M.; Cupboard, L. *A Dictionary of Culinary Curiosities;* Insomniac Press: Toronto, Canada, 2004.

Moussavi-Nik, S. M.; Salari, M.; Mobasser, H. R.; Keshavarzi, M. H. B. The Effect of Different Irrigation Intervals and Mineral Nutrition on Seed Yield of Ajowan (*Trachyspermum ammi*). *Ann. Bio. Res.* **2011,** *2,* 692–698.

Murtagh, G. J.; Southwell, I. A.; Curtis, A.; Stiff, I. A. Essential Oil and Pharmaceutical Crops for North-Eastern Australia. Rural Credits Development Fund Projects No. NSWDA/8637 and NSWDA/8933, 1990.

Murthy, G.; Umesha, K.; Smitha, G. R.; Krishnamanohar, R. Effect of Growth Regulators and Bio-inoculants on Rooting and Growth of Vanilla Stem Cuttings. *Indian J. Hortic.* **2010,** *67*(1), 90–93.

Muvel, R.; Naruka, I. S.; Chundawat, R. S.; Shaktawat, R. P. S.; Rathore, S. S.; Verma, K. S. Production, Productivity and Quality of Ajwain (*Trachyspermum ammi* L. Sprague) as Affected by Plant Geometry and Fertilizer Levels. *Int. J. Seed Spice* **2015,** *5*(2), 32–37.

Mylavarapu, R. S.; Zinati, G. M. Improvement of Soil Properties Using Compost for Optimum Parsley Production in Sandy Soils. *Sci. Hortic.* **2009,** *120*(3), 426–430.

Nabizadeh, E.; Habibi, H.; Hosainpour, M. The Effect of Fertilizers and Biological Nitrogen and Planting Density on Yield Quality and Quantity of *Pimpinella anisum* L. *Eur. J. Exp. Biol.* **2012,** *2,* 1326–1336.

Nadjafi, F.; Bannayan, M.; Tabrizi, L.; Rastgoo, M. Seed Germination and Dormancy Breaking Techniques for *Ferula gummosa* and *Teucrium polium*. *J. Arid. Environ.* **2006,** *64,* 542–547.

Naik, B. G.; Saifulla, M.; Manjunath, B.; Prasad, P. S. Status of Stem Rot of Vanilla Incited by *Fusarium oxysporium* f. sp. Vanillae in Karnataka. *Trends Biosci.* **2010,** *3*(2), 206–207.

Naik, H. P.; Balasubrahmanyam, N.; Dhanaraj, S.; Gurudutt, K. N. Packaging and Storage Studies in Flue-Cured Large Cardamom (*Amomum subulatum* Roxb.). *J. Food Sci. Technol.* **2000,** *37*(6), 577–581.

Naik, J. P.; Ramesh, B. S.; Gurudutt, K. N. Fumigation Studies on Cured Large Cardamom (*Amomum subulatum* Roxb.) Capsules. *J. Food Sci. Technol.* **2005,** *42*(6), 531–533.

Najla, S.; Sanoubar, R.; Murshed, R. Morphological and Biochemical Changes in Two Parsley Varieties upon Water Stress. *Physiol. Mol. Biol. Plants* **2012,** *18*(2), 133–139.

Naruka, I. S.; Singh, P. P.; Megha, B.; Rathore, S. S. Effect of Spacing and Nitrogen Levels on Growth, Yield and Quality of Ajwain (*Trachyspermum ammi* L. Sprague). *Int. J. Seed Spices* **2012,** *2*(1), 12–17.

Nawrocki, J.; Mazur, S. Effectiveness of Some Means Using Against Root Rot on Parsley Seedling Root. *Commun. Agric. Appl. Biol. Sci.* **2004,** *69*(4), 693–696.

Negbi, M.; Dagan, B.; Dror, A.; Basker, D. Growth, Flowering, Vegetative Reproduction and Dormancy in the Saffron Crocus (*Crocus sativus* L.). *Israel J. Bot.* **1989,** *38*, 95–113.

NHB, Indian Horticulture Database-2010. National Horticulture Board, Ministry of Agriculture, Government of India, 2011.

NHB, Indian Horticulture Database-2014. National Horticulture Board, Ministry of Agriculture, Government of India, 2014.

Nickavar, B.; Mojab, F.; Dolatbadi, R. Analysis of the Essential Oils of two Thymus species from Iran. *Food Chem.* **2005,** *90*, 609–611.

Nicola, S.; Fontana, E.; Hoeberechts, J. Cultural Techniques to Optimize the Thyme (*Thymus vulgaris*) Propagation. *Acta Hortic.* **2004,** *631*, 187–192.

Nielsen, L. R.; Siegismund, H. R. Interspecific Differentiation and Hybridization in Vanilla Species (Orchidaceae). *Heredity* **1999,** *83*(5), 560–567.

Nimet, K. Yield, Quality, and Growing Degree Days of Anise (*Pimpinella anisum* L.) Under Different Agronomic Practices. *Turk. J. Agric. For.* **2015,** *39*, 1–9.

Nirmal Babu, K.; Minoo, D.; Geetha, S. P.; Samsudeen, K.; Rema, J.; Ravindran, P. N.; Peter, K. V. Plant Biotechnology—It's Role in Improvement of Spices. *Indian J. Agric. Sci.* **1998,** *68*(Special Issue.), 533–547.

NISCHAIR. *Wealth of India: Raw Material*, CSIR; NISCHAIR Press: New Delhi, 2003; vol. 8, p 98.

Nissar, V. A. M.; Hrideek, T. K.; Kuruvilla, K. M.; Madhusoodanan, K. J.; Thomas J Studies on Pollination, Inter Specific Hybridization and Fruit Development in Vanilla. *J. Plant. Crops* **2006,** *34*(3), 167–170.

Odoux, E.; Brillouet, J. M. Anatomy, Histochemistry and Biochemistry of Glucovanillin, Oleoresin and Mucilage Accumulation Sites in Green Mature Vanilla Pod (*Vanilla planifolia*; Orchidaceae): A Comprehensive and Critical Reexamination. *Fruits Paris* **2009,** *64*(4), 221–241.

Ogbuchiekwe, E. J.; McGiffen, M. E. Jr. Efficacy and Economic Value of Weed Control for Drip and Sprinkler Irrigated Celery. *Hortic. Sci.* **2001,** *36*(7), 1278–1282.

Omidbaigi, R. Effect of Corms Weight on Quality of Saffron (*Crocus sativus* Linn.). *Nat. Prod. Radiance* **2005,** *4*, 193–194.

Omidbaigi, R.; Rezaei, N. A. The Influence of Nitrogen-Fertilizer and Harvest Time on the Productivity of *Thymus vulgaris* L. *Int. J. Hortic. Sci.* **2000,** *6*(4), 43–46.

Orav, A.; Raal, A.; Arak, E. Essential Oil Composition of *Pimpinella anisum* L. Fruits from Various European Countries. *Nat. Prod. Res.* **2008,** *22*(3), 227–232.

Ortiz, E. L. (Ed.) *The Encyclopedia of Herbs, Spices, and Flavorings;* Sterling Publishing Co., Inc.: New York, 1996.

Osbaldeston, T. A.; Wood, R. P. A. (Transl.) *Dioscorides, De Materia Medica;* Ibidis Press: Johannesburg, 2000.

Osinska, E.; Roslon, W.; Drzewiecka, M. The Evaluation of Quality of Selected Cultivars of Parsley (*Petroselinum sativum* L. ssp. crispum). *Acta Scientiarum Polonorum Hortorum Cultus* **2012,** *11*(4), 47–57.

Osorio, A. I.; Osorio Vega, N. W.; Diez, M. C.; Moreno, F. H. Nutrient Status and Vegetative Growth of *Vanilla planifolia* Jacks Plants as Affected by Fertilization and Organic Substrate Composition. *Acta Agron. Univ. Nacional Colombia* **2014,** *63*(4), 326–334.

Osorio, A. I.; Osorio, N. W.; Diez, M. C.; Moreno, F. H. Effects of Organic Substrate Composition, Fertilizer Dose, and Microbial Inoculation on Vanilla Plant Nutrient Uptake and Growth. *Acta Hortic.* **2012,** *964,* 135–142.

Oya, T.; Osawa, T.; Kawakishi, S. Spice Constituents Scavenging Free Radicals and Inhibiting Pentosidine Formation in a Model System. *Biosci. Biotechnol. Biochem.* **1997,** *61,* 263–266.

Özcan, M. Inhibitory Effects of Spice Extracts on the Growth of *Aspergillus parasiticus* NRRL 2999 strain. Z. Leb.-Mitt. Unters. u- Forsch. A 207, 253–255.

Ozcan, M.; Chalchat, J. Essential Oil Composition of *Ocimum basilicum* L. and *Ocimum minimum* L. in Turkey. *Czech J. Food Sci.* **2002,** *20*(6), 223–228.

Ozcan, M.; Erkmen, O. Antimicrobial Activity of the Essential Oils of Turkish Plant Species. *Eur. Food Res. Technol.* **2001,** *212,* 658–660.

Ozcan, M. M.; Chalchat, J. C. Chemical Composition and Antifungal Effect of Anise (*Pimpinella anisum* L.) Fruit Oil at Ripening Stage. *Ann. Microbiol.* **2006,** *56*(4), 353–358.

Ozudogru, E. A.; Kaya, E.; Kirdok, E.; Issever-Ozturk,S. In Vitro Propagation from Young and Mature Explants of Thyme (*Thymus vulgaris* and *T. longicaulis*) Resulting in Genetically Stable Shoots. *In Vitro Cell. Dev. Biol. Plant* **2011,** *47*(2), 309–320.

Ozudogru, E. A.; Kaya, E.; Kirdok, E. Development of Protocols for Short-, Medium- and Long-Term Conservation of Thyme. *Acta Hortic.* **2011,** *918,* 43–50.

Palama, T. L.; Khatib, A.; Hae, C. Y.; Payet, B.; Fock, I.; Verpoorte, R.; Kodja, H. Metabolic Changes in Different Developmental Stages of Vanilla planifolia Pods. *J. Agric. Food Chem.* **2009,** *57*(17), 7651–7658.

Palama, T. L.; Grisoni, M.; Fock-Bastide, I.; Jade, K.; Bartet, L.; Hae, C. Y.; Verpoorte, R.; Kodja, H. Metabolome of *Vanilla planifolia* (Orchidaceae) and Related Species Under Cymbidium mosaic virus (CymMV) Infection. *Plant Physiol. Biochem.* **2012,** *60,* 25–34.

Palumbo, A. D.; Amore, D. R. Parsley Cultivation (*Petroselinum sativum* Hoffm.) in Cold Tunnel. 13th International Congress on Plastics in Agriculture [Congresso-internazionale-del-CIPA]. Proceedings of a Conference Held in Verona, Italy, March 8–11, 1994, 1, 18.

Pank, F.; Kruger, H. Sources of Variability of Thyme Populations (*Thymus vulgaris* L.) and Conclusions for Breeding. *Zeitschrift fur Arznei and Gewurzpflanzen* **2003,** *8*(3), 117–124.

Pankaja, H. K.; Krishnamurthy, B.; Kumar, R. V. Correlates of Knowledge Level of Vanilla Growers on Cultivation Practices of Vanilla. *J. Agric. Sci.* **2009,** *43*(1), 143–146.

Parker, J. The 9 Billion-People Question. A Special Report on Feeding the World. *The Economist,* 26 February, 2011.

Patra, D. D.; Singh, A.; Shasany, A. K.; Kalra, A.; Khanuja, S. P. S. Agro-package for Commercial Cultivation of Menthol Mint (*Mentha arvensis*). *J. Med. Aromat. Plant Sci.* **2003,** *25,* 1016–1023.

Patrakar, R.; Mansuriya, M.; Patil, P. Phytochemical and Pharmacological Review on *Laurus nobilis*. *Int. J. Pharm. Chem. Sci.* **2012,** *1* (2), 595–692.

Paul, R.; Datta, A. K.; Majumder, D. Quantitative Evaluation of Induced Macromutants in Celery (*Apium graveolens* L.) and Ajowan (*Trachyspermum ammi* L.). *Int. J. Plant Sci.* **2006,** *1*(1), 24–26.

Peter, K. V. Spices Research in India. *Indian J. Agric. Sci.* **1998,** *68,* 527–532.

Peter, K. V.; Nirmal Babu, K.; Minoo, D. Spices Biotechnology. *J. Hortic. Sci.* **2006,** *1*(1), 1–14.

Petropoulos, S. A.; Akoumianakis, C. A.; Passam, H. C. Effect of Sowing Date and Cultivar on Yield and Quality of Turnip Rooted Parsley (*Petroselinum crispum* ssp. tuberosum). *J. Food Agric. Environ.* **2005,** *3*(2), 205–207.

Petropoulos, S. A.; Daferera, D.; Polissiou, M. G.; Passam, H. C. Effect of Nitrogen-Application Rate on the Biomass, Concentration, and Composition of Essential Oils in the Leaves and Roots of Three Types of Parsley. *J. Plant Nutr. Soil Sci.* **2009,** *172*(2), 210–215.

Petropoulosa, S. A.; Dafererab, D.; Polissioub, M. G.; Passama H. C. The Effect of Water Deficit Stress on the Growth, Yield and Composition of Essential Oils of Parsley. *Sci. Hortic.* **2008,** *115*(4),393–397.

Petrovic, S. S.; Ristic, M. S.; Petrovic, N. V.; Lazic, M. L.; Franciskovic, M.; Petrovic, S. D.; Chemical Composition and Antioxidative Activity of Essential Oil of *Thymus serpyllum* L. *Hemijska Industrija* **2014,** *68*(3), 389–397.

Pill, W.; Kilian, E. A. Germination and Emergence of Parsley in Response to Osmotic or Matric SeedP and Treatment with Gibberellin. *Hortic. Sci.* **2000,** *35*(5), 907–909.

Pinaria, A. G.; Liew, E. C. Y.; Burgess, L. W. Fusarium Species Associated with Vanilla Stem Rot in Indonesia. *Australas. Plant Pathol.* **2010,** *39*(2), 176–183.

Pino, J. A.; Rosado, A.; Fuentes, V. Composition of the Essential Oil from the Leaves and Flowers of *Ocimum gratissimum* L. Grown in Cuba. *J. Essent. Oil Res.* **1996,** *8*, 139–141.

Pirbalouti, A. G.; Samani, M. R.; Hashemi, M.; Zeinali, H. Salicylic Acid Affects Growth, Essential Oil and Chemical Compositions of Thyme (*Thymus daenensis* Celak.) Under Reduced Irrigation. *Plant Growth Regul.* **2014,** *72*(3), 289–301.

Pirbalouti, A. G.; Rahimmalek, M.; Malekpoor, F.; Karimi, A. Variation in Antibacterial Activity, Thymol and Carvacrol Contents of Wild Populations of *Thymus daenensis* subsp. daenensis Celak. *Plant Omics J.* **2011,** *4*(4),209–214.

Platel, K.; Srinivasan, K. Digestive Stimulant Action of Spices: A Myth or Reality? *Indian J. Med. Res.* **2004,** *119*, 167–179.

Plessner, O.; Ziv, M.; Negbi, M. In vitro Corm Production in the Saffron Crocus (*Crocus sativus* L.). *Plant Cell Tiss. Organ Cult.* **1990,** *20*, 89–94.

Pliny. Natural History.Vol.iv:LibriXII–XVI.(Loeb Classical Library,370). Rackham, H., Ed.; Harvard University: Cambridge, Mass/London, 1945.

Poggi, L. M.; Portela, A. J.; Pontin, M. A.; Molina, R. V. Corm Size and Incubation Effects on Time to Flowering and Threads Yield and Quality in Saffron Production in Argentina. *Acta Hortic.* **2010,** *850*, 193–198.

Prajakta, J.; Kiran, B.; Vijayalakshmi, P. A Phytopharmacological Review on Garciniaindica. *Int. J. Herb. Med.* **2015,** *3*(4), 2–7.

Prakash, K. V.; Sudarshan, M. R.; Radhika, N. S. *Studies on Insect Pests of Vanilla in Karnataka*, In Proceedings of the 15th Plantation Crops Symposium Placrosym XV, Mysore, India, December 10–13, 2002, 592–595.

Prakash, V. *Leafy Spices;* CRC Press Inc.: Boca Raton, Florida, USA, 1990; pp 31–32.

Premnath J, R. C.; Verma, R. B.; Yadav, G. C. Effect of Date of Sowing, Nitrogen Levels and Spacing on Growth and Yield of Ajwain (*Trachyspermum ammi* (L.) Sprague). *J. Spices Aromat. Crops.* **2008,** *17*(1), 1–4.

Pruthi, J. S. *Minor Spices and Condiments—Crop Management and Post Harvest Technology;* Indian Council of Agricultural Research: New Delhi, 2001.

Pruthi, J. S. *Major Spices of India—Crop Management and Post Harvest Technology;* ICAR Publications: New Delhi, 1993; pp 114–179.

Pruthi, J. S. N. *Minor Spices and Condiments;* ICAR: New Delhi, 2001.

Puthur, J. T.; Kumar, V. K. Studies on the Role of Nutrients and Metabolites on Flowering of Vanilla (*Vanilla planifolia* Andrews). J. Plant. Crops **2006,** *34*(2), 90–93.

Quesada Moraga, E.; Lopez Diaz, C.; Landa, B. B. The Hidden Habit of the Entomopathogenic Fungus *Beauveria bassiana*: First Demonstration of Vertical Plant Transmission. *PLoS One* **2014,** *9*(2), e 89278.

Radhakrishnan, V. V.; Mohanan, K. V. Cardamom Breeding in India—An Overview. *Indian J. Arecanut Spices Med. Plants* **2012,** *14*(1), 17–23.

Radjacommare, R.; Venkatesan, S.; Samiyappan, R. Biological Control of Phytopathogenic Fungi of Vanilla Through Lytic Action of *Trichoderma* Species and *Pseudomonas fluorescens*. *Arch. Phytopathol. Plant Prot.* **2010,** *43*(1/3), 1–17.

Radulescu, V.; Popescu, M. L.; Ilies, D. C. Chemical Composition of the Volatile Oil from Different Plant Parts of *Anethum graveolens* L. (Umbelliferae). *Farmacia* **2010,** *58*, 594–600.

Rahimi, S.; Rahimi, H.; Hasanshahi, S.; Ahmadi, M.; Hoseyni, N. Floristic Evaluation of Saffron Field's Weed and Ways to Improve their Control. *Adv. Environ. Biol.* **2014,** *8*, 2542–2547.

Rahman, L. U.; Kukreja, A. K.; Singh, S. K.; Singh, A.; Yadav, A.; Khanuja, S. P. S. Qualitative Analysis of Essential Oil of *Rosmarinus officinalis* L. Cultivated in Uttaranchal Hills, India. *J. Spices Aromat. Crops* **2007,** *16*, 55–57.

Rajapakse, R. H. S.; Kumara Wasantha, K. L. A Review of Identification and Management of Pests and Diseases of Cinnamon (*Cinnamomum zeylanicum* Blume). *Trop. Agric. Res. Ext.* **2007,** *10*, 1–10.

Ram, M.; Kumar, S. Yield improvement in the regenerated and transplanted mint Mentha arvensis by recycling the organic wastes and manures. Bioresour Technol 1997, 59, 141–149.

Ram, M.; Kumar, S. Optimization of Interplant Space and Harvesting Time for High Essential Oil Yield in Different Varieties of Mint (*Mentha arvensis*). *J. Med. Aromat. Plant Sci.* **1999,** *21*, 38–45.

Ramin, A. A.; Atherton, J. G. Manipulation of Bolting and Flowering in Celery (*Apium graveolens* L. var. dulce). Effects of Chilling During Germination and Seed Development. *J. Hortic. Sci.* **1991,** *66*(4), 435–441.

Ramos Castella, A.; Iglesias Andreu, L. G.; Bello Bello, J.; Lee Espinosa, H. Improved Propagation of Vanilla (*Vanilla planifolia* Jacks. ex Andrews) Using a Temporary Immersion System. *In-vitro Cell. Dev. Biol. Plant* **2014,** *50*(5), 576–581.

Ranadive, A. S. Vanilla Cultivation, Curing, Chemistry, Technology and Commercial Products. In *Spices, Herbs and Edible Fungi;* Charalambous, G., Ed.; Elsevier: Amsterdam, 1994; pp 517–577.

Rands, R. D. Stripe Canker of Cinnamon Caused by *Phytophthora cinnamomi* n. sp. *Meded. Inst. Voor Plantenziekten* **1922,** *54*, 50–53.

Rao, B. R.; Kothari, S. K.; Rajput, D. K.; Patel, R. P.; Sastry, K. P. Composition of Rosemary (*Rosmarinus officinalis* L.) Oil Produced from the Semi-arid Tropical Climate of South India. *J. Med. Aromat. Plant Sci.* **2010,** *32*, 20–23.

Rao, B. R. Species of Alternaria on Some cruciferae. *Geobios* **1977,** *4*, 163–166.

Rao, L. J. M. Spices and Plantation Crops: Bioactive Compounds, Colourants and Flavourants, 2008. http://www.scitopics.com/Spices_and_Plantation_Crops_Bioactive_ Compounds_ Colourants_and_Flavourants.html.

Rao, V. G.; Sanjay, M.; Kishore, V. V. N. Study of Drying Characteristics of Large Cardamom. *Biomass Bioenergy* **2001,** 20(1), 37–43.

Rao, Y. S.; Kumar, A.; Chatterjee, S.; Naidu, R.; George, C. K. Large Cardamom (*Amomum subulatum* Roxb.). *J. Spices Aromat. Crops* **1993,** *2*, 1–15.

Rao, Y. S.; Mathew, K. M.; Lakshmanan, R.; Kumar, K. P.; Thomas, J. Generation of Inter Specific Hybrids Between *Vanilla planifolia* Andrews and *V. wightiana* Lindl and Their Validation Through Molecular Biology Techniques. Adv. Plant Sci. **2012,** *25*(2), 433–438.

Rao, X. L.; Krom, N.; Tang, Y. H.; Widiez, T.; Havkin-Frenkel, D.; Belanger, F. C.; Dixon, R. A.; Chen, F. A Deep Transcriptomic Analysis of Pod Development in the Vanilla Orchid (*Vanilla planifolia*). *BMC Genomics* **2014,** *15*, 964.

Rasoul, R. A.; Behrooz, K. T.; Abdollah, S.; Raham, A.; Askar, G. O. Changes in Essential Oil Content and Yield Components of Anise (*Pimpinella anisum* L.) Under Different Irrigation regimes. *Int. J. Agric. Crop Sci.* **2013,** *6*(7), 364–369.

Rastogi, R. P.; Malhotra B, N. *Compendium of Indian Medicinal Plants;* CDRI: Lucknow and New Delhi, India, Nisc 1993; pp 504–857.

Ravikumar, P. Rosemary. *Spice India* **2002,** *15*(10), 16–19.

Ravindran, P. N.; Johny, A. K.; Nirmei, B. K. *Spices in Our Daily Life*. Satabdi Smaramha, 2, 2002.

Reddy, N. S.; Anjanappa, M.; Murali, K.; Krishnappa, K. S.; Nagaraj, H. Effect of Biochemical Constituents on Rooting of Curry Leaf (*Murraya koenigii* S.) Stem Cuttings. *South Indian Hortic.* **1998,** *46*, 326–329.

Reghunath, B. R.; Priyadarshan, P. M. Somaclonal Variation in Cardamom (*Elettaria cardamomum* Maton) Derived from Axenic Culture of Juvenile Shoot Primordia. *Acta Hortic.* **1993,** *330*, 235–242.

Reis, C. A. M.; Brondani, G. E.; Almeida, M. de. Floral Biology, Reproductive Biology and Vegetative Propagation of Vanilla. *Scientia Agraria Paranaensis* **2011,** *10*(1), 69–82.

Rema, J.; Krishnamoorthy, B. Effect of Packing Materials and Storage of Scions on Graft Success in Nutmeg (*Myristica fragrans* Houtt.). *J. Spices Aromat. Crops* **1998,** *7*, 147–148.

Rema, J.; Krishnamoorthy, B.; Mathew, P. A. High Yielding Varieties of Cinnamon and Nutmeg. *Spice India* **2003,** *16*, 7–16.

Retheesh, S. T.; Bhat, A. I. Simultaneous Elimination of Cucumber Mosaic Virus and Cymbidium Mosaic Virus Infecting *Vanilla planifolia* Through Meristem Culture. *Crop Prot.* **2010,** *29*(10), 1214–1217.

Richard, A.; Farreyrol, K.; Rodier, B.; Leoce, M.; . San, K.; Wong, M.; Pearson, M.; Grisoni, M. Control of Virus Diseases in Intensively Cultivated Vanilla Plots of French Polynesia. *Crop Prot.* **2009,** *28*(10), 870–877.

Rodinson M, Arberry, A. J., Perry, C. (Eds.) *Medieval Arab Cookery;* Prospect Books: Totnes, 2001.

Rometsch, S. Ecology and Cultivation of Thyme (*Thymus vulgaris* L.) in the Canton Valais, Switzerland. *Acta Hortic.* **1993,** *344*, 411–415.

Ross, I. A. *Medicinalplants of the World: Chemical Constituents, Traditional and Medicinal Uses*; Humana Press: Totowa, NJ, 2001; vol. 1.

Rout, G. R. Direct Plant Regeneration of Curry Leaf Tree (*Murraya koenigii* Koenig.), an Aromatic Plant. *In Vitro Cell. Dev. Biol. Plant* **2005,** *41*, 133–136.

Royandezagh, S. D.; Khawar, K. M.; Osalou, A. R.; Ozcan, S. Agrobacterium Mediated Genetic Transformation of *Papaver somniferum* L. Using Semi Solid Agar Gelled Primed Seeds as Explant. *Bulg. J. Agric. Sci.* **2013,** *19*, 222–227.

Rozek, E. Reaction of Leaf Celery (*Apium graveolens* L. var. Secalinum) to Planting Density and Irrigation. *Veg. Crops Res. Bull.* **2007,** *66*, 69–77.

Rozek, E. Yielding of Leaf Celery (*Apium graveolens* L. var. secalinum Alef.) Depending on the Number of Harvests and Irrigation. *Mod. Phytomorphol.* 2013, 3, 83–86.

Rozek, E.; Nurzynska Wierdak, R.; Kosior, M. Efficiency of Some Agrotechnical Treatments in Quantity and Quality Yield Modification of Leaf Celery (*Apium graveolens* L. var. secalinum Alef.). *Acta Scientiarum Polonorum Hortorum Cultus* **2013,** *12*(6), 227–239.

Ruangamnart, A.; Buranaphalin, S.; Temsiririrkkul, R.; Chuakul, W.; Pratuangdejkul, J. Chemical Compositions and Antibacterial Activity of Essential Oil from Dill Fruits (*Anethum graveolens* L.) Cultivated in Thailand, *J. Pharm. Sci.* **2015,** *42*(3), 135–143.

Ruf, F.; Yoddang; Syarifuddin. Vanilla on Coffee Farms. From Slash and Burn to Replanting: Green Revolutions in the Indonesian Uplands, 2004, pp 113–119.

Rumasz Rudnicka, E.; Koszanski, Z.; Podsiado, C. The Effect of Various Salinity of Irrigation Water on the Onion and Celery Yield. *Woda Srodowisko Obszary Wiejskie* **2005,** *5*(14), 275–285.

Rumasz, E.; Koszanski, Z.; Wronkowska, H. Influence of Saline Water Irrigation on Celery Yield. *Folia Univ. Agric. Stetinensis, Agric.* **1999,** *73*, 207–211.

Sadeghi, S.; Rahnavard, A.; Ashrafi, Z. Y. The Effect of Plant-Density and Sowing Date on Yield of Basil (*Ocimum basilicum* L.). *Iran J. Agric. Technol.* **2009,** *5*(2), 413–422.

SaeidZehtab, S.; Kazem Ghassemi, G.; Sajad, M. Effect of Sowing Date and Limited Irrigation on the Seed Yield and Quality of Dill (*Anethum graveolens* L.). *Turk. J. Agric. For. Sci.* **2006,** *30*(4), 281–286.

Safaii, L.; Ashoorabadi, E. S.; Emami, S. D.; Afiuni, D. The Effect of Harvesting Stages on Quantitative and Qualitative Characters of Essential Oil and Phenolic Yield Composition in Two Thyme Species (*Thymus daenensis* Celak & *T. vulgaris* L.) in Iran. *Int. J. Agric. Crop Sci.* **2014,** *7*(13), 1346–1351.

Saffari, A.; Jamnejad, M.; Echi, R. M.; Abdollahi, M.; Kashani, Z. F. The Effect of Phosphorus Fertilizer Changes on *Thymus vulgaris* L. Yield and Essence in Different Irrigation Levels. *Int. J. Biosci.* **2013,** *3*(8), 110–115.

Saidj, F.; Bentahar, F.; Boutekedjiret, C. Kinetic Study and Optimization of the Operating Conditions of the Extraction by Steam Distillation of the Essential Oil of *Thymus numidicus* (Poiret) of Algeria. *J. Essent. Oil Bear. Plants* **2009,** *12*(2), 144–154.

Saini, S. S.; Cheema, S. S.; Randhawa, G. S.; Mahey, R. K. Effect of Irrigation and Nitrogen on Celery. *Indian J. Agron.* **1989,** *34*(3), 341–345.

Saju, K. A.; Kuldeep, R.; Deka, T. N.; Gupta, U.; Biswas, A. K.; Sudharshan, M. R. Colletotrichum Blight Control in Large Cardamom (*Amomum subulatum* Roxb.) Nursery. *J. Spices Aromat. Crops* **2013,** *22*, 219–221.

Sakaguchi, I.; Ishimoto, H.; Matsuo, M.; Ikeda, N.; Minamino, M.; Kato, Y. The Water-Soluble Extract of *Illicium anisatum* Stimulates Mouse Vibrissae Follicles in Organ Culture. *Exp. Dermatol.* **2004,** *13*(8), 499–504.

Salehi, S.; Golparvar, A. R.; Hadipanah, A. Effect of Harvest Time on Yield and Quality of *Thymus vulgaris* L. Essential Oil in Isfahan Province, Iran. *Agric. Conspectus Scientificus Poljopr. Znan. Smotra* **2014,** *79*(2), 115–118.

Salwee, Y.; Nehvi, F. A. Seasonal Bud Sprouting: A Challenge for In vitro Micropropagation in Saffron. *Vegetos* **2013,** *26*, 154–158.

Sandheep, A. R.; Asok, A. K.; Jisha, M. S. Combined Inoculation of *Pseudomonas fluorescens* and *Trichoderma harzianum* for Enhancing Plant Growth of Vanilla (*Vanilla planifolia*). *Pak. J. Biol. Sci.* **2013a,** *16*(12), 580–584.

Sandheep, A. R.; Asok, A. K.; Jisha, M. S. Colonization Study of Antagonistic *Pseudomonas* sp. in *Vanilla planifolia* Using Green Fluorescent Protein (GFP) as a Marker. *Afr. J. Microbiol. Res.* **2013b,** *7*(48), 5417–5423.

Sandheep, A. R.; Asok, A. K.; Fathima, P. A.; Jisha, M. S. Exploiting the Biocontrol Potential of Indigenous *Trichoderma* sp. Against Major Phytopathogens of Vanilla (*Vanilla planifolia*). *South Asian J. Exp. Biol.* **2014,** *4*(1), 1–7.

Santos, V. M. C. S.; Schneider, T. R.; Bizzo, H. R.; Deschamps, C. Propagation Alternatives for the Production of *Mentha canadensis* L. Essential Oil at the Santa Catarina State Northern Coast. *Rev. Bras. Plant. Med.* **2012,** *14*, 97–102.

Santosa, E.; Sugiyama, N.; Nakata, M.; Kawabata, S. Profitability of Vanilla Intercropping in Pine Forests in West Java, Indonesia. *Jpn. J. Tropic. Agric.* **2005,** *49*(3), 207–214.

Sathyanarayana, E.; Hadole, S. S.; Laharia. G. S.; Ghawade. S. M. Effect on Nutrient Levels on Fertility Status of Soil and Quality of Ajwain (*Trachyspermum ammi* L. Sprague). *Int. J. Trop. Agric.* **2015,** *33*(4), 3727–3729.

Sathyanarayana, E.; Hadole, S. S.; Ghawade, S. M.; Laharia, G. S. Effect of Nutrient Levels on Growth and Yield of Ajwain (*Trachyspermum ammi* L. Sprague). Asian Reson. **2015,** *4*(4), 7–10.

Satyavati, G. V.; Gupta, A. K. *Medicinal Plants of India*; ICMR: New Delhi, 1987; vol. II.

Sawant, D. S.; Haldankar, P. M.; Nagwekar, D. D.; Rajput, J. C. Screening of Kokam (*Garcinia indica* Choisy) Genotypes. *India J. Arecanut Spices Med. Plants* **1999,** *1*(2), 55–59.

Score, C.; Lorenzi, R.; Ranall, P. The Effect of (S)-(+)-Carvonetreatments on Seed Potato Tuber Dormancy and Sprouting. *Potato Res.* **1997,** *40*, 155–161.

Seidler ozykowska, K.; Golcz, A.; Kozik, E.; Wojcik, J. *Quality of Thyme Herb (Thymus vulgaris L.) from Organic Cultivation. Cultivating the Future Based on Science. Volume-1.* Organic Crop Production Proceedings of the Second Scientific Conference of the International Society of Organic Agriculture Research ISOFAR held at the 16th IFOAM Organic World Conference in Cooperation with the International Federation of Organic Agriculture Movements IFOAM and the Consorzio Modena. Bio in Modena, Italy, June 18–20, 2008, 1, 600–603.

Sekeroglu, N.; Ozguven, M.; Erden, U. Effects of the Drying Temperature on Essential Oil Content of Bay Leaf (*Laurus nobilis* l.) Harvested at Different Times. *Acta Hortic.* **2007,** *756*, 315–320.

Selvi, A. T.; Joseph, G. S.; Jayaprakasha, G. K. Inhibition of Growth and Aflatoxin Production in *Aspergillus flavus* by *Garcinia indica* Extract and its Antioxidant Activity. *Food Microbiol.* **2003,** *20*(4), 455–460.

Shadakshari, Y. G.; Madaiah, D.; Kumar, M. D.; Shivakumar, K. V.; Goudra, K. H. B. Pollen Viability and Stigma Receptivity in Vanilla (*Vanilla planifolia* Andrews). *J. Spices Aromat. Crops* **2003,** *12*(2), 194–196.

Shaddad, R. A. M.; El, B. R. E. I.; El-Asdoudi, A. M.; Abdallah, M. M. F. Effect of Solarization and Organic Fertilizer on Yield and Quality of Rocket and Parsley Fresh Herbs. *Ann. Agric. Sci.* **2009,** *54(*1), 151–164.

Shahida, K.; Gopal, K. S.; Mathew, S. K. Efficacy of Native Bioagents Against *Phytophthora meadii* Causing Phytophthora Rot in Vanilla and its Compatibility with Fungicides. *SAARC J. Agric.* **2010,** *8*(1), 103–111.

Shal, C.; Puthur, J. T. Assorted Response of Mutated Variants of *Vanilla planifolia* Andr. Towards Drought. *Acta Physiol. Plant.* **2009,** *31*(5), 1023–1029.

Sharafzadeh, S.; Alizadeh, A.; Sarvestani, M. M. S. Comparison of Essential Oil Yield and Components in Two Parts of Garden Thyme Shoot. *Adv. Environ. Biol.* **2011,** *5*(10), 3179–3182.

Sharafzadeh, S. Effect of Nitrogen, Phosphorous and Potassium on Growth, Essential Oil and Total Phenolic Content of Garden Thyme (*Thymus vulgaris* L.). *Adv. Environ. Biol.* **2011,** *5*(4), 699–703.

Sharma, A.; Verma, S. C.; Saxena, N.; Chadda, N.; Singh, N. P.; Sinha, A. K. Microwave and Ultrasound Assisted Extraction of Vanillin and its Quantification by High Performance Liquid Chromatography in *Vanilla planifolia*. *J. Sep. Sci.* **2006,** *29*, 613–619.

Sharma, E.; Sharma, R.; Singh, K. K.; Sharma, G. A Boon for Mountain Populations: Large Cardamom Farming in the Sikkim Himalaya. *Mountain Res. Dev*. **2000,** *20*(2), 108–111.

Sharma, G.; Sharma, R.; Sharma, E. Traditional Knowledge Systems in Large Cardamom Farming: Biophysical and Management Diversity in Indian Mountainous Regions. *Indian J. Tradit. Knowl.* **2009,** *8*(1),17–22.

Sharma, R. *Agro-techniques of Medicinal Plants;* Daya Publishing House: New Delhi, 2004; pp 3–10.

Shetty, A. A.; Rana, M. K. Effect of Gibberellic Acid on Yield and Seed Quality of Ajowain (*Trachyspermum ammi* L.). *BIOINFOLET: Q. J. Life Sci.* **2012,** *9*(2), 190–194.

Shinde, S. R.; Haldankar, P. M. Effect of Post Flowering Foliar Sprays of Nutrients on Physico-Chemical Properties of Kokam (*Garcinia indica* Choisy). *Asian J. Hortic.* **2010,** *5*(1), 177–179.

Shivakumar, K. V.; Kumar, M. D.; Devaraju, K. M. Effect of Plant Growth Regulators on Growth and Development of Vanilla (*Vanilla planifolia* Andrews) Fruits. *J. Plant. Crops* **2008,** *36*(2), 108–111.

Shukla, G.; Sharma, N. *Biodiversity in Medicinal and Aromatic Plants in India.* Conservation and Utilization, National Bureau of Plant Genitie Resources. PUSA Campus: New Delhi, 1996.

Siddagangaiah.; Koshy.; John.; Sudharshan, M. R Influence of Sowing Methods on Production of Seedlings in Cardamom (*Elettaria cardamomum* Maton). *J. Spices Aromat. Crops* **1997,** *6*, 137–139.

Siddappa, R.; Hegde, N. K. Effect of Foliar Spray of Vermiwash and Nutritional Treatments on Growth and Yield Performance of Curry Leaf var. Suvasini. *Asian J. Hortic.* **2011,** *6*(1), 68–70.

Siddiqui, Y.; Meon, S. Fungi Associated with Leaf and Stem Blight of Vanilla. *J. Agric. Res. Dev.* **2009,** *7*(1). http://dx.doi.org/10.4314/jard.v7i1.49661.

Sies, H. (Ed.) *Antioxidants in Disease, Mechanisms and Therapy,* 1st ed.; Academic Press: New York, 1996; vol. 38, 707.

Silva, M.; Das, D. D.; Pasqual, M.; Silva, F. P.; Dias, J. M. M.; Araujo, A. G. de. Rooting of Cuttings of Vanilla Plant (Orchidaceae). *Semina: Ciencias Agrarias Londrina* **2009,** *30*(1), 71–80.

Simon, J. E.; Quinn, J. Characterization of Essential Oil of Parsley. *J. Agric. Food Chem.* **1988,** *36*, 467.

Singh, A. I.; Pothula, A. K. Postharvest Processing of Large Cardamom in the Eastern Himalaya: A Review and Recommendations for Increasing the Sustainability of a Niche Crop. *Mountain Res. Dev.* **2013,** *33*, 453–462.

Singh, A. K.; Khanuja SPSCIMAP Initiatives for Menthol Mint. *Spice India* **2007,** *December*, 14–17.

Singh, R.; Pandey, R. M. Combining Ability and Heterosis in Opium Poppy (*Papaversom niferum* L.). *Curr. Adv. Agric. Sci.* **2011,** *3*, 130–134.

Singh, R.; Singh, M. K.; Pal, S. B.; Singh, D. P.; Rajiv. and Dharmendra, Y. Role of *Bacillus circulans*—A Bacterial Fertilizer on Yield Quality and Economics of Aniseed (*Pimpinella anisum*). *Int. J. For. Crop Improv.* **2013,** *4*(1), 44–46.

Singh, S.; Omreb, P. K.; Sandhya, M. M. Curry Leaves (*Murraya koenigii* Linn. Sprengal)—A Mircale Plant. *Indian J. Sci. Res.* **2014,** *4*(1), 46–52.

Smith-Palmer, A.; Stewart, J.; Fyfe, L. The Potential Application of Plant Essential Oils as Natural Food Preservatives in Soft Cheese. *Food Microbiol.* **2001,** *18*, 463–470.

Sokhangoy, S. H.; Ansari Kh; Eradatmand, A. D. Effect of Bio-fertilizers on Performance of Dill (*Anethum graveolens* L.). *Iran. J. Plant Physiol.* **2011,** *2*, 547–552.

Soufizadeh, S.; Zand, E.; Baghestani, M. A.; Kashani, F. B.; Nezamabadi, N.; Sheibany, K. Integrated Weed Management in Saffron (*Crocus sativus*). *Acta Hortic.* **2007,** *739*, 133–137.

Spirling, L. I.; Daniels, I. R. Botanical Perspectives on Health Peppermint: More than Just an After-Dinner Mint. *J. R. Soc.* **2001,** *121*, 62.

Sreedhar, R. V.; Lakshmanan, V.; Neelwame, B. Genetic Fidelity of Long Term Micropropagated Shoot Cultures of Vanilla (*Vanilla planifolia* Andrews) as Assessed by Molecular Markers. *Biotechnol. J.* **2007,** *2*(8), 1007–1013.

Sridharan, S.; Nagarajan, N.; Thamburaj, S.; Mohideen, M. K. Effect of Planting Density on the Capsule Damage by *Sciothrips cardamomi* in Cardamom. *South Indian Hortic.* **1990,** *38*, 120–121.

Srinivasan, K. Plant Foods in the Management of Diabetes Mellitus: Spices as Beneficial Antidiabetic Food Adjuncts. *Int. J. Food Sci. Nutr.* **2005,** *56*(6), 399–414.

Srivastava, P. N. Removal of Acrosomes of Ram and Rabbit Spermatozoa. *J. Reprod. Fert.* **1973,** *33*, 323.

Stace, C. A. *New Flora of the British Isles,* 3rd ed.; Cambridge University Press: Cambridge, U.K., 2010.

Stahl-Biskup, E. Essential Oil Chemistry of the Genus Thymus—A Global View. In *Thyme—The Genus Thymus;* Taylor and Francis: London, 2002; pp 75–124.

Staugaitis, G.; Starkute, R. Impact of Planting Time and Agrofilm Cover on Yield, Quality and Profitability of Plain and Curly Leaved Parsley. *Sodininkyste ir Darzininkyste* **2000,** *19*(2), 98–106.

Stavri, M.; Gibbons, S. The Ant Mycobacterial Constituents of Dill (*Anethum graveolens*). *Phytother. Res.* **2005,** *19*, 938–941.

Stephen, R.; Jayabalan, N. In vitro Flowering and Seed Setting Formation of Coriander (*Coriandrum sativum*). *Curr. Sci.* **1998,** *74*(3),195–197.

Straubinger, M.; Bau, B.; Eckstein, S.; Fink, M.; Winterhalter, P. Identification of Novel Glycosidic Aroma Precursors in Saffron (*Crocus sativus* L). *J. Agric. Food Chem.* **1998,** *46*, 3228–3243.

Subha, R.; Jansirani, P.; Babu, C. R. Studies on Crop Regulation in Curry Leaf (*Murraya koenigii* Spreng.) During off Season. *Int. J. Plant Sci.* **2010,** *5*, 269–273.

SubhasChandran, M. D. Conservation and Sustainable Use of Cultivated and Wild Tropical Diversity. Introduction to Good Practices. Good Practices Workshop, Bangkok, Thailand, April 4–5, 2005.

Sudan, R.; Bhagat, M.; Gupta, S.; Devi, C. T. Comparative Analysis of Cytotoxic and Antioxidant Potential of Edible *Cinnamomum verum* (bark) and *Cinnamomum tamala* (Indian Bay Leaf). *Free Rad. Antioxid.* **2013,** *3*, 70–73.

Suganthy, M.; Kalyanasundaram, M. Insect Pests of Tree Spices and Their Management. In: Proc. Nat. Sem. Tree spices, HRS, Kanyakumari, March 5–7, 73–82.

Sujatha, S.; Bhat, R. Response of Vanilla (*Vanilla planifolia* A.) Intercropped in Arecanut to Irrigation and Nutrition in Humid Tropics of India. *Agric. Water Manage.* **2010,** *97*(7), 988–994.

Sulieman, A. E.; Abdelrahman, S. E.; Abdel Rahim, A. M. Phytochemical Analysis of Local Spearmint (*Mentha spicata*) Leaves and Detection of the Antimicrobial Activity of its Oil. *J. Microbiol. Res.* **2011,** *1*(1), 1–4.

Sulochanamma, G.; Ramalakshmi, K.; Kumar, T. M. M.; Indiramma, A. R.; Ramesh, B. S.; Sampathu, S. R.; Naik, J. P. Storage Characteristics of Large Cardamom (*Amomum subulatum* Roxb.) and Seeds in Different Packages. *J. Food Sci. Technol.* **2008,** *45*(2), 183–186.

Sumy, O.; Ved, D. K.; Krishnan, R. *Tropical Indian Medicinal Plants, Propagation Method;* Foundation for Revitalisation of Local Health Traditions: Bangalore, 2000; pp 268–269.

Kumar, S.; Kamboj, J.; Suman; Sharma, S. Overview for Various Aspects of the Health Benefits of *Piper longum* L. Fruit. *J. Acupunct. Meridian Stud.* **2011,** *4*(2),134–140.

Suryanarayana, B. C. A Complete Guide to Vanilla Cultivation. *Vanilla India* **2004,** 17–27.

Taheri, A. M.; Daneshian, J.; Valadabadi, S. A. R.; Aliabadi, F. H. Effects of Water Deficit and Plant Density on Morphological Characteristics of Chicory (*Cichorium intybus* L.). Abstracts Book of 5th International Crop Science Congress & Exhibition, 2008, p 26.

Takeoka, G. Volatile Constituents of Asafoetida. In *Aroma Active Compounds in Foods;* Takeoka, G. R., Guntert, M., Engel, K. H., Eds.; American Chemical Society, Washington, DC, 2001; pp 33–44.

Talubnak, C.; Soytong, K. Biological Control of Vanilla Anthracnose Using *Emericella nidulans*. *Int. J. Agric. Technol.* **2010,** *6*(1), 47–55.

Talwari, G.; Ghuman, B. S. Optimization of Microwave Assisted Process for Extraction of Celery Seed Essential Oil. *J. Agric. Eng.* **2014,** *51*(2), 9–18.

Tammaro, F. *Crocus sativus* L.—cv. Piano di Navelli (L'Aquila saffron): Environment, Cultivation, Morphometric Characteristics, Active Principles, Uses. In *Proceedings of the international conference on saffron (Crocus sativus L.), L'Aquila;* Tammaro, F, Marra, L., Eds.; Università Degli Studi L'Aquila e Accademia Italiana della Cucina: L'Aquila, 1990; pp 47–97.

Tammaro, F. Saffron (*Crocus sativus* L.) in Italy. In *Saffron: Crocus sativus L.;* Negbi M., Ed.; Harwood Academic Publishers: Australia, 1999; pp 53–62.

Tan, B. C.; Chin, C. F.; Alderson, P. An Improved Plant Regeneration of *Vanilla planifolia* Andrews. *J. Plant Tissue Cult. Biotechnol. PTC B* **2011,** *21*(1), 27–33.

Tanne, I.; Cantliffe, D. J. *Seed Treatments to Improve Rate and Uniformity of Celery Seed Germination.* Proceedings of the Florida State Horticultural Society, 1989, 102, 319–322.

Tanwar, A.; Yadav, K.; Prasad, K.; Aggarwal, A. Biological Amendments on Growth, Nutritional Quality, and Yield of Celery. *Int. J. Veg. Sci.* **2013,** *19*(3),228–239.

Telci, I.; Bayram, E.; Yilmaz, G.; Avci, B. Variability in Essential oil Composition of Turkish Basils (*Ocimum basilicum* L.). *Biochem. Syst. Ecol.* **2006,** *34*, 489–497.

Temperini, O.; Colla, G.; Saccardo, F.; Brancaleone, M. Fertilizer Application and Choice of Variety as the Basis for Improving Celery Yield. *Informatore Agrario* **2000,** *56*(8), 95–98.

Tepe, B.; Akpulat, A. H.; Sokmen, M.; Daferera, D.; Yumrutas, O.; Aydin, E.; Polissiou, M.; Sokmen, M. Screening of the Antioxidtaive and Antimicrobial Properties of the Essential Oil of *Pimpinella anisum* and *Pimpinella flabellifolia* from Turkey. *Food Chem.* **2006,** *97*, 719–724.

Tesfaendrias, M. T.; Trueman, C. L.; Gossen, B. D.; McKeown, A. W.; McDonald, M. R. The Influence of Nitrogen and Calcium Fertilizers on Septoria Late Blight and Yield of Celery. *Can. J. Plant Sci.* **2014,** *94*(8), 1391–1399.

Thangaselvabai, T.; Sudha, K. R.; Selvakumar, T.; Balakumbahan, R. Nutmeg (*Myristica fragrans* Houtt.). *Agric. Rev.* **2011,** *32*, 283–293.

Thankamani, C. K.; Sivaraman, K.; Kandiannan, K.; Peter, K. V. Agronomy of Tree Spices (Clove, Nutmeg, Cinnamon and Allspice)—A Review. *J. Spices Aromat. Crops* **1994,** *3*, 105–123.

The Economic Times. Spices Board to explore global market for large cardamom . The Economic Times, January25,2016.http://economictimes.indiatimes.com/articleshow/50717200.cms?utm_source=contentofinterest&utm_medium=text&utm_campaign=cppst (accessed April 28, 2017).

The Hindu. Business Line, Economy, March 7, 2017. http://www.thehindubusinessline.com/economy/agri-business/india-temporarily-suspends-import-of-agricommodities-from-vietnam/article9574295.ece (accessed April 29, 2017).

The World Bank. Saffron: A Major Source of Income and an Alternative to Poppy, January 19, 2015. http://www.worldbank.org/en/news/feature/2015/01/20/saffron-major-source-income-alternative-poppy (accessed May 01, 2017).

Thimmarayappa, M.; Shivashankar, K. T.; Shanthaveerabhadraiah, S. M. Effect of Organic Manure and Inorganic Fertilizers on Growth, Yield Attributes and Yield of Cardamom (*Elettaria cardamomum* Maton). *J. Spices Aromat. Crops* **2000,** *9*(1), 57–59.

Thomas, J.; Bhai, R. S. Sclerotium Rot—A New Disease of Vanilla (*Vanilla planifolia* Andrews) in India. *J. Spices Aromat. Crops* **2000,** *9*(2), 175–176.

Thomas, T. H. Hormonal Involvement in Photoregulation of Celery Seed Dormancy (Monograph).*Br. Soc. Plant Growth Regul.* **1990,** *20*, 51–58.

Tirapelli, C. R.; Andrade, C. R.; de Cassano, A. O.; de Souza, F. A.; Ambrosio, S. R.; Costa, F. B.; da Oliveria, A. M. Antispasmodic and Relaxant Effects of the Hydroalcoholic Extract of *Pimpinella anisum* (Apiaceae) on Rat Anococcygeous Smooth Muscle. *J. Enthopharmacol.* **2007,** *110*(1), 23–29.

Tiwari, R. K. S.; Das, K.; Pandey, D.; Tiwari, R. B.; Dubey, J. Rhizome Yield of Sweet Flag (*Acorus calamus* L.) as Influenced by Planting Season, Harvest Time, and Spacing. *Int. J. Agron.* 2012, ID 731375.

Torres, A.; Nand Hoy, C. W. Relationship Between Carrot Weevil Infestation and Parsley Yield. *J. Econ. Entomol.* **2005,** *98*(4), 1213–1220.

Tort, N.; Demiray, H.; Dereboylu, A. E. Effects of Different Growth Media, Temperature and Salt (NaCl) Concentrations on Germination and Proline Content of Aniseed (*Pimpinella anisum* L. cv. Burdur). *Turk. J. Field Crops* **2005,** *10*(1), 1–7.

Toth, S.; Lee, K. J.; Havkin Frenkel, D.; Belanger, F. C.; Hartman, T. G. *Volatile Compounds in Vanilla. Handbook of Vanilla Science and Technology;* 2011, pp 183–219.

Tuteja, S. S.; Lakpale, R.; Singh, A. P.; Tripathi, R. S. Effect of Harvesting Intervals on Herbage, Oil Yield and Economics of Different Varieties of Japanese Mint (*Mentha arvensis*). *Indian J. Agron.* **2006,** *51*, 245–246.

Tuttolomondo, T.; Bella, S. la.; Garbo, D. di.; Virga, G.; Leto, C. A Study of Thyme Germplasm in Sicily. *Acta Hortic.* **2009,** *826*, 73–80.

Tyler, V. E.; Brady, L. R.; Robbers, J. E. *Pharmacognosy,* 7th ed.; Lea & Febiger: Philadelphia, 1976.

Umamaheswari, R.; Mohanan, K. V. A Study of Vegetative and Floral Morphology of Vanilla planifolia and Vanilla tahitensis. *J. Non-Timber Forest Prod.* **2011,** *18*(2), 95–100.

Umamaheswari, R.; Mohanan, K. V. A study of the Association of Agronomic Characters in *Vanilla planifolia* Andrews. *Int. J. Plant Breed. Genet.* **2011,** *5*(1), 53–58.

Umesha, K.; Murthy, G.; Smitha, G. R. Environmental Conditions and Type of Cuttings on Rooting and Growth of Vanilla (*Vanilla planifolia* Andrews). *J. Trop. Agric.* **2011,** *49*(1/2), 121–123.

Umetsu, H.; Wake, H.; Saitoh, M.; Yamaguchi, H.; Shimomura, K. Characteristics of Cold Preserved Embryogenic Suspension cells in Fennel *Foeniculum vulgare* Miller. *J. Plant Physiol.* **1995,** *146*, 337–342.

USDA National Nutrient Database for Standard Reference Release 28, slightly revised May, 2016. https://www.google.co.in/?gws_rd=ssl#q=USDA+National+Nutrient+data+base,+poppy.

Vakili, M.; Sharafzadeh, S. Growth and Volatile Oil Yield of Garden Thyme as Affected by Nitrogen Source and Level. *Indian J. Fundam. Appl. Life Sci.* **2014,** *4*(1), 205–208.

Vanitha, K.; Karuppuchamy, P.; Sivasubramanian, P. Occurrence and Management of White Grubs on Vanilla. *Ann. Plant Prot. Sci.* **2011,** *19*(1), 233–235.

Vanitha, K.; Karuppuchamy, P.; Sivasubramanian, P. Feeding Preference of Achatina fulica Attacking Vanilla and its Management Through Barrier Substances. *Pest Manage. Hortic. Ecosys.* **2011,** *17*(1), 38–41.

Vanitha, K.; Karuppuchamy, P.; Sivasubramanian, P. Pests of Vanilla (*Vanilla planifolia* Andrews) and Their Natural Enemies in Tamil Nadu, India. *Int. J. Biodiversity Conserv.* **2011,** *3*(4), 116–120.

Varadarasan, S.; Ali, M. A. A.; Chandrasekar, S. S. *Vanilla Vine Weevil, a New Insect Pest on Vanilla (Vanilla planifolia)*, In Proceedings of the 15th Plantation Crops Symposium Placrosym XV, Mysore, India, December 10–13, 2002, 620–622.

Varadarasan, S.; Gopakumar, B.; Chandrasekar, S. S.; Ansar, Ali, M. A.; Prakash, K. V. Pests and Their Management. In *Vanilla—The Prince of Spices;* Thomas, J., Rao, Y. S., Eds.; Spices Board: Cochin, Kerala, 2003; pp 42–54.

Veitch, N. C. Horseradish Peroxidase: A Modern View of a Classic Enzyme. *Phytochemistry* **2004,** *65*, 249–259.

Venugopal, M. N.; Mathew, M. J. Integrated Management of Viral Diseases of Cardamom. In *Diseases of Plantation Crops, Spices, Betelvine and Mulberry;* Dasgupta, M. K., Ed.; ICAR—Indian Institute of Spices Research: Kozhikode 2000; pp 106–110.

Vercruysse, P.; Meert, F.; Tirry, L.; Hofte, M. Evaluation of Insecticides for Control of *Cavariella aegopodii* and Carrot Motley Dwarf Disease in Parsley. *Mededelingen Faculteit Landbouwkundige en Toegepaste Biologische Wetenschappen, Universiteit Gent.* **2000,** *65*(1), 9–18.

Verma, P. C.; Debasis, C.; Jena, S. N.; Mishra, D. K.; Singh, P. K.; Sawant, S. V.; Rakesh, T. The Extent of Genetic Diversity Among Vanilla Species: Comparative Results for RAPD and ISSR. *Ind. Crops Prod.* **2009,** *29*(2/3), 581–589.

Vijayan, A. K.; Joseph, T.; Dhanapal, K.; Thomas, J. Management of Rot Diseases of Vanilla (*Vanilla planifolia* Andrews) Using Bioagents. *J. Plant. Crops* **2009,** *37*(3), 229–231.

Vijayan, A. K.; Sithara, L.; Sreelakshmi, K. P.; Joseph, T.; Thomas, J.; Misra, R. S.; Saju, K. A. Molecular Diversity of *Fusarium oxysporum* Causing Rot Diseases of Vanilla in South India. *Arch. Phytopathol. Plant Prot.* **2012,** *45*(11), 1319–1326.

Vikrant, V. R.; Singh, B. P.; Singh, R. P. Oil Production and Returns of Japanese Mint as Influenced by Nitrogen and Cutting Management. *Ann. Biol.* **2004,** *20*, 27–29.

Vina, A.; Murillo, E. Essential Oil Composition from Twelve Varieties of Basil (*Ocimum* spp) Grown in Colombia. *J. Braz. Chem Soc.* **2003,** *14*(5), 744–749.

Viswanathan, T. V. Medicinal and Aromatic Plants. In *Advances in Horticulture;* Chadha, K. L., Gupta, R., Eds.; Malhotra Publishing House: New Delhi, 1995; 373–383.

Vokk, R.; Lõugas, T.; Mets, K.; Kravets, M. Dill (*Anethum graveolens* L.) and Parsley (*Petroselinum crispum* (Mill.) Fuss) from Estonia: Seasonal Differences in Essential Oil Composition. *Agron. Res.* **2011,** *9*(Special Issue II), 515–520.

Vouillamoz, J. F.; Schaller, M.; Rossinelli, M.; Carron, C. A.; Carlen, C. 'Varico 3', a New Hybrid for Thyme (*Thymus vulgaris* L.) Production in Switzerland. *Rev. Suisse Vitic. Arboric. Hortic.* **2011,** *43*(6), 370–376.

Wagner, H.; Bladt, S. *Plant Drug Analysis;* Berlin-Heidelberg: Springer-Verlag, 1996; pp 154–175.

Waliszewski, K. N.; Pardio, V. T.; Ovando, S. L. A Simple and Rapid HPLC Technique for Vanillin Determination in Alcohol Extract. *Food Chem.* **2007,** *101*(3), 1059–1062.

Wander, J. G. N.; Bouwmeester, H. J. Effects of Nitrogen Fertilization on Dill (*Anethum graveolens* L.) Seed and Carvone Production. *Ind. Crops Prod.* **1998,** 7(2–3), 211–216.

Wang, H. J.; Liu, L. F.; Zhang, J. S. Effect of Ultra High Pressure Treatment on the Extraction Efficiency of Vanillin in Vanilla. *China Condiment* **2011,** *9*, 113–115.

Weber, H. Technologies for Safer Products. 1. Survey of Possibilities for Sterilizing of Spices. *Fleischwirtschaft* **2003a,** *83*, 33–36.

Weber, H. Technologies for Safer Products. 2. Sterilization Methods Which have not Succeeded. *Fleischwirtschaft* **2003b,** *83*, 37–38.

Wei, Q. Y.; Liu, G. Q.; Wei, X. M.; Ma, X. X.; Xu, D. L.; Dong, R. J. Influence of Wood Vinegar as Leaves Fertilizer on Yield and Quality of Celery. *J. China Agric. Univ.* **2009,** *14*(1), 89–92.

Westcott, R. J.; Cheetham, P. S. J.; Arraclough, A. J. B. Use of Organized Viable Vanilla Plant Aerial Roots for the Production of Natural Vanillin. *Phytochemistry* **1994,** *35*, 135–138.

Wiethaler, A.; Fischer, P. Damage in Parsley—Nutrient Deficiencies as Causes. *Gemuse-Munchen* **2001,** *37*(12), 8–13.

Wildman, R. E. C. *Handbook of Nutraceuticals & Functional Foods;* CRC Press: New York, Washington D.C.

Winterhalter, P.; Straubinger, M. Saffron-renewed Interest in an Ancient Spice. *Food Rev. Int.* **2000,** *16*, 39–59.

Wisler, G.; Zettler, G.; Mu, L. Virus Infections of Vanilla and Other Orchids in French Polynesia. *Plant Dis.* **1987,** *71*, 1125–1129.

Xiong, Y. S.; Chen, M. L.; He, Y. Q.; Xiong, G.; Yun,; Yu, Y. X. Influence of Coated Urea on the Yield and Quality of Celery and Nitrogen Balance. *Plant Nutr. Fert. Sci.* **2005,** *11*(1), 104–109.

Yadav, H. K.; Singh, S. P. Heterosis and Inbreeding Depression for Seed and Opium Yield in Opium Poppy (*Papaversomni ferum* L.). *J. Plant Dev. Sci.* **2011,** *3*, 81–86.

Yadav, R.; Chandan, K. P.; Deepika, G.; Rahul, K. Health Benefits of Indian Aromatic Plant Ajwain (*Trachycpermum ammi*). *Int. J. Pharm. Technol.* **2011,** *3*(3), 1356–1366.

Yadegari, M. Effect of Foliar Application of Micronutrients on Growth, Yield and Essential Oil Content of Thyme (*Thymus vulgaris* L.). *Crop Res.* **2014,** *47*(1/3), 56–65.

Yadegari, M.; Farahani, G. H. N.; Mosadeghzad, Z. Biofertilizers Effects on Quantitative and Qualitative Yield of Thyme (*Thymus vulgaris*). *Afr. J. Agric. Res.* **2012,** *7*(34), 4716–4723.

Yan, X.; Hou, J.; Xie, G. *Traditional Chinese Medicines Molecular Structure, Natural Sources and Application;* Ashgate Publishing Ltd.: Gower House craft road, England, 1998; pp. 124–148.

Yogita, R.; Nikam. T. D.; Dhumal, K. N. Effect of Foliar Application of Plant Growth Regulators on Growth, Yield and Essential Oil Components of Ajwain (*Trachyspermum ammi* L.). *Int. J. Seed Spices* **2013,** *3*(2), 34–41.

Yoldas, F.; Ceylan, S.; Mordogan, N.; Cakc, H. Effects of Sowing Dates and Nitrogen Application on Phosphorus and Potassium Contents in Parsley. *Asian J. Chem.* **2007,** *19*(5), 4059–4064.

Yoshikawa, T. (Ed.). *Free Radicals in Chemistry, Biology and Medicine;* OICA International: London, 2000.

Yousef, T. R.; Tawil, G. C. Antimicrobial Activity of Volatile Oils. *Pharmaize* **1980,** *35*, 698–701.

Yu, T.; Su, L.; Niu, S. W.; He, Z. G. Effects of Reduction and Optimization of Nitrogen Fertilizer Application on the Yield and Quality of Celery and Soil Nitrate Leaching Under Greenhouse Condition. *Soil Fertilizer Sci. China* **2012,** *3*, 55–58.

Yurembam, G. S. Performance Evaluation of Cardamom Polisher as Large Cardamom Detailing Machine [B. Tech. thesis], Sikkim, India, College of Agricultural Engineering and Post Harvest Technology, Ranipool, 2010.

Zahedi, H.; Jahanshahi, S. Effect of Planting Date and Vermicompost on Seed and Essence Production of Dill (*Anethum graveolens* L.). *Biol. Forum* **2014,** *6*(2), 357–361.

Zaidi, M. A.; Crow, S. A. Biologically Active Traditional Medicinal Herbs from Balochistan. *J. Ethnopharmacol.* **2005,** *96*, 331–334.

Zanzucchi, C. *Research Carried Out by the Consorzio Comunale Parmense on Saffron (Crocus sativus L.)*, In Proceedings of the Conference on the "coltivazione delle piante officinali", Trento ottobre 9–10, 1986, 1987(Publ.), 347–395.

Zaouali, L. *Medieval Cuisine of the Islamic World: A Concise History with 174 Recipes* (Translator M. B. DeBevoise); University of California Press: Berkley, 2007.

Zare, A. R.; Solouki, M.; Omidi, M.; Irvani, N.; Oladzad Abasabadi, A.; Mahdi Nezad, N. Effect of Various Treatments on Seed Germination and Dormancy Breaking in Ferula assafoetida L.(Asafoetida), a Threatened Medicinal Herb. *Trakia J. Sci.* **2011,** *9*(2), 57–61.

Zargari, A. *Medicinal Plants;* Tehran University Press: Tehran, 1995; Vol. 4, pp 325–328.

Zeinali, N.; Kashi, A.; Moghadam, M. R. F. A Study of the Effects of Planting Density and Pattern on Cucumber Yield and Yield Components in its Intercropping with Celery. *Iran. J. Hortic. Sci.* **2010,** *41*(1), 55–62.

Zhang, H.; Chen, F.; Wang, X.; Yao, H. Y. Evaluation of Antioxidant Activity of Parsley (Petroselinum crispum) Essential Oil and Identification of its Antioxidant Constituents. *Food Res. Int.* **2006,** *39*(8), 833–839.

Zhang, Y. L.; Sun, S. R.; Wang, L. J.; Qiao, B. J.; Lu, X. Y. Effects of Low Temperature Treatments on Germination of Celery Seeds. *Acta Agric. Univ. Henanensis* **1998,** *32*(1), 73–75.

Zheljazkov, V. D.; Cantrell, C. L.; Astatkie, T.; Hristov, A. Yield, Content, and Composition of Peppermint and Spearmints as a Function of Harvesting Time and Drying. *J. Agric. Food Chem.* **2010,** *58*, 11400–11407.

INDEX

E

F

G

H

N

T

U

V